Analysis on Manifolds

Analysis on Manifolds

James R. Munkres

Massachusetts Institute of Technology
Cambridge, Massachusetts

The Advanced Book Program

Westview Press

A Member of the Perseus Books Group

Library of Congress Cataloging-in-Publication Data
Munkres, James R., 1930-
Analysis on manifolds/James R. Munkres.
p. cm.
Includes bibliographical references.
1. Mathematical analysis. 2. Manifolds (Mathematics)
QA300.M75 1990
516.3'6'20—dc20 91-39786
ISBN 0-201-51035-9 CIP
ISBN 0-201-31596-3 (pbk.)

Find us on the World Wide Web at
http://www.westviewpress.com

This book was prepared using the TEX typesetting language.

20

Preface

This book is intended as a text for a course in analysis, at the senior or first-year graduate level.

A year-long course in real analysis is an essential part of the preparation of any potential mathematician. For the first half of such a course, there is substantial agreement as to what the syllabus should be. Standard topics include: sequences and series, the topology of metric spaces, and the derivative and the Riemann integral for functions of a single variable. There are a number of excellent texts for such a course, including books by Apostol [A], Rudin [Ru], Goldberg [Go], and Royden [Ro], among others.

There is no such universal agreement as to what the syllabus of the second half of such a course should be. Part of the problem is that there are simply too many topics that belong in such a course for one to be able to treat them all within the confines of a single semester, at more than a superficial level.

At M.I.T., we have dealt with the problem by offering two independent second-term courses in analysis. One of these deals with the derivative and the Riemann integral for functions of several variables, followed by a treatment of differential forms and a proof of Stokes' theorem for manifolds in euclidean space. The present book has resulted from my years of teaching this course. The other deals with the Lebesgue integral in euclidean space and its applications to Fourier analysis.

Prerequisites

As indicated, we assume the reader has completed a one-term course in analysis that included a study of metric spaces and of functions of a single variable. We also assume the reader has some background in linear algebra, including vector spaces and linear transformations, matrix algebra, and determinants.

The first chapter of the book is devoted to reviewing the basic results from linear algebra and analysis that we shall need. Results that are truly basic are stated without proof, but proofs are provided for those that are sometimes omitted in a first course. The student may determine from a perusal of this chapter whether his or her background is sufficient for the rest of the book.

How much time the instructor will wish to spend on this chapter will depend on the experience and preparation of the students. I usually assign Sections 1 and 3 as reading material, and discuss the remainder in class.

How the book is organized

The main part of the book falls into two parts. The first, consisting of Chapters 2 through 4, covers material that is fairly standard: derivatives, the inverse function theorem, the Riemann integral, and the change of variables theorem for multiple integrals. The second part of the book is a bit more sophisticated. It introduces manifolds and differential forms in $\mathbf{R}^n$, providing the framework for proofs of the n-dimensional version of Stokes' theorem and of the Poincaré lemma.

A final chapter is devoted to a discussion of abstract manifolds; it is intended as a transition to more advanced texts on the subject.

The dependence among the chapters of the book is expressed in the following diagram:

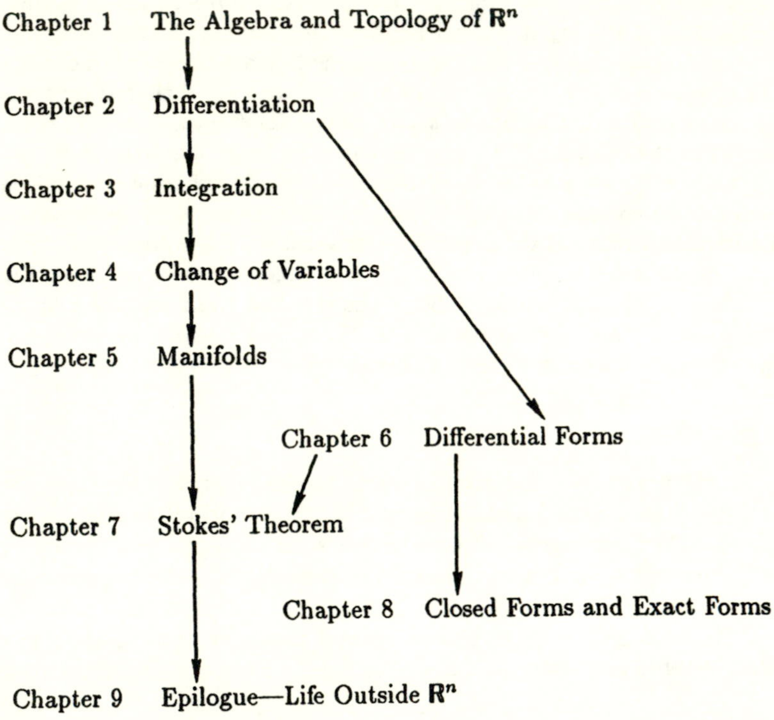

Certain sections of the books are marked with an asterisk; these sections may be omitted without loss of continuity. Similarly, certain theorems that may be omitted are marked with asterisks. When I use the book in our undergraduate analysis sequence, I usually omit Chapter 8, and assign Chapter 9 as reading. With graduate students, it should be possible to cover the entire book.

At the end of each section is a set of exercises. Some are computational in nature; students find it illuminating to know that one can compute the volume of a five-dimensional ball, even if the practical applications are limited! Other exercises are theoretical in nature, requiring that the student analyze carefully the theorems and proofs of the preceding section. The more difficult exercises are marked with asterisks, but none is unreasonably hard.

Acknowledgements

Two pioneering works in this subject demonstrated that such topics as manifolds and differential forms could be discussed with undergraduates. One is the set of notes used at Princeton c. 1960, written by Nickerson, Spencer, and Steenrod [N-S-S]. The second is the book by Spivak [S]. Our indebtedness to these sources is obvious. A more recent book on these topics is the one by Guillemin and Pollack [G-P]. A number of texts treat this material at a more advanced level. They include books by Boothby [B], Abraham, Mardsen, and Raitu [A-M-R], Berger and Gostiaux [B-G], and Fleming [F]. Any of them would be suitable reading for the student who wishes to pursue these topics further.

I am indebted to Sigurdur Helgason and Andrew Browder for helpful comments. To Ms. Viola Wiley go my thanks for typing the original set of lecture notes on which the book is based. Finally, thanks is due to my students at M.I.T., who endured my struggles with this material, as I tried to learn how to make it understandable (and palatable) to them!

J.R.M.

Contents

Analysis on Manifolds

1

The Algebra and Topology of $\mathbb{R}^n$

§1. REVIEW OF LINEAR ALGEBRA

Vector spaces

Suppose one is given a set V of objects, called **vectors**. And suppose there is given an operation called **vector addition**, such that the sum of the vectors $\mathbf{x}$ and $\mathbf{y}$ is a vector denoted $\mathbf{x} + \mathbf{y}$. Finally, suppose there is given an operation called **scalar multiplication**, such that the product of the scalar (i.e., real number) c and the vector $\mathbf{x}$ is a vector denoted $c\mathbf{x}$.

The set V, together with these two operations, is called a **vector space** (or **linear space**) if the following properties hold for all vectors $\mathbf{x}$, $\mathbf{y}$, $\mathbf{z}$ and all scalars c, d:

(1) $\mathbf{x} + \mathbf{y} = \mathbf{y} + \mathbf{x}$.

(2) $\mathbf{x} + (\mathbf{y} + \mathbf{z}) = (\mathbf{x} + \mathbf{y}) + \mathbf{z}$.

(3) There is a unique vector $\mathbf{0}$ such that $\mathbf{x} + \mathbf{0} = \mathbf{x}$ for all $\mathbf{x}$.

(4) $\mathbf{x} + (-1)\mathbf{x} = \mathbf{0}$.

(5) $1\mathbf{x} = \mathbf{x}$.

(6) $c(d\mathbf{x}) = (cd)\mathbf{x}$.

(7) $(c + d)\mathbf{x} = c\mathbf{x} + d\mathbf{x}$.

(8) $c(\mathbf{x} + \mathbf{y}) = c\mathbf{x} + c\mathbf{y}$.

One example of a vector space is the set $\mathbb{R}^n$ of all n-tuples of real numbers, with component-wise addition and multiplication by scalars. That is, if $\mathbf{x} = (x_1, \ldots, x_n)$ and $\mathbf{y} = (y_1, \ldots, y_n)$, then

$$\mathbf{x} + \mathbf{y} = (x_1 + y_1, \ldots, x_n + y_n),$$
$$c\mathbf{x} = (cx_1, \ldots, cx_n).$$

The vector space properties are easy to check.

If V is a vector space, then a subset W of V is called a **linear subspace** (or simply, a **subspace**) of V if for every pair $\mathbf{x},\mathbf{y}$ of elements of W and every scalar c, the vectors $\mathbf{x}+\mathbf{y}$ and $c\mathbf{x}$ belong to W. In this case, W itself satisfies properties (1)–(8) if we use the operations that W inherits from V, so that W is a vector space in its own right.

In the first part of this book, $\mathbb{R}^n$ and its subspaces are the only vector spaces with which we shall be concerned. In later chapters we shall deal with more general vector spaces.

Let V be a vector space. A set $\mathbf{a}_1, \ldots, \mathbf{a}_m$ of vectors in V is said to **span** V if to each $\mathbf{x}$ in V, there corresponds *at least* one m-tuple of scalars $c_1, \ldots, c_m$ such that

$$\mathbf{x} = c_1\mathbf{a}_1 + \cdots + c_m\mathbf{a}_m.$$

In this case, we say that $\mathbf{x}$ can be written as a **linear combination** of the vectors $\mathbf{a}_1, \ldots, \mathbf{a}_m$.

The set $\mathbf{a}_1, \ldots, \mathbf{a}_m$ of vectors is said to be **independent** if to each $\mathbf{x}$ in V there corresponds *at most* one m-tuple of scalars $c_1, \ldots, c_m$ such that

$$\mathbf{x} = c_1\mathbf{a}_1 + \cdots + c_m\mathbf{a}_m.$$

Equivalently, $\{\mathbf{a}_1, \ldots, \mathbf{a}_m\}$ is independent if to the zero vector $\mathbf{0}$ there corresponds only one m-tuple of scalars $d_1, \ldots, d_m$ such that

$$\mathbf{0} = d_1\mathbf{a}_1 + \cdots + d_m\mathbf{a}_m,$$

namely the scalars $d_1 = d_2 = \cdots = d_m = 0$.

If the set of vectors $\mathbf{a}_1, \ldots, \mathbf{a}_m$ both spans V and is independent, it is said to be a **basis** for V.

One has the following result:

Theorem 1.1. *Suppose V has a basis consisting of m vectors. Then any set of vectors that spans V has at least m vectors, and any set of vectors of V that is independent has at most m vectors. In particular, any basis for V has exactly m vectors.* □

If V has a basis consisting of m vectors, we say that m is the **dimension** of V. We make the convention that the vector space consisting of the zero vector alone has dimension zero.

It is easy to see that $\mathbb{R}^n$ has dimension n. (Surprise!) The following set of vectors is called the **standard basis for $\mathbb{R}^n$**:

$$\begin{aligned} \mathbf{e}_1 &= (1,0,0,\ldots,0), \\ \mathbf{e}_2 &= (0,1,0,\ldots,0), \\ &\ldots \\ \mathbf{e}_n &= (0,0,0,\ldots,1). \end{aligned}$$

The vector space $\mathbb{R}^n$ has many other bases, but any basis for $\mathbb{R}^n$ must consist of precisely n vectors.

One can extend the definitions of *spanning, independence,* and *basis* to allow for infinite sets of vectors; then it is possible for a vector space to have an infinite basis. (See the exercises.) However, we shall not be concerned with this situation.

Because $\mathbb{R}^n$ has a finite basis, so does every subspace of $\mathbb{R}^n$. This fact is a consequence of the following theorem:

Theorem 1.2. *Let V be a vector space of dimension m. If W is a linear subspace of V (different from V), then W has dimension less than m. Furthermore, any basis $\mathbf{a}_1,\ldots,\mathbf{a}_k$ for W may be extended to a basis $\mathbf{a}_1,\ldots,\mathbf{a}_k$, $\mathbf{a}_{k+1},\ldots,\mathbf{a}_m$ for V.* □

Inner products

If V is a vector space, an **inner product** on V is a function assigning, to each pair $\mathbf{x}$, $\mathbf{y}$ of vectors of V, a real number denoted $\langle \mathbf{x},\mathbf{y}\rangle$, such that the following properties hold for all $\mathbf{x}$, $\mathbf{y}$, $\mathbf{z}$ in V and all scalars c:

(1) $\langle \mathbf{x},\mathbf{y}\rangle = \langle \mathbf{y},\mathbf{x}\rangle$.

(2) $\langle \mathbf{x}+\mathbf{y},\mathbf{z}\rangle = \langle \mathbf{x},\mathbf{z}\rangle + \langle \mathbf{y},\mathbf{z}\rangle$.

(3) $\langle c\mathbf{x},\mathbf{y}\rangle = c\langle \mathbf{x},\mathbf{y}\rangle = \langle \mathbf{x},c\mathbf{y}\rangle$.

(4) $\langle \mathbf{x},\mathbf{x}\rangle > 0$ if $\mathbf{x} \neq \mathbf{0}$.

A vector space V together with an inner product on V is called an **inner product space**.

A given vector space may have many different inner products. One particularly useful inner product on $\mathbb{R}^n$ is defined as follows: If $\mathbf{x} = (x_1,\ldots,x_n)$ and $\mathbf{y} = (y_1,\ldots,y_n)$, we define

$$\langle \mathbf{x},\mathbf{y}\rangle = x_1y_1 + \cdots + x_ny_n.$$

The properties of an inner product are easy to verify. This is the inner product we shall commonly use in $\mathbb{R}^n$. It is sometimes called the **dot product**; we denote it by $\langle \mathbf{x},\mathbf{y}\rangle$ rather than $\mathbf{x}\cdot\mathbf{y}$ to avoid confusion with the matrix product, which we shall define shortly.

If V is an inner product space, one defines the **length** (or **norm**) of a vector of V by the equation

$$\|\mathbf{x}\| = \langle \mathbf{x}, \mathbf{x} \rangle^{1/2}.$$

The norm function has the following properties:

(1) $\|\mathbf{x}\| > 0$ if $\mathbf{x} \neq \mathbf{0}$.

(2) $\|c\mathbf{x}\| = |c|\,\|\mathbf{x}\|$.

(3) $\|\mathbf{x} + \mathbf{y}\| \leq \|\mathbf{x}\| + \|\mathbf{y}\|$.

The third of these properties is the only one whose proof requires some work; it is called the **triangle inequality**. (See the exercises.) An equivalent form of this inequality, which we shall frequently find useful, is the inequality

(3′) $\|\mathbf{x} - \mathbf{y}\| \geq \|\mathbf{x}\| - \|\mathbf{y}\|$.

Any function from V to the reals $\mathbf{R}$ that satisfies properties (1)–(3) just listed is called a **norm** on V. The length function derived from an inner product is one example of a norm, but there are other norms that are not derived from inner products. On $\mathbf{R}^n$, for example, one has not only the familiar norm derived from the dot product, which is called the **euclidean norm,** but one has also the **sup norm,** which is defined by the equation

$$|\mathbf{x}| = \max\{|x_1|, \ldots, |x_n|\}.$$

The sup norm is often more convenient to use than the euclidean norm. We note that these two norms on $\mathbf{R}^n$ satisfy the inequalities

$$|\mathbf{x}| \leq \|\mathbf{x}\| \leq \sqrt{n}|\mathbf{x}|.$$

Matrices

A **matrix** A is a rectangular array of numbers. The general number appearing in the array is called an **entry** of A. If the array has n rows and m columns, we say that A has **size** n by m, or that A is "an n by m matrix." We usually denote the entry of A appearing in the ith row and j^{th} column by a_{ij}; we call i the **row index** and j the **column index** of this entry.

If A and B are matrices of size n by m, with general entries a_{ij} and b_{ij}, respectively, we define $A + B$ to be the n by m matrix whose general entry is $a_{ij} + b_{ij}$, and we define cA to be the n by m matrix whose general entry is ca_{ij}. With these operations, the set of all n by m matrices is a vector space; the eight vector space properties are easy to verify. This fact is hardly surprising, for an n by m matrix is very much like an nm-tuple; the only difference is that the numbers are written in a rectangular array instead of a linear array.

The set of matrices has, however, an additional operation, called **matrix multiplication.** If A is a matrix of size n by m, and if B is a matrix of size m by p, then the **product** $A \cdot B$ is defined to be the matrix C of size n by p whose general entry c_{ij} is given by the equation

$$c_{ij} = \sum_{k=1}^{m} a_{ik} b_{kj}.$$

This product operation satisfies the following properties, which are straightforward to verify:

(1) $A \cdot (B \cdot C) = (A \cdot B) \cdot C$.

(2) $A \cdot (B + C) = A \cdot B + A \cdot C$.

(3) $(A + B) \cdot C = A \cdot C + B \cdot C$.

(4) $(cA) \cdot B = c(A \cdot B) = A \cdot (cB)$.

(5) For each k, there is a k by k matrix I_k such that if A is any n by m matrix,

$$I_n \cdot A = A \qquad \text{and} \qquad A \cdot I_m = A.$$

In each of these statements, we assume that the matrices involved are of appropriate sizes, so that the indicated operations may be performed.

The matrix I_k is the matrix of size k by k whose general entry δ_{ij} is defined as follows: $\delta_{ij} = 0$ if $i \neq j$, and $\delta_{ij} = 1$ if $i = j$. The matrix I_k is called the **identity matrix** of size k by k; it has the form

$$I_k = \begin{bmatrix} 1 & 0 & \dots & 0 \\ 0 & 1 & \dots & 0 \\ & \dots & \dots & \\ 0 & 0 & \dots & 1 \end{bmatrix},$$

with entries of 1 on the "main diagonal" and entries of 0 elsewhere.

We extend to matrices the sup norm defined for n-tuples. That is, if A is a matrix of size n by m with general entry a_{ij}, we define

$$|A| = \max\{|a_{ij}|;\ i = 1, \dots, n \text{ and } j = 1, \dots, m\}.$$

The three properties of a norm are immediate, as is the following useful result:

Theorem 1.3. *If A has size n by m, and B has size m by p, then*

$$|A \cdot B| \leq m|A|\,|B|. \quad \square$$

Linear transformations

If V and W are vector spaces, a function $T : V \to W$ is called a **linear transformation** if it satisfies the following properties, for all $\mathbf{x}$, $\mathbf{y}$ in V and all scalars c:

(1) $T(\mathbf{x}+\mathbf{y}) = T(\mathbf{x}) + T(\mathbf{y})$.

(2) $T(c\mathbf{x}) = cT(\mathbf{x})$.

If, in addition, T carries V onto W in a one-to-one fashion, then T is called a **linear isomorphism.**

One checks readily that if $T : V \to W$ is a linear transformation, and if $S : W \to X$ is a linear transformation, then the composite $S \circ T : V \to X$ is a linear transformation. Furthermore, if $T : V \to W$ is a linear isomorphism, then $T^{-1} : W \to V$ is also a linear isomorphism.

A linear transformation is uniquely determined by its values on basis elements, and these values may be specified arbitrarily. That is the substance of the following theorem:

Theorem 1.4. *Let V be a vector space with basis $\mathbf{a}_1, \ldots, \mathbf{a}_m$. Let W be a vector space. Given any m vectors $\mathbf{b}_1, \ldots, \mathbf{b}_m$ in W, there is exactly one linear transformation $T : V \to W$ such that, for all i, $T(\mathbf{a}_i) = \mathbf{b}_i$.* □

In the special case where V and W are "tuple spaces" such as $\mathbb{R}^m$ and $\mathbb{R}^n$, matrix notation gives us a convenient way of specifying a linear transformation, as we now show.

First we discuss row matrices and column matrices. A matrix of size 1 by n is called a **row matrix;** the set of all such matrices bears an obvious resemblance to $\mathbb{R}^n$. Indeed, under the one-to-one correspondence

$$(x_1, \ldots, x_n) \longrightarrow [x_1 \cdots x_n]$$

the vector space operations also correspond. Thus this correspondence is a linear isomorphism. Similarly, a matrix of size n by 1 is called a **column matrix;** the set of all such matrices also bears an obvious resemblance to $\mathbb{R}^n$. Indeed, the correspondence

$$(x_1, \ldots, x_n) \longrightarrow \begin{bmatrix} x_1 \\ \vdots \\ x_n \end{bmatrix}$$

is a linear isomorphism.

The second of these isomorphisms is particularly useful when studying linear transformations. Suppose for the moment that we represent elements of $\mathbf{R}^m$ and $\mathbf{R}^n$ by column matrices rather than by tuples. If A is a fixed n by m matrix, let us define a function $T : \mathbf{R}^m \longrightarrow \mathbf{R}^n$ by the equation

$$T(\mathbf{x}) = A \cdot \mathbf{x}.$$

The properties of matrix product imply immediately that T is a linear transformation.

In fact, every linear transformation of $\mathbf{R}^m$ to $\mathbf{R}^n$ has this form. The proof is easy. Given T, let $\mathbf{b}_1, \ldots, \mathbf{b}_m$ be the vectors of $\mathbf{R}^n$such that $T(\mathbf{e}_j) = \mathbf{b}_j$. Then let A be the n by m matrix $A = [\mathbf{b}_1 \cdots \mathbf{b}_m]$ with successive columns $\mathbf{b}_1, \ldots, \mathbf{b}_m$. Since the identity matrix has columns $\mathbf{e}_1, \ldots, \mathbf{e}_m$, the equation $A \cdot I_m = A$ implies that $A \cdot \mathbf{e}_j = \mathbf{b}_j$ for all j. Then $A \cdot \mathbf{e}_j = T(\mathbf{e}_j)$ for all j; it follows from the preceding theorem that $A \cdot \mathbf{x} = T(\mathbf{x})$ for all $\mathbf{x}$.

The convenience of this notation leads us to make the following convention:

Convention. *Throughout, we shall represent the elements of* $\mathbf{R}^n$ *by column matrices, unless we specifically state otherwise.*

Rank of a matrix

Given a matrix A of size n by m, there are several important linear spaces associated with A. One is the space spanned by the columns of A, looked at as column matrices (equivalently, as elements of $\mathbf{R}^n$). This space is called the **column space** of A, and its dimension is called the **column rank** of A. Because the column space of A is spanned by m vectors, its dimension can be no larger than m; because it is a subspace of $\mathbf{R}^n$, its dimension can be no larger than n.

Similarly, the space spanned by the rows of A, looked at as row matrices (or as elements of $\mathbf{R}^m$) is called the **row space** of A, and its dimension is called the **row rank** of A.

The following theorem is of fundamental importance:

Theorem 1.5. *For any matrix A, the row rank of A equals the column rank of A.* □

Once one has this theorem, one can speak merely of the **rank** of a matrix A, by which one means the number that equals both the row rank of A and the column rank of A.

The rank of a matrix A is an important number associated with A. One cannot in general determine what this number is by inspection. However, there is a relatively simple procedure called **Gauss-Jordan reduction** that can be used for finding the rank of a matrix. (It is used for other purposes as well.) We assume you have seen it before, so we merely review its major features here.

One considers certain operations, called **elementary row operations,** that are applied to a matrix A to obtain a new matrix B of the same size. They are the following:

(1) Exchange rows i_1 and i_2 of A (where $i_1 \neq i_2$).

(2) Replace row i_1 of A by itself plus the scalar c times row i_2 (where $i_1 \neq i_2$).

(3) Multiply row i of A by the non-zero scalar λ.

Each of these operations is invertible; in fact, the inverse of an elementary operation is an elementary operation of the same type, as you can check. One has the following result:

Theorem 1.6. *If B is the matrix obtained by applying an elementary row operation to A, then*

$$\text{rank } B = \text{rank } A. \quad \square$$

Gauss-Jordan reduction is the process of applying elementary operations to A to reduce it to a special form called **echelon form** (or **stairstep form**), for which the rank is obvious. An example of a matrix in this form is the following:

$$B = \begin{bmatrix} \circledast & * & * & * & * & * \\ 0 & \circledast & * & * & * & * \\ 0 & 0 & 0 & \circledast & * & * \\ 0 & 0 & 0 & 0 & 0 & 0 \end{bmatrix}.$$

Here the entries beneath the "stairsteps" are 0; the entries marked $*$ may be zero or non-zero, and the "corner entries," marked $\circledast$, are *non-zero.* (The corner entries are sometimes called "pivots.") One in fact needs only operations of types (1) and (2) to reduce A to echelon form.

Now it is easy to see that, for a matrix B in echelon form, the non-zero rows are independent. It follows that they form a basis for the row space of B, so the rank of B equals the number of its non-zero rows.

For some purposes it is convenient to reduce B to an even more special form, called **reduced echelon form.** Using elementary operations of type (2), one can make all the entries lying directly above each of the corner entries into 0's. Then by using operations of type (3), one can make all the corner entries into 1's. The reduced echelon form of the matrix B considered previously has the form:

$$C = \begin{bmatrix} 1 & 0 & * & 0 & * & * \\ 0 & 1 & * & 0 & * & * \\ 0 & 0 & 0 & 1 & * & * \\ 0 & 0 & 0 & 0 & 0 & 0 \end{bmatrix}.$$

It is even easier to see that, for the matrix C, its rank equals the number of its non-zero rows.

Transpose of a matrix

Given a matrix A of size n by m, we define the **transpose** of A to be the matrix D of size m by n whose general entry in row i and column j is defined by the equation $d_{ij} = a_{ji}$. The matrix D is often denoted A^{tr}.

The following properties of the transpose operation are readily verified:

(1) $(A^{tr})^{tr} = A$.

(2) $(A + B)^{tr} = A^{tr} + B^{tr}$.

(3) $(A \cdot C)^{tr} = C^{tr} \cdot A^{tr}$.

(4) rank A^{tr} = rank A.

The first three follow by direct computation, and the last from the fact that the row rank of A^{tr} is obviously the same as the column rank of A.

EXERCISES

1. Let V be a vector space with inner product $\langle \mathbf{x}, \mathbf{y} \rangle$ and norm $\|\mathbf{x}\| = \langle \mathbf{x}, \mathbf{x} \rangle^{1/2}$.
   (a) Prove the **Cauchy-Schwarz inequality** $\langle \mathbf{x}, \mathbf{y} \rangle \leq \|\mathbf{x}\| \, \|\mathbf{y}\|$. [*Hint:* If $\mathbf{x}, \mathbf{y} \neq \mathbf{0}$, set $c = 1/\|\mathbf{x}\|$ and $d = 1/\|\mathbf{y}\|$ and use the fact that $\|c\mathbf{x} \pm d\mathbf{y}\| \geq 0$.]
   (b) Prove that $\|\mathbf{x} + \mathbf{y}\| \leq \|\mathbf{x}\| + \|\mathbf{y}\|$. [*Hint:* Compute $\langle \mathbf{x} + \mathbf{y}, \mathbf{x} + \mathbf{y} \rangle$ and apply (a).]
   (c) Prove that $\|\mathbf{x} - \mathbf{y}\| \geq \|\mathbf{x}\| - \|\mathbf{y}\|$.
2. If A is an n by m matrix and B is an m by p matrix, show that
$$|A \cdot B| \leq m|A| \, |B|.$$
3. Show that the sup norm on $\mathbf{R}^2$ is not derived from an inner product on $\mathbf{R}^2$. [*Hint:* Suppose $\langle \mathbf{x}, \mathbf{y} \rangle$ is an inner product on $\mathbf{R}^2$ (*not* the dot product) having the property that $|\mathbf{x}| = \langle \mathbf{x}, \mathbf{x} \rangle^{1/2}$. Compute $\langle \mathbf{x} \pm \mathbf{y}, \mathbf{x} \pm \mathbf{y} \rangle$ and apply to the case $\mathbf{x} = \mathbf{e}_1$ and $\mathbf{y} = \mathbf{e}_2$.]
4. (a) If $\mathbf{x} = (x_1, x_2)$ and $\mathbf{y} = (y_1, y_2)$, show that the function
$$\langle \mathbf{x}, \mathbf{y} \rangle = [x_1 \; x_2] \begin{bmatrix} 2 & -1 \\ -1 & 1 \end{bmatrix} \begin{bmatrix} y_1 \\ y_2 \end{bmatrix}$$
   is an inner product on $\mathbf{R}^2$.
   *(b) Show that the function
$$\langle \mathbf{x}, \mathbf{y} \rangle = [x_1 \; x_2] \begin{bmatrix} a & b \\ b & c \end{bmatrix} \begin{bmatrix} y_1 \\ y_2 \end{bmatrix}$$
   is an inner product on $\mathbf{R}^2$ if and only if $b^2 - ac < 0$ and $a > 0$.

*5. Let V be a vector space; let $\{\mathbf{a}_\alpha\}$ be a set of vectors of V, as α ranges over some index set J (which may be infinite). We say that the set $\{\mathbf{a}_\alpha\}$ **spans** V if every vector $\mathbf{x}$ in V can be written as a finite linear combination

$$\mathbf{x} = c_{\alpha_1}\mathbf{a}_{\alpha_1} + \cdots + c_{\alpha_k}\mathbf{a}_{\alpha_k}$$

of vectors from this set. The set $\{\mathbf{a}_\alpha\}$ is **independent** if the scalars are uniquely determined by $\mathbf{x}$. The set $\{\mathbf{a}_\alpha\}$ is a **basis** for V if it both spans V and is independent.

(a) Check that the set $\mathbb{R}^\omega$ of all "infinite-tuples" of real numbers

$$\mathbf{x} = (x_1, x_2, \ldots)$$

is a vector space under component-wise addition and scalar multiplication.

(b) Let $\mathbb{R}^\infty$ denote the subset of $\mathbb{R}^\omega$ consisting of all $\mathbf{x} = (x_1, x_2, \ldots)$ such that $x_i = 0$ for all but finitely many values of i. Show $\mathbb{R}^\infty$ is a subspace of $\mathbb{R}^\omega$; find a basis for $\mathbb{R}^\infty$.

(c) Let $\mathcal{F}$ be the set of all real-valued functions $f\colon [a,b] \to \mathbb{R}$. Show that $\mathcal{F}$ is a vector space if addition and scalar multiplication are defined in the natural way:

$$(f+g)(x) = f(x) + g(x),$$
$$(cf)(x) = cf(x).$$

(d) Let $\mathcal{F}_{\mathrm{B}}$ be the subset of $\mathcal{F}$ consisting of all bounded functions. Let $\mathcal{F}_{\mathrm{I}}$ consist of all integrable functions. Let $\mathcal{F}_{\mathrm{C}}$ consist of all continuous functions. Let $\mathcal{F}_{\mathrm{D}}$ consist of all continuously differentiable functions. Let $\mathcal{F}_{\mathrm{P}}$ consist of all polynomial functions. Show that each of these is a subspace of the preceding one, and find a basis for $\mathcal{F}_{\mathrm{P}}$.

There is a theorem to the effect that *every* vector space has a basis. The proof is non-constructive. No one has ever exhibited specific bases for the vector spaces $\mathbb{R}^\omega, \mathcal{F}, \mathcal{F}_{\mathrm{B}}, \mathcal{F}_{\mathrm{I}}, \mathcal{F}_{\mathrm{C}}, \mathcal{F}_{\mathrm{D}}$.

(e) Show that the integral operator and the differentiation operator,

$$(If)(x) = \int_a^x f(t)\,dt \qquad \text{and} \qquad (Df)(x) = f'(x),$$

are linear transformations. What are possible domains and ranges of these transformations, among those listed in (d)?

§2. MATRIX INVERSION AND DETERMINANTS

We now treat several further aspects of linear algebra. They are the following: elementary matrices, matrix inversion, and determinants. Proofs are included, in case some of these results are new to you.

Elementary matrices

Definition. An **elementary matrix** of size n by n is the matrix obtained by applying one of the elementary row operations to the identity matrix I_n.

The elementary matrices are of three basic types, depending on which of the three operations is used. The elementary matrix corresponding to the first elementary operation has the form

$$E = \begin{bmatrix} 1 & & & & & & \\ & 1 & & & & & \\ & & 0 & \dots & 1 & & \\ & & \vdots & & \vdots & & \\ & & 1 & \dots & 0 & & \\ & & & & & 1 & \\ & & & & & & 1 \end{bmatrix} \begin{matrix} \\ \\ \nwarrow \text{row } i_1 \\ \swarrow \text{row } i_2 \\ \\ \\ \\ \end{matrix}.$$

The elementary matrix corresponding to the second elementary row operation has the form

$$E' = \begin{bmatrix} 1 & & & & & & \\ & 1 & & & & & \\ & & 1 & \dots & c & & \\ & & \vdots & & \vdots & & \\ & & 0 & \dots & 1 & & \\ & & & & & 1 & \\ & & & & & & 1 \end{bmatrix} \begin{matrix} \\ \\ \nwarrow \text{row } i_1 \\ \swarrow \text{row } i_2 \\ \\ \\ \\ \end{matrix}.$$

And the elementary matrix corresponding to the third elementary row operation has the form

$$E'' = \begin{bmatrix} 1 & & & & & \\ & 1 & & & & \\ & & \lambda & & & \\ & & & \ddots & & \\ & & & & 1 & \\ & & & & & 1 \end{bmatrix} \nwarrow \text{row } i.$$

One has the following basic result:

Theorem 2.1. *Let A be an n by m matrix. Any elementary row operation on A may be carried out by premultiplying A by the corresponding elementary matrix.*

Proof. One proceeds by direct computation. The effect of multiplying A on the left by the matrix E is to interchange rows i_1 and i_2 of A. Similarly, multiplying A by E' has the effect of replacing row i_1 by itself plus c times row i_2. And multiplying A by E'' has the effect of multiplying row i by λ. □

We will use this result later on when we prove the change of variables theorem for a multiple integral, as well as in the present section.

The inverse of a matrix

Definition. Let A be a matrix of size n by m; let B and C be matrices of size m by n. We say that B is a **left inverse** for A if $B \cdot A = I_m$, and we say that C is a **right inverse** for A if $A \cdot C = I_n$.

Theorem 2.2. *If A has both a left inverse B and a right inverse C, then they are unique and equal.*

Proof. Equality follows from the computation

$$C = I_m \cdot C = (B \cdot A) \cdot C = B \cdot (A \cdot C) = B \cdot I_n = B.$$

If B_1 is another left inverse for A, we apply this same computation with B_1 replacing B. We conclude that $C = B_1$; thus B_1 and B are equal. Hence B is unique. A similar computation shows that C is unique. □

Definition. If A has both a right inverse and a left inverse, then A is said to be **invertible.** The unique matrix that is both a right inverse and a left inverse for A is called the **inverse of** A, and is denoted A^{-1}.

A necessary and sufficient condition for A to be invertible is that A be square and of maximal rank. That is the substance of the following two theorems:

Theorem 2.3. *Let A be a matrix of size n by m. If A is invertible, then*

$$n = m = \text{ rank } A.$$

Proof. *Step 1.* We show that for any k by n matrix D,

$$\text{rank } (D \cdot A) \leq \text{ rank } A.$$

The proof is easy. If R is a row matrix of size 1 by n, then $R \cdot A$ is a row matrix that equals a linear combination of the rows of A, so it is an element of the row space of A. The rows of $D \cdot A$ are obtained by multiplying the rows of D by A. Therefore each row of $D \cdot A$ is an element of the row space of A. Thus the row space of $D \cdot A$ is contained in the row space of A and our inequality follows.

Step 2. We show that if A has a left inverse B, then the rank of A equals the number of columns of A.

The equation $I_m = B \cdot A$ implies by Step 1 that $m = \text{ rank } (B \cdot A) \leq$ rank A. On the other hand, the row space of A is a subspace of m-tuple space, so that rank $A \leq m$.

Step 3. We prove the theorem. Let B be the inverse of A. The fact that B is a left inverse for A implies by Step 2 that rank $A = m$. The fact that B is a right inverse for A implies that

$$B^{\text{tr}} \cdot A^{\text{tr}} = I_n^{\text{tr}} = I_n,$$

whence by Step 2, rank $A = n$. □

We prove the converse of this theorem in a slightly strengthened version:

Theorem 2.4. *Let A be a matrix of size n by m. Suppose*

$$n = m = \text{rank } A.$$

Then A is invertible; and furthermore, A equals a product of elementary matrices.

Proof. *Step 1.* We note first that every elementary matrix is invertible, and that its inverse is an elementary matrix. This follows from the fact that elementary operations are invertible. Alternatively, you can check directly that the matrix E corresponding to an operation of the first type is its own inverse, that an inverse for E' can be obtained by replacing c by $-c$ in the formula for E', and that an inverse for E'' can be obtained by replacing λ by $1/\lambda$ in the formula for E''.

Step 2. We prove the theorem. Let A be an n by n matrix of rank n. Let us reduce A to reduced echelon form C by applying elementary row operations. Because C is square and its rank equals the number of its rows, C must equal the identity matrix I_n. It follows from Theorem 2.1 that there is a sequence $E_1, \ldots, E_k$ of elementary matrices such that

$$E_k(E_{k-1}(\cdots(E_2(E_1 \cdot A))\cdots)) = I_n.$$

If we multiply both sides of this equation on the left by E_k^{-1}, then by E_{k-1}^{-1}, and so on, we obtain the equation

$$A = E_1^{-1} \cdot E_2^{-1} \cdots E_k^{-1};$$

thus A equals a product of elementary matrices. Direct computation shows that the matrix

$$B = E_k \cdot E_{k-1} \cdots E_1$$

is both a right and a left inverse for A. □

One very useful consequence of this theorem is the following:

Theorem 2.5. *If A is a square matrix and if B is a left inverse for A, then B is also a right inverse for A.*

Proof. Since A has a left inverse, Step 2 of the proof of Theorem 2.3 implies that the rank of A equals the number of columns of A. Since A is square, this is the same as the number of rows of A, so the preceding theorem implies that A has an inverse. By Theorem 2.2, this inverse must be B. □

An n by n matrix A is said to be **singular** if rank $A < n$; otherwise, it is said to be **non-singular.** The theorems just proved imply that A is invertible if and only if A is non-singular.

Determinants

The determinant is a function that assigns, to each square matrix A, a number called the *determinant* of A and denoted $\det A$.

The notation $|A|$ is often used for the determinant of A, but we are using this notation to denote the sup norm of A. So we shall use "det A" to denote the determinant instead.

In this section, we state three axioms for the determinant function, and we assume the existence of a function satisfying these axioms. The actual construction of the general determinant function will be postponed to a later chapter.

Definition. A function that assigns, to each n by n matrix A, a real number denoted $\det A$, is called a **determinant function** if it satisfies the following axioms:

(1) If B is the matrix obtained by exchanging any two rows of A, then $\det B = -\det A$.

(2) Given i, the function $\det A$ is linear as a function of the ith row alone.

(3) $\det I_n = 1$.

Condition (2) can be formulated as follows: Let i be fixed. Given an n-tuple $\mathbf{x}$, let $A_i(\mathbf{x})$ denote the matrix obtained from A by replacing the ith row by $\mathbf{x}$. Then condition (2) states that

$$\det A_i(a\mathbf{x} + b\mathbf{y}) = a \det A_i(\mathbf{x}) + b \det A_i(\mathbf{y}).$$

These three axioms characterize the determinant function uniquely, as we shall see.

EXAMPLE 1. In low dimensions, it is easy to construct the determinant function. For 1 by 1 matrices, the function

$$\det [a] = a$$

will do. For 2 by 2 matrices, the function

$$\det \begin{bmatrix} a & b \\ c & d \end{bmatrix} = ad - bc$$

suffices. And for 3 by 3 matrices, the function

$$\det \begin{bmatrix} a_1 & a_2 & a_3 \\ b_1 & b_2 & b_3 \\ c_1 & c_2 & c_3 \end{bmatrix} = \begin{matrix} a_1b_2c_3 + a_2b_3c_1 + a_3b_1c_2 \\ -a_3b_2c_1 - a_1b_3c_2 - a_2b_1c_3 \end{matrix}$$

will do, as you can readily check. For matrices of larger size, the definition is more complicated. For example, the expression for the determinant of a 4 by 4 matrix involves 24 terms; and for a 5 by 5 matrix, it involves 120 terms! Obviously, a less direct approach is needed. We shall return to this matter in Chapter 6.

Using the axioms, one can determine how the elementary row operations affect the value of the determinant. One has the following result:

Theorem 2.6. *Let A be an n by n matrix.*

(a) *If E is the elementary matrix corresponding to the operation that exchanges rows i_1 and i_2, then* $\det(E \cdot A) = -\det A$.

(b) *If E' is the elementary matrix corresponding to the operation that replaces row i_1 of A by itself plus c times row i_2, then* $\det(E' \cdot A) = \det A$.

(c) *If E'' is the elementary matrix corresponding to the operation that multiplies row i of A by the non-zero scalar λ, then* $\det(E'' \cdot A) = \lambda(\det A)$.

(d) *If A is the identity matrix I_n, then* $\det A = 1$.

Proof. Property (a) is a restatement of Axiom 1, and (d) is a restatement of Axiom 3. Property (c) follows directly from linearity (Axiom 2); it states merely that

$$\det A_i(\lambda \mathbf{x}) = \lambda(\det A_i(\mathbf{x})).$$

Now we verify (b). Note first that if A has two equal rows, then $\det A = 0$. For exchanging these rows does not change the matrix A, but by Axiom 1 it changes the sign of the determinant. Now let E' be the elementary operation that replaces row $i = i_1$ by itself plus c times row i_2. Let $\mathbf{x}$ equal row i_1 and let $\mathbf{y}$ equal row i_2. We compute

$$\begin{aligned}
\det(E' \cdot A) &= \det A_i(\mathbf{x} + c\mathbf{y}) \\
&= \det A_i(\mathbf{x}) + c \det A_i(\mathbf{y}) \\
&= \det A_i(\mathbf{x}), \text{ since } A_i(\mathbf{y}) \text{ has two equal rows,} \\
&= \det A, \text{ since } A_i(\mathbf{x}) = A. \quad \square
\end{aligned}$$

The four properties of the determinant function stated in this theorem are what one usually uses in practice rather than the axioms themselves. They also characterize the determinant completely, as we shall see.

One can use these properties to compute the determinants of the elementary matrices. Setting $A = I_n$ in Theorem 2.6, we have

$$\det E = -1 \quad \text{and} \quad \det E' = 1 \quad \text{and} \quad \det E'' = \lambda.$$

We shall see later how they can be used to compute the determinant in general.

Now we derive the further properties of the determinant function that we shall need.

Theorem 2.7. *Let A be a square matrix. If the rows of A are independent, then* $\det A \neq 0$; *if the rows are dependent, then* $\det A = 0$. *Thus an n by n matrix A has rank n if and only if* $\det A \neq 0$.

Proof. First, we note that if the ith row of A is the zero row, then $\det A = 0$. For multiplying row i by 2 leaves A unchanged; on the other hand, it must multiply the value of the determinant by 2.

Second, we note that applying one of the elementary row operations to A does not affect the vanishing or non-vanishing of the determinant, for it alters the value of the determinant by a factor of either -1 or 1 or λ (where $\lambda \neq 0$).

Now by means of elementary row operations, let us reduce A to a matrix B in echelon form. (Elementary operations of types (1) and (2) will suffice.) If the rows of A are dependent, rank $A < n$; then rank $B < n$, so that B must have a zero row. Then $\det B = 0$, as just noted; it follows that $\det A = 0$.

If the rows of A are independent, let us reduce B further to echelon form C. Since C is square and has rank n, C must equal the identity matrix I_n. Then $\det C \neq 0$; it follows that $\det A \neq 0$. □

The proof just given can be refined so as to provide a method for calculating the determinant function:

Theorem 2.8. *Given a square matrix A, let us reduce it to echelon form B by elementary row operations of types (1) and (2). If B has a zero row, then $\det A = 0$. Otherwise, let k be the number of row exchanges involved in the reduction process. Then $\det A$ equals $(-1)^k$ times the product of the diagonal entries of B.*

Proof. If B has a zero row, then rank $A < n$ and $\det A = 0$. So suppose that B has no zero row. We know from (a) and (b) of Theorem 2.6 that $\det A = (-1)^k \det B$. Furthermore, B must have the form

$$B = \begin{bmatrix} b_{11} & * & \dots & * \\ 0 & b_{22} & \dots & * \\ & \vdots & & \vdots \\ 0 & 0 & \dots & b_{nn} \end{bmatrix},$$

where the diagonal entries are non-zero. It remains to show that

$$\det B = b_{11} b_{22} \cdots b_{nn}.$$

For that purpose, let us apply elementary operations of type (2) to make the entries above the diagonal into zeros. The diagonal entries are unaffected by the process; therefore the resulting matrix has the form

$$C = \begin{bmatrix} b_{11} & 0 & \dots & 0 \\ 0 & b_{22} & \dots & 0 \\ & \vdots & & \vdots \\ 0 & 0 & \dots & b_{nn} \end{bmatrix}.$$

Since only operations of type (2) are involved, we have $\det B = \det C$. Now let us multiply row 1 of C by $1/b_{11}$, row 2 by $1/b_{22}$, and so on, obtaining as our end result the identity matrix I_n. Property (c) of Theorem 2.6 implies that

$$\det I_n = (1/b_{11})(1/b_{22})\cdots(1/b_{nn})\det C,$$

so that (using property (d))

$$\det C = b_{11}b_{22}\cdots b_{nn},$$

as desired. □

Corollary 2.9. *The determinant function is uniquely characterized by its three axioms. It is also characterized by the four properties listed in Theorem 2.6.*

Proof. The calculation of $\det A$ just given uses only properties (a)–(d) of Theorem 2.6. These in turn follow from the three axioms. □

Theorem 2.10. *Let A and B be n by n matrices. Then*

$$\det(A \cdot B) = (\det A) \cdot (\det B).$$

Proof. *Step 1.* The theorem holds when A is an elementary matrix. Indeed:

$$\begin{aligned}\det(E \cdot B) &= -\det B = (\det E)(\det B),\\ \det(E' \cdot B) &= \det B = (\det E')(\det B),\\ \det(E'' \cdot B) &= \lambda \cdot \det B = (\det E'')(\det B).\end{aligned}$$

Step 2. The theorem holds when rank $A = n$. For in that case, A is a product of elementary matrices, and one merely applies Step 1 repeatedly. Specifically, if $A = E_1 \cdots E_k$, then

$$\begin{aligned}\det(A \cdot B) &= \det(E_1 \cdots E_k \cdot B)\\ &= (\det E_1)\det(E_2 \cdots E_k \cdot B)\\ &= \quad \cdots\\ &= (\det E_1)(\det E_2)\cdots(\det E_k)(\det B).\end{aligned}$$

This equation holds for all B. In the case $B = I_n$, it tells us that

$$\det A = (\det E_1)(\det E_2)\cdots(\det E_k).$$

The theorem follows.

Step 3. We complete the proof by showing that the theorem holds if rank $A < n$. We have in general,

$$\text{rank } (A \cdot B) = \text{rank } (A \cdot B)^{\text{tr}} = \text{rank } (B^{\text{tr}} \cdot A^{\text{tr}}) \le \text{rank } A^{\text{tr}},$$

where the inequality follows from Step 1 of Theorem 2.3. Thus if rank $A < n$, the theorem holds because both sides of the equation vanish. □

Even in low dimensions, this theorem would be very unpleasant to prove by direct computation. You might try it in the 2 by 2 case!

Theorem 2.11. $\det A^{\text{tr}} = \det A$.

Proof. *Step 1.* We show the theorem holds when A is an elementary matrix.

Let E, E', and E'' be elementary matrices of the three basic types. Direct inspection shows that $E^{\text{tr}} = E$ and $(E'')^{\text{tr}} = E''$, so the theorem is trivial in these cases. For the matrix E' of type (2), we note that its transpose is another elementary matrix of type (2), so that both have determinant 1.

Step 2. We verify the theorem when A has rank n. In that case, A is a product of elementary matrices, say

$$A = E_1 \cdot E_2 \cdots E_k.$$

Then

$$\begin{aligned}
\det A^{\text{tr}} &= \det(E_k^{\text{tr}} \cdots E_2^{\text{tr}} \cdot E_1^{\text{tr}}) \\
&= (\det E_k^{\text{tr}}) \cdots (\det E_2^{\text{tr}})(\det E_1^{\text{tr}}) && \text{by Theorem 2.10,} \\
&= (\det E_k) \cdots (\det E_2)(\det E_1) && \text{by Step 1,} \\
&= (\det E_1)(\det E_2) \cdots (\det E_k) \\
&= \det(E_1 \cdot E_2 \cdots E_k) \\
&= \det A.
\end{aligned}$$

Step 3. The theorem holds if rank $A < n$. In this case, rank $A^{\text{tr}} < n$, so that $\det A^{\text{tr}} = 0 = \det A$. □

A formula for A^{-1}

We know that A is invertible if and only if $\det A \neq 0$. Now we derive a formula for A^{-1} that involves determinants explicitly.

Definition. Let A be an n by n matrix. The matrix of size $n-1$ by $n-1$ that is obtained from A by deleting the ith row and the jth column of A is called the (i,j)-minor of A. It is denoted A_{ij}. The number

$$(-1)^{i+j} \det A_{ij}$$

is called the (i,j)-cofactor of A.

Lemma 2.12. *Let A be an n by n matrix; let b denote its entry in row i and column j.*

(a) *If all the entries in row i other than b vanish, then*

$$\det A = b(-1)^{i+j} \det A_{ij}.$$

(b) *The same equation holds if all the entries in column j other than the entry b vanish.*

Proof. *Step 1.* We verify a special case of the theorem. Let $b, a_2, \dots, a_n$ be fixed numbers. Given an $n-1$ by $n-1$ matrix D, let $A(D)$ denote the n by n matrix

$$A(D) = \begin{bmatrix} b & a_2 & \dots & a_n \\ 0 & & & \\ \vdots & & D & \\ 0 & & & \end{bmatrix}.$$

We show that $\det A(D) = b(\det D)$.

If $b = 0$, this result is obvious, since in that case rank $A(D) < n$. So assume $b \neq 0$. Define a function f by the equation

$$f(D) = (1/b) \det A(D).$$

We show that f satisfies the four properties stated in Theorem 2.6, so that $f(D) = \det D$.

Exchanging two rows of D has the effect of exchanging two rows of $A(D)$, which changes the value of f by a factor -1. Replacing row i_1 of D by itself plus c times row i_2 of D has the effect of replacing row $(i_1 + 1)$ of $A(D)$ by itself plus row $(i_2 + 1)$ of $A(D)$, which leaves the value of f unchanged. Multiplying row i of D by λ has the effect of multiplying row $(i+1)$ of $A(D)$ by λ, which changes the value of f by a factor of λ. Finally, if $D = I_{n-1}$, then $A(D)$ is in echelon form, so $\det A(D) = b \cdot 1 \cdots 1$ by Theorem 2.8, and $f(D) = 1$.

Step 2. It follows by taking transposes that

$$\det \begin{bmatrix} b & 0 & \dots & 0 \\ a_2 & & & \\ \vdots & & D & \\ a_n & & & \end{bmatrix} = b(\det D).$$

Step 3. We prove the theorem. Let A be a matrix satisfying the hypotheses of our theorem. One can by a sequence of $i-1$ exchanges of adjacent rows bring the ith row of A up to the top of matrix, without affecting the order of the remaining rows. Then by a sequence of $j-1$ exchanges of adjacent columns, one can bring the jth column of this matrix to the left edge of the matrix, without affecting the order of the remaining columns. The matrix C that results has the form of one of the matrices considered in Steps 1 and 2. Furthermore, the (1,1)-minor $C_{1,1}$ of the matrix C is identical with the (i,j)-minor A_{ij} of the original matrix A.

Now each row exchange changes the sign of the determinant. So does each column exchange, by Theorem 2.11. Therefore

$$\det C = (-1)^{(i-1)+(j-1)} \det A = (-1)^{i+j} \det A.$$

Thus

$$\begin{aligned} \det A &= (-1)^{i+j} \det C, \\ &= (-1)^{i+j} b \det C_{1,1} \qquad \text{by Steps 1 and 2,} \\ &= (-1)^{i+j} b \det A_{ij}. \quad \square \end{aligned}$$

Theorem 2.13 (Cramer's rule). *Let A be an n by n matrix with successive columns $\mathbf{a}_1, \ldots, \mathbf{a}_n$. Let*

$$\mathbf{x} = \begin{bmatrix} x_1 \\ \vdots \\ x_n \end{bmatrix} \quad \textit{and} \quad \mathbf{c} = \begin{bmatrix} c_1 \\ \vdots \\ c_n \end{bmatrix}$$

be column matrices. If $A \cdot \mathbf{x} = \mathbf{c}$, then

$$(\det A) \cdot x_i = \det\, [\mathbf{a}_1 \cdots \mathbf{a}_{i-1}\ \mathbf{c}\ \mathbf{a}_{i+1} \cdots \mathbf{a}_n].$$

Proof. Let $\mathbf{e}_1, \ldots, \mathbf{e}_n$ be the standard basis for $\mathbb{R}^n$, where each $\mathbf{e}_i$ is written as a column matrix. Let C be the matrix

$$C = [\mathbf{e}_1 \cdots \mathbf{e}_{i-1}\ \mathbf{x}\ \mathbf{e}_{i+1} \cdots \mathbf{e}_n].$$

The equations $A \cdot \mathbf{e}_j = \mathbf{a}_j$ and $A \cdot \mathbf{x} = \mathbf{c}$ imply that

$$A \cdot C = [\mathbf{a}_1 \cdots \mathbf{a}_{i-1}\ \mathbf{c}\ \mathbf{a}_{i+1} \cdots \mathbf{a}_n].$$

By Theorem 2.10,

$$(\det A) \cdot (\det C) = \det\, [\mathbf{a}_1 \cdots \mathbf{a}_{i-1}\ \mathbf{c}\ \mathbf{a}_{i+1} \cdots \mathbf{a}_n].$$

Now C has the form

$$C = \begin{bmatrix} 1 & \cdots & x_1 & \cdots & 0 \\ \vdots & & \vdots & & \vdots \\ 0 & \cdots & x_i & \cdots & 0 \\ \vdots & & \vdots & & \vdots \\ 0 & \cdots & x_n & \cdots & 1 \end{bmatrix},$$

where the entry x_i appears in row i and column i. Hence by the preceding lemma,

$$\det C = x_i(-1)^{i+i} \det I_{n-1} = x_i.$$

The theorem follows. □

Here now is the formula we have been seeking:

Theorem 2.14. *Let A be an n by n matrix of rank n; let $B = A^{-1}$. Then*

$$b_{ij} = \frac{(-1)^{j+i} \det A_{ji}}{\det A}.$$

Proof. Let j be fixed throughout this argument. Let

$$\mathbf{x} = \begin{bmatrix} x_1 \\ \vdots \\ x_n \end{bmatrix}$$

denote the jth column of the matrix B. The fact that $A \cdot B = I_n$ implies in particular that $A \cdot \mathbf{x} = \mathbf{e}_j$. Cramer's rule tells us that

$$(\det A) \cdot x_i = \det [\mathbf{a}_1 \cdots \mathbf{a}_{i-1}\ \mathbf{e}_j\ \mathbf{a}_{i+1} \cdots \mathbf{a}_n].$$

We conclude from Lemma 2.12 that

$$(\det A) \cdot x_i = 1 \cdot (-1)^{j+i} \det A_{ji}.$$

Since $x_i = b_{ij}$, our theorem follows. □

This theorem gives us an algorithm for computing the inverse of a matrix A. One proceeds as follows:

(1) First, form the matrix whose entry in row i and column j is $(-1)^{i+j} \det A_{ij}$; this matrix is called the **matrix of cofactors** of A.

(2) Second, take the transpose of this matrix.

(3) Third, divide each entry of this matrix by $\det A$.

This algorithm is in fact not very useful for practical purposes; computing determinants is simply too time-consuming. The importance of this formula for A^{-1} is theoretical, as we shall see. If one wishes actually to compute A^{-1}, there is an algorithm based on Gauss-Jordan reduction that is more efficient. It is outlined in the exercises.

Expansion by cofactors

We now derive a final formula for evaluating the determinant. This is the one place we actually need the axioms for the determinant function rather than the properties stated in Theorem 2.6.

Theorem 2.15. *Let A be an n by n matrix. Let i be fixed. Then*

$$\det A = \sum_{k=1}^{n} (-1)^{i+k} a_{ik} \cdot \det A_{ik}.$$

Here A_{ik} is, as usual, the (i,k)-minor of A. This rule is called the "rule for expansion of the determinant by cofactors of the ith row." There is a similar rule for expansion by cofactors of the jth column, proved by taking transposes.

Proof. Let $A_i(\mathbf{x})$, as usual, denote the matrix obtained from A by replacing the ith row by the n-tuple $\mathbf{x}$. If $\mathbf{e}_1, \ldots, \mathbf{e}_n$ denote the usual basis vectors in $\mathbb{R}^n$ (written as *row* matrices in this case), then the ith row of A can be written in the form

$$\sum_{k=1}^{n} a_{ik} \mathbf{e}_k.$$

Then

$$\begin{aligned} \det A &= \sum_{k=1}^{n} a_{ik} \cdot \det A_i(\mathbf{e}_k) && \text{by linearity (Axiom 2),} \\ &= \sum_{k=1}^{n} a_{ik} (-1)^{i+k} \det A_{ik} && \text{by Lemma 2.12. } \square \end{aligned}$$

EXERCISES

1. Consider the matrix

$$A = \begin{bmatrix} 1 & 2 \\ 1 & -1 \\ 0 & 1 \end{bmatrix}.$$

   (a) Find two different left inverses for A.

   (b) Show that A has no right inverse.

2. Let A be an n by m matrix with $n \neq m$.

   (a) If rank $A = m$, show there exists a matrix D that is a product of elementary matrices such that

$$D \cdot A = \begin{bmatrix} I_m \\ 0 \end{bmatrix}.$$

   (b) Show that A has a left inverse if and only if rank $A = m$.

   (c) Show that A has a right inverse if and only if rank $A = n$.

3. Verify that the functions defined in Example 1 satisfy the axioms for the determinant function.

4. (a) Let A be an n by n matrix of rank n. By applying elementary row operations to A, one can reduce A to the identity matrix. Show that by applying the same operations, in the same order, to I_n, one obtains the matrix A^{-1}.

   (b) Let

$$A = \begin{bmatrix} 1 & 2 & 3 \\ 0 & 1 & 2 \\ 1 & 2 & 1 \end{bmatrix}.$$

   Calculate A^{-1} by using the algorithm suggested in (a). [*Hint:* An easy way to do this is to reduce the 3 by 6 matrix $[A\ I_3]$ to reduced echelon form.]

   (c) Calculate A^{-1} using the formula involving determinants.

5. Let

$$A = \begin{bmatrix} a & b \\ c & d \end{bmatrix},$$

   where $ad - bc \neq 0$. Find A^{-1}.

*6. Prove the following:

   Theorem. *Let A be a k by k matrix, let D have size n by n and let C have size n by k. Then*

$$\det \begin{bmatrix} A & 0 \\ C & D \end{bmatrix} = (\det A) \cdot (\det D).$$

Proof. First show that

$$\begin{bmatrix} A & 0 \\ 0 & I_n \end{bmatrix} \cdot \begin{bmatrix} I_k & 0 \\ C & D \end{bmatrix} = \begin{bmatrix} A & 0 \\ C & D \end{bmatrix}.$$

Then use Lemma 2.12.

§3. REVIEW OF TOPOLOGY IN $\mathbf{R}^n$

Metric spaces

Recall that if A and B are sets, then $A \times B$ denotes the set of all ordered pairs (a, b) for which $a \in A$ and $b \in B$.

Given a set X, a **metric** on X is a function $d : X \times X \to \mathbf{R}$ such that the following properties hold for all $x, y, z \in X$:

(1) $d(x, y) = d(y, x)$.

(2) $d(x, y) \geq 0$, and equality holds if and only if $x = y$.

(3) $d(x, z) \leq d(x, y) + d(y, z)$.

A **metric space** is a set X together with a specific metric on X. We often suppress mention of the metric, and speak simply of "the metric space X."

If X is a metric space with metric d, and if Y is a subset of X, then the restriction of d to the set $Y \times Y$ is a metric on Y; thus Y is a metric space in its own right. It is called a **subspace** of X.

For example, $\mathbf{R}^n$ has the metrics

$$d(\mathbf{x}, \mathbf{y}) = \|\mathbf{x} - \mathbf{y}\| \qquad \text{and} \qquad d(\mathbf{x}, \mathbf{y}) = |\mathbf{x} - \mathbf{y}|;$$

they are called the **euclidean metric** and the **sup metric**, respectively. It follows immediately from the properties of a norm that they are metrics. For many purposes, these two metrics on $\mathbf{R}^n$ are equivalent, as we shall see.

We shall in this book be concerned only with the metric space $\mathbf{R}^n$ and its subspaces, except for the expository final section, in which we deal with general metric spaces. The space $\mathbf{R}^n$ is commonly called ***n*-dimensional euclidean space.**

If X is a metric space with metric d, then given $x_0 \in X$ and given $\epsilon > 0$, the set

$$U(x_0; \epsilon) = \{x \mid d(x, x_0) < \epsilon\}$$

is called the ϵ-**neighborhood** of x_0, or the ϵ-**neighborhood centered at** x_0. A subset U of X is said to be **open** in X if for each $x_0 \in U$ there is a corresponding $\epsilon > 0$ such that $U(x_0;\epsilon)$ is contained in U. A subset C of X is said to be **closed** in X if its complement $X - C$ is open in X. It follows from the triangle inequality that an ϵ-neighborhood is itself an open set.

If U is *any* open set containing x_0, we commonly refer to U simply as a **neighborhood** of x_0.

Theorem 3.1. *Let (X,d) be a metric space. Then finite intersections and arbitrary unions of open sets of X are open in X. Similarly, finite unions and arbitrary intersections of closed sets of X are closed in X.* □

Theorem 3.2. *Let X be a metric space; let Y be a subspace. A subset A of Y is open in Y if and only if it has the form*

$$A = U \cap Y,$$

where U is open in X. Similarly, a subset A of Y is closed in Y if and only if it has the form

$$A = C \cap Y,$$

where C is closed in X. □

It follows that if A is open in Y and Y is open in X, then A is open in X. Similarly, if A is closed in Y and Y is closed in X, then A is closed in X.

If X is a metric space, a point x_0 of X is said to be a **limit point** of the subset A of X if every ϵ-neighborhood of x_0 intersects A in at least one point different from x_0. An equivalent condition is to require that every neighborhood of x_0 contain infinitely many points of A.

Theorem 3.3. *If A is a subset of X, then the set $\bar{A}$ consisting of A and all its limit points is a closed set of X. A subset of X is closed if and only if it contains all its limit points.* □

The set $\bar{A}$ is called the **closure** of A.

In $\mathbf{R}^n$, the ϵ-neighborhoods in our two standard metrics are given special names. If $\mathbf{a} \in \mathbf{R}^n$, the ϵ-neighborhood of $\mathbf{a}$ in the euclidean metric is called the **open ball** of radius ϵ centered at $\mathbf{a}$, and denoted $B(\mathbf{a};\epsilon)$. The ϵ-neighborhood of $\mathbf{a}$ in the sup metric is called the **open cube** of radius ϵ centered at $\mathbf{a}$, and denoted $C(\mathbf{a};\epsilon)$. The inequalities $|\mathbf{x}| \leq \|\mathbf{x}\| \leq \sqrt{n}\,|\mathbf{x}|$ lead to the following inclusions:

$$B(\mathbf{a};\epsilon) \subset C(\mathbf{a};\epsilon) \subset B(\mathbf{a};\sqrt{n}\,\epsilon).$$

These inclusions in turn imply the following:

Theorem 3.4. *If X is a subspace of $\mathbf{R}^n$, the collection of open sets of X is the same whether one uses the euclidean metric or the sup metric on X. The same is true for the collection of closed sets of X.* □

In general, any property of a metric space X that depends only on the collection of open sets of X, rather than on the specific metric involved, is called a **topological property** of X. Limits, continuity, and compactness are examples of such, as we shall see.

Limits and Continuity

Let X and Y be metric spaces, with metrics d_X and d_Y, respectively.

We say that a function $f : X \to Y$ is **continuous at the point x_0** of X if for each open set V of Y containing $f(x_0)$, there is an open set U of X containing x_0 such that $f(U) \subset V$. We say f is **continuous** if it is continuous at each point x_0 of X. Continuity of f is equivalent to the requirement that for each open set V of Y, the set

$$f^{-1}(V) = \{x \mid f(x) \in V\}$$

is open in X, or alternatively, the requirement that for each closed set D of Y, the set $f^{-1}(D)$ is closed in X.

Continuity may be formulated in a way that involves the metrics specifically. The function f is continuous at x_0 if and only if the following holds: *For each $\epsilon > 0$, there is a corresponding $\delta > 0$ such that*

$$d_Y(f(x), f(x_0)) < \epsilon \quad \textit{whenever} \quad d_X(x, x_0) < \delta.$$

This is the classical "ϵ–δ formulation of continuity."

Note that given $x_0 \in X$ it may happen that for some $\delta > 0$, the δ-neighborhood of x_0 consists of the point x_0 alone. In that case, x_0 is called an **isolated point** of X, and any function $f : X \to Y$ is automatically continuous at x_0!

A constant function from X to Y is continuous, and so is the identity function $i_X : X \to X$. So are restrictions and composites of continuous functions:

Theorem 3.5. (a) *Let $x_0 \in A$, where A is a subspace of X. If $f : X \to Y$ is continuous at x_0, then the restricted function $f \mid A : A \to Y$ is continuous at x_0.*

(b) *Let $f : X \to Y$ and $g : Y \to Z$. If f is continuous at x_0 and g is continuous at $y_0 = f(x_0)$, then $g \circ f : X \to Z$ is continuous at x_0.* □

Theorem 3.6. (a) *Let X be a metric space. Let $f : X \to \mathbf{R}^n$ have the form*

$$f(x) = (f_1(x), \ldots, f_n(x)).$$

Then f is continuous at x_0 if and only if each function $f_i : X \to \mathbb{R}$ is continuous at x_0. The functions f_i are called the **component functions** *of f.*

(b) *Let $f, g : X \to \mathbb{R}$ be continuous at x_0. Then $f+g$ and $f-g$ and $f \cdot g$ are continuous at x_0; and f/g is continuous at x_0 if $g(x_0) \neq 0$.*

(c) *The projection function $\pi_i : \mathbb{R}^n \to \mathbb{R}$ given by $\pi_i(\mathbf{x}) = x_i$ is continuous.* □

These theorems imply that functions formed from the familiar real-valued continuous functions of calculus, using algebraic operations and composites, are continuous in $\mathbb{R}^n$. For instance, since one knows that the functions e^x and $\sin x$ are continuous in $\mathbb{R}$, it follows that such a function as

$$f(s,t,u,v) = (\sin(s+t))/e^{uv}$$

is continuous in $\mathbb{R}^4$.

Now we define the notion of *limit.* Let X be a metric space. Let $A \subset X$ and let $f : A \to Y$. Let x_0 be a limit point of the domain A of f. (The point x_0 may or may not belong to A.) We say that **$f(x)$ approaches y_0 as x approaches x_0** if for each open set V of Y containing y_0, there is an open set U of X containing x_0 such that $f(x) \in V$ whenever x is in $U \cap A$ and $x \neq x_0$. This statement is expressed symbolically in the form

$$f(x) \to y_0 \quad \text{as} \quad x \to x_0.$$

We also say in this situation that the **limit** of $f(x)$, as x approaches x_0, is y_0. This statement is expressed symbolically by the equation

$$\lim_{x \to x_0} f(x) = y_0.$$

Note that the requirement that x_0 be a limit point of A guarantees that there *exist* points x different from x_0 belonging to the set $U \cap A$. *We do not attempt to define the limit of f if x_0 is not a limit point of the domain of f.*

Note also that the value of f at x_0 (provided f is even *defined* at x_0) is not involved in the definition of the limit.

The notion of limit can be formulated in a way that involves the metrics specifically. One shows readily that $f(x)$ approaches y_0 as x approaches x_0 if and only if the following condition holds: *For each $\epsilon > 0$, there is a corresponding $\delta > 0$ such that*

$$d_Y(f(x), y_0) < \epsilon \quad \textit{whenever} \quad x \in A \quad \textit{and} \quad 0 < d_X(x, x_0) < \delta.$$

There is a direct relation between limits and continuity; it is the following:

Theorem 3.7. *Let $f : X \to Y$. If x_0 is an isolated point of X, then f is continuous at x_0. Otherwise, f is continuous at x_0 if and only if $f(x) \to f(x_0)$ as $x \to x_0$.* □

Most of the theorems dealing with continuity have counterparts that deal with limits:

Theorem 3.8. (a) *Let $A \subset X$; let $f : A \to \mathbf{R}^n$ have the form*

$$f(x) = (f_1(x), \ldots, f_n(x)).$$

Let $\mathbf{a} = (a_1, \ldots, a_n)$. Then $f(x) \to \mathbf{a}$ as $x \to x_0$ if and only if $f_i(x) \to a_i$ as $x \to x_0$, for each i.

(b) *Let $f, g : A \to \mathbf{R}$. If $f(x) \to a$ and $g(x) \to b$ as $x \to x_0$, then as $x \to x_0$,*

$$f(x) + g(x) \to a + b,$$
$$f(x) - g(x) \to a - b,$$
$$f(x) \cdot g(x) \to a \cdot b;$$

also, $f(x)/g(x) \to a/b$ if $b \neq 0$. □

Interior and Exterior

The following concepts make sense in an arbitrary metric space. Since we shall use them only for $\mathbf{R}^n$, we define them only in that case.

Definition. Let A be a subset of $\mathbf{R}^n$. The **interior** of A, as a subset of $\mathbf{R}^n$, is defined to be the union of all open sets of $\mathbf{R}^n$ that are contained in A; it is denoted Int A. The **exterior** of A is defined to be the union of all open sets of $\mathbf{R}^n$ that are disjoint from A; it is denoted Ext A. The **boundary** of A consists of those points of $\mathbf{R}^n$ that belong neither to Int A nor to Ext A; it is denoted Bd A.

A point $\mathbf{x}$ is in Bd A if and only if every open set containing $\mathbf{x}$ intersects both A and the complement $\mathbf{R}^n - A$ of A. The space $\mathbf{R}^n$ is the union of the disjoint sets Int A, Ext A, and Bd A; the first two are open in $\mathbf{R}^n$ and the third is closed in $\mathbf{R}^n$.

For example, suppose Q is the **rectangle**

$$Q = [a_1, b_1] \times \cdots \times [a_n, b_n],$$

consisting of all points $\mathbf{x}$ of $\mathbf{R}^n$ such that $a_i \leq x_i \leq b_i$ for all i. You can check that

$$\text{Int } Q = (a_1, b_1) \times \cdots \times (a_n, b_n).$$

We often call Int Q an **open rectangle.** Furthermore, Ext $Q = \mathbf{R}^n - Q$ and Bd $Q = Q - \text{Int } Q$.

An open cube is a special case of an open rectangle; indeed,

$$C(\mathbf{a};\epsilon) = (a_1 - \epsilon, a_1 + \epsilon) \times \cdots \times (a_n - \epsilon, a_n + \epsilon).$$

The corresponding (closed) rectangle

$$C = [a_1 - \epsilon, a_1 + \epsilon] \times \cdots \times [a_n - \epsilon, a_n + \epsilon]$$

is often called a **closed cube,** or simply a **cube,** centered at **a.**

EXERCISES

Throughout, let X be a metric space with metric d.

1. Show that $U(x_0;\epsilon)$ is an open set.
2. Let $Y \subset X$. Give an example where A is open in Y but not open in X. Give an example where A is closed in Y but not closed in X.
3. Let $A \subset X$. Show that if C is a closed set of X and C contains A, then C contains $\bar{A}$.
4. (a) Show that if Q is a rectangle, then Q equals the closure of Int Q.
   (b) If D is a closed set, what is the relation in general between the set D and the closure of Int D?
   (c) If U is an open set, what is the relation in general between the set U and the interior of $\bar{U}$?
5. Let $f: X \to Y$. Show that f is continuous if and only if for each $x \in X$ there is a neighborhood U of x such that $f \mid U$ is continuous.
6. Let $X = A \cup B$, where A and B are subspaces of X. Let $f: X \to Y$; suppose that the restricted functions

$$f \mid A : A \to Y \quad \text{and} \quad f \mid B : B \to Y$$

   are continuous. Show that if both A and B are closed in X, then f is continuous.
7. Finding the limit of a composite function $g \circ f$ is easy if both f and g are continuous; see Theorem 3.5. Otherwise, it can be a bit tricky:

   Let $f: X \to Y$ and $g: Y \to Z$. Let x_0 be a limit point of X and let y_0 be a limit point of Y. See Figure 3.1. Consider the following three conditions:

   (i) $f(x) \to y_0$ as $x \to x_0$.

   (ii) $g(y) \to z_0$ as $y \to y_0$.

   (iii) $g(f(x)) \to z_0$ as $x \to x_0$.

   (a) Give an example where (i) and (ii) hold, but (iii) does not.

   (b) Show that if (i) and (ii) hold and if $g(y_0) = z_0$, then (iii) holds.

8. Let $f : \mathbf{R} \to \mathbf{R}$ be defined by setting $f(x) = \sin x$ if x is rational, and $f(x) = 0$ otherwise. At what points is f continuous?

9. If we denote the general point of $\mathbf{R}^2$ by (x, y), determine Int A, Ext A, and Bd A for the subset A of $\mathbf{R}^2$ specified by each of the following conditions:
   (a) $x = 0$.
   (b) $0 \le x < 1$.
   (c) $0 \le x < 1$ and $0 \le y < 1$.
   (d) x is rational and $y > 0$.
   (e) x and y are rational.
   (f) $0 < x^2 + y^2 < 1$.
   (g) $y < x^2$.
   (h) $y \le x^2$.

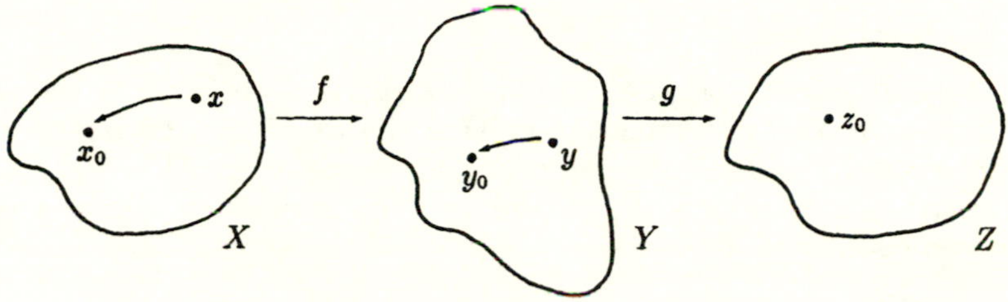

Figure 3.1

§4. COMPACT SUBSPACES AND CONNECTED SUBSPACES OF $\mathbf{R}^n$

An important class of subspaces of $\mathbf{R}^n$ is the class of compact spaces. We shall use the basic properties of such spaces constantly. The properties we shall need are summarized in the theorems of this section. Proofs are included, since some of these results you may not have seen before.

A second useful class of spaces is the class of connected spaces; we summarize here those few properties we shall need.

We do not attempt to deal here with compactness and connectedness in arbitrary metric spaces, but comment that many of our proofs do hold in that more general situation.

Compact spaces

Definition. Let X be a subspace of $\mathbf{R}^n$. A **covering** of X is a collection of subsets of $\mathbf{R}^n$ whose union contains X; if each of the subsets is open in $\mathbf{R}^n$, it is called an **open covering** of X. The space X is said to be **compact** if every open covering of X contains a finite subcollection that also forms an open covering of X.

While this definition of compactness involves open sets of $\mathbf{R}^n$, it can be reformulated in a manner that involves only open sets of the space X:

Theorem 4.1. *A subspace X of $\mathbf{R}^n$ is compact if and only if for every collection of sets open in X whose union is X, there is a finite subcollection whose union equals X.*

Proof. Suppose X is compact. Let $\{A_\alpha\}$ be a collection of sets open in X whose union is X. Choose, for each α, an open set U_α of $\mathbf{R}^n$ such that $A_\alpha = U_\alpha \cap X$. Since X is compact, some finite subcollection of the collection $\{U_\alpha\}$ covers X, say for $\alpha = \alpha_1, \ldots, \alpha_k$. Then the sets A_α, for $\alpha = \alpha_1, \ldots, \alpha_k$, have X as their union.

The proof of the converse is similar. □

The following result is always proved in a first course in analysis, so the proof will be omitted here:

Theorem 4.2. *The subspace $[a, b]$ of $\mathbf{R}$ is compact.* □

Definition. A subspace X of $\mathbf{R}^n$ is said to be **bounded** if there is an M such that $|\mathbf{x}| \le M$ for all $\mathbf{x} \in X$.

We shall eventually show that a subspace of $\mathbf{R}^n$ is compact if and only if it is closed and bounded. Half of that theorem is easy; we prove it now:

Theorem 4.3. *If X is a compact subspace of $\mathbb{R}^n$, then X is closed and bounded.*

Proof. Step 1. We show that X is bounded. For each positive integer N, let U_N denote the open cube $U_N = C(0; N)$. Then U_N is an open set; and $U_1 \subset U_2 \subset \cdots$; and the sets U_N cover all of $\mathbb{R}^n$ (so in particular they cover X). Some finite subcollection also covers X, say for $N = N_1, \ldots, N_k$. If M is the largest of the numbers $N_1, \ldots, N_k$, then X is contained in U_M; thus X is bounded.

Step 2. We show that X is closed by showing that the complement of X is open. Let $\mathbf{a}$ be a point of $\mathbb{R}^n$ not in X; we find an ϵ-neighborhood of $\mathbf{a}$ that lies in the complement of X.

For each positive integer N, consider the cube

$$C_N = \{\mathbf{x}; |\mathbf{x} - \mathbf{a}| \leq 1/N\}.$$

Then $C_1 \supset C_2 \supset \cdots$, and the intersection of the sets C_N consists of the point $\mathbf{a}$ alone. Let V_N be the complement of C_N; then V_N is an open set; and $V_1 \subset V_2 \subset \cdots$; and the sets V_N cover all of $\mathbb{R}^n$ except for the point $\mathbf{a}$ (so they cover X). Some finite subcollection covers X, say for $N = N_1, \ldots, N_k$. If M is the largest of the numbers $N_1, \ldots, N_k$, then X is contained in V_M. Then the set C_M is disjoint from X, so that in particular the open cube $C(\mathbf{a}; 1/M)$ lies in the complement of X. See Figure 4.1. □

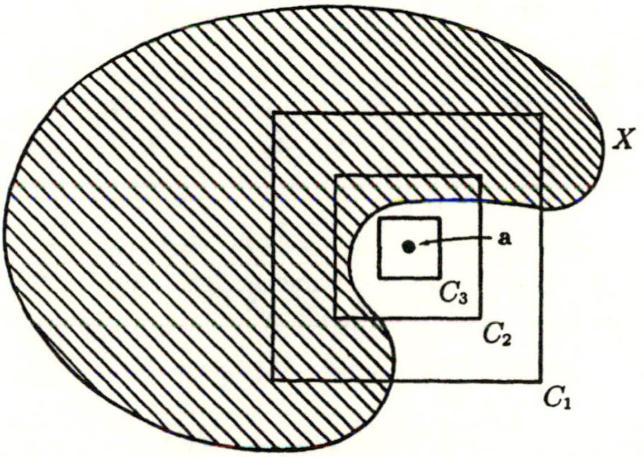

Figure 4.1

Corollary 4.4. *Let X be a compact subspace of $\mathbb{R}$. Then X has a largest element and a smallest element.*

Proof. Since X is bounded, it has a greatest lower bound and a least upper bound. Since X is closed, these elements must belong to X. □

Here is a basic (and familiar) result that is used constantly:

Theorem 4.5 (Extreme-value theorem). *Let X be a compact subspace of $\mathbf{R}^m$. If $f : X \to \mathbf{R}^n$ is continuous, then $f(X)$ is a compact subspace of $\mathbf{R}^n$.*

In particular, if $\phi : X \to \mathbf{R}$ is continuous, then ϕ has a maximum value and a minimum value.

Proof. Let $\{V_\alpha\}$ be a collection of open sets of $\mathbf{R}^n$ that covers $f(X)$. The sets $f^{-1}(V_\alpha)$ form an open covering of X. Hence some finitely many of them cover X, say for $\alpha = \alpha_1, \ldots, \alpha_k$. Then the sets V_α for $\alpha = \alpha_1, \ldots, \alpha_k$ cover $f(X)$. Thus $f(X)$ is compact.

Now if $\phi : X \to \mathbf{R}$ is continuous, $\phi(X)$ is compact, so it has a largest element and a smallest element. These are the maximum and minimum values of ϕ. □

Now we prove a result that may not be so familiar.

Definition. Let X be a subset of $\mathbf{R}^n$. Given $\epsilon > 0$, the union of the sets $B(\mathbf{a};\epsilon)$, as $\mathbf{a}$ ranges over all points of X, is called the **ϵ-neighborhood** of X in the euclidean metric. Similarly, the union of the sets $C(\mathbf{a};\epsilon)$ is called the ϵ-neighborhood of X in the sup metric.

Theorem 4.6 (The ϵ-neighborhood theorem). *Let X be a compact subspace of $\mathbf{R}^n$; let U be an open set of $\mathbf{R}^n$ containing X. Then there is an $\epsilon > 0$ such that the ϵ-neighborhood of X (in either metric) is contained in U.*

Proof. The ϵ-neighborhood of X in the euclidean metric is contained in the ϵ-neighborhood of X in the sup metric. Therefore it suffices to deal only with the latter case.

Step 1. Let C be a fixed subset of $\mathbf{R}^n$. For each $\mathbf{x} \in \mathbf{R}^n$, we define

$$d(\mathbf{x}, C) = \inf\{|\mathbf{x} - \mathbf{c}|; \mathbf{c} \in C\}.$$

We call $d(\mathbf{x}, C)$ the **distance from x to C**. We show it is continuous as a function of $\mathbf{x}$:

Let $\mathbf{c} \in C$; let $\mathbf{x}, \mathbf{y} \in \mathbf{R}^n$. The triangle inequality implies that

$$d(\mathbf{x}, C) - |\mathbf{x} - \mathbf{y}| \leq |\mathbf{x} - \mathbf{c}| - |\mathbf{x} - \mathbf{y}| \leq |\mathbf{y} - \mathbf{c}|.$$

This inequality holds for all $\mathbf{c} \in C$; therefore

$$d(\mathbf{x}, C) - |\mathbf{x} - \mathbf{y}| \leq d(\mathbf{y}, C),$$

so that

$$d(\mathbf{x}, C) - d(\mathbf{y}, C) \leq |\mathbf{x} - \mathbf{y}|.$$

The same inequality holds if $\mathbf{x}$ and $\mathbf{y}$ are interchanged; continuity of $d(\mathbf{x}, C)$ follows.

Step 2. We prove the theorem. Given U, define $f : X \to \mathbf{R}$ by the equation

$$f(\mathbf{x}) = d(\mathbf{x}, \mathbf{R}^n - U).$$

Then f is a continuous function. Furthermore, $f(\mathbf{x}) > 0$ for all $\mathbf{x} \in X$. For if $\mathbf{x} \in X$, then some δ-neighborhood of $\mathbf{x}$ is contained in U, whence $f(\mathbf{x}) \geq \delta$. Because X is compact, f has a minimum value ϵ. Because f takes on only positive values, this minimum value is positive. Then the ϵ-neighborhood of X is contained in U. □

This theorem does not hold without some hypothesis on the set X. If X is the x-axis in $\mathbf{R}^2$, for example, and U is the open set

$$U = \{ (x, y) \mid y^2 < 1/(1 + x^2) \},$$

then there is no ϵ such that the ϵ-neighborhood of X is contained in U. See Figure 4.2.

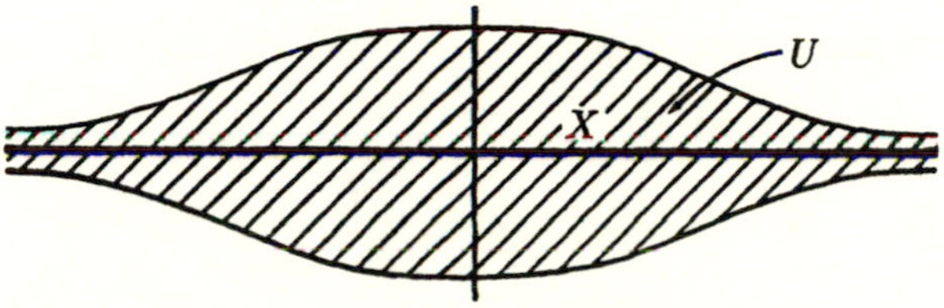

Figure 4.2

Here is another familiar result.

Theorem 4.7 (Uniform continuity). *Let X be a compact subspace of $\mathbf{R}^m$; let $f : X \to \mathbf{R}^n$ be continuous. Given $\epsilon > 0$, there is a $\delta > 0$ such that whenever $\mathbf{x}, \mathbf{y} \in X$,*

$$|\mathbf{x} - \mathbf{y}| < \delta \quad \text{implies} \quad |f(\mathbf{x}) - f(\mathbf{y})| < \epsilon.$$

This result also holds if one uses the euclidean metric instead of the sup metric.

The condition stated in the conclusion of the theorem is called the condition of **uniform continuity.**

Proof. Consider the subspace $X \times X$ of $\mathbf{R}^m \times \mathbf{R}^m$; and within this, consider the space

$$\Delta = \{(\mathbf{x},\mathbf{x}) \mid \mathbf{x} \in X\},$$

which is called the **diagonal** of $X \times X$. The diagonal is a compact subspace of $\mathbf{R}^{2m}$, since it is the image of the compact space X under the continuous map $d(\mathbf{x}) = (\mathbf{x},\mathbf{x})$.

We prove the theorem first for the euclidean metric. Consider the function $g : X \times X \to \mathbf{R}$ defined by the equation

$$g(\mathbf{x},\mathbf{y}) = \|f(\mathbf{x}) - f(\mathbf{y})\|.$$

Then consider the set of points $(\mathbf{x},\mathbf{y})$ of $X \times X$ for which $g(\mathbf{x},\mathbf{y}) < \epsilon$. Because g is continuous, this set is an open set of $X \times X$. Also, it contains the diagonal Δ, since $g(\mathbf{x},\mathbf{x}) = 0$. Therefore, it equals the intersection with $X \times X$ of an open set U of $\mathbf{R}^m \times \mathbf{R}^m$ that contains Δ. See Figure 4.3.

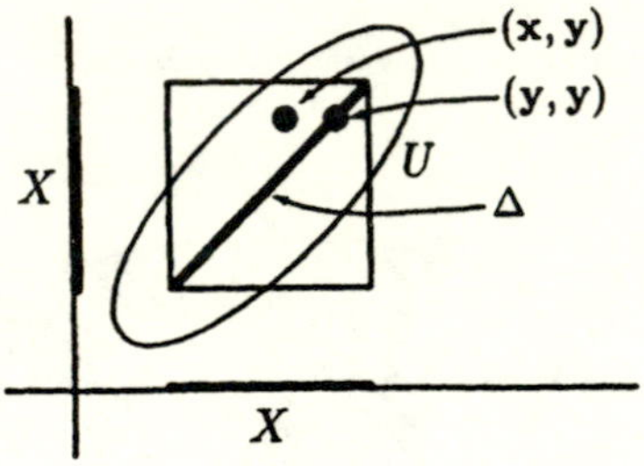

Figure 4.3

Compactness of Δ implies that for some δ, the δ-neighborhood of Δ is contained in U. This is the δ required by our theorem. For if $\mathbf{x},\mathbf{y} \in X$ with $\|\mathbf{x} - \mathbf{y}\| < \delta$, then

$$\|(\mathbf{x},\mathbf{y}) - (\mathbf{y},\mathbf{y})\| = \|(\mathbf{x} - \mathbf{y}, 0)\| = \|\mathbf{x} - \mathbf{y}\| < \delta,$$

so that $(\mathbf{x},\mathbf{y})$ belongs to the δ-neighborhood of the diagonal Δ. Then $(\mathbf{x},\mathbf{y})$ belongs to U, so that $g(\mathbf{x},\mathbf{y}) < \epsilon$, as desired.

The corresponding result for the sup metric can be derived by a similar proof, or simply by noting that if $|\mathbf{x}-\mathbf{y}| < \delta/\sqrt{n}$, then $\|\mathbf{x}-\mathbf{y}\| < \delta$, whence

$$|f(\mathbf{x}) - f(\mathbf{y})| \leq \|f(\mathbf{x}) - f(\mathbf{y})\| < \epsilon. \quad \square$$

To complete our characterization of the compact subspaces of $\mathbf{R}^n$, we need the following lemma:

Lemma 4.8. *The rectangle*

$$Q = [a_1, b_1] \times \cdots \times [a_n, b_n]$$

in $\mathbf{R}^n$ *is compact.*

Proof. We proceed by induction on n. The lemma is true for $n = 1$; we suppose it true for $n - 1$ and prove it true for n. We can write

$$Q = X \times [a_n, b_n],$$

where X is a rectangle in $\mathbf{R}^{n-1}$. Then X is compact by the induction hypothesis. Let $\mathcal{A}$ be an open covering of Q.

Step 1. We show that given $t \in [a_n, b_n]$, there is an $\epsilon > 0$ such that the set

$$X \times (t - \epsilon, t + \epsilon)$$

can be covered by finitely many elements of $\mathcal{A}$.

The set $X \times t$ is a compact subspace of $\mathbf{R}^n$, for it is the image of X under the continuous map $f : X \to \mathbf{R}^n$ given by $f(\mathbf{x}) = (\mathbf{x}, t)$. Therefore it may be covered by finitely many elements of $\mathcal{A}$, say by $A_1, \ldots, A_k$.

Let U be the union of these sets; then U is open and contains $X \times t$. See Figure 4.4.

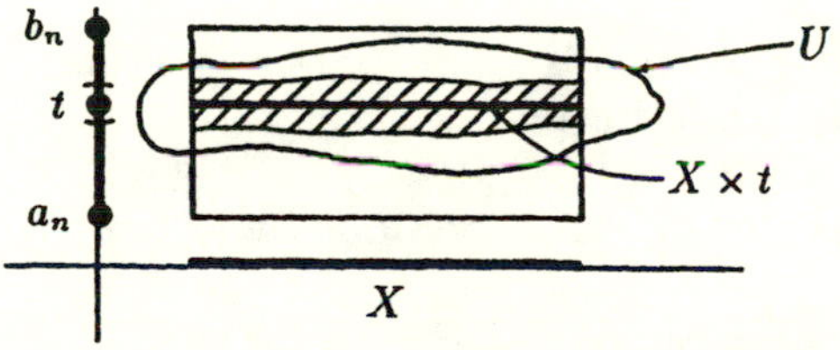

Figure 4.4

Because $X \times t$ is compact, there is an $\epsilon > 0$ such that the ϵ-neighborhood of $X \times t$ is contained in U. Then in particular, the set $X \times (t - \epsilon, t + \epsilon)$ is contained in U, and hence is covered by $A_1, \ldots, A_k$.

Step 2. By the result of Step 1, we may for each $t \in [a_n, b_n]$ choose an open interval V_t about t, such that the set $X \times V_t$ can be covered by finitely many elements of the collection $\mathcal{A}$.

Now the open intervals V_t in $\mathbf{R}$ cover the interval $[a_n, b_n]$; hence finitely many of them cover this interval, say for $t = t_1, \ldots, t_m$.

Then $Q = X \times [a_n, b_n]$ is contained in the union of the sets $X \times V_t$ for $t = t_1, \ldots, t_m$; since each of these sets can be covered by finitely many elements of $\mathcal{A}$, so may Q be covered. □

Theorem 4.9. *If X is a closed and bounded subspace of $\mathbf{R}^n$, then X is compact.*

Proof. Let $\mathcal{A}$ be a collection of open sets that covers X. Let us adjoin to this collection the single set $\mathbf{R}^n - X$, which is open in $\mathbf{R}^n$ because X is closed. Then we have an open covering of all of $\mathbf{R}^n$. Because X is bounded, we can choose a rectangle Q that contains X; our collection then in particular covers Q.

Since Q is compact, some finite subcollection covers Q. If this finite subcollection contains the set $\mathbf{R}^n - X$, we discard it from the collection. We then have a finite subcollection of the collection $\mathcal{A}$; it may not cover all of Q, but it certainly covers X, since the set $\mathbf{R}^n - X$ we discarded contains no point of X. □

All the theorems of this section hold if $\mathbf{R}^n$ and $\mathbf{R}^m$ are replaced by arbitrary metric spaces, *except for the theorem just proved.* That theorem does not hold in an arbitrary metric space; see the exercises.

Connected spaces

If X is a metric space, then X is said to be **connected** if X cannot be written as the union of two disjoint non-empty sets A and B, each of which is open in X.

The following theorem is always proved in a first course in analysis, so the proof will be omitted here:

Theorem 4.10. *The closed interval $[a, b]$ of $\mathbf{R}$ is connected.* □

The basic fact about connected spaces that we shall use is the following:

Theorem 4.11 (Intermediate-value theorem). *Let X be connected. If $f : X \to Y$ is continuous, then $f(X)$ is a connected subspace of Y.*

In particular, if $f : X \to \mathbf{R}$ is continuous and if $f(x_0) < r < f(x_1)$ for some points x_0, x_1 of X, then $f(x) = r$ for some point x of X.

Proof. Suppose $f(X) = A \cup B$, where A and B are disjoint sets open in $f(X)$. Then $f^{-1}(A)$ and $f^{-1}(B)$ are disjoint sets whose union is X, and each is open in X because f is continuous. This contradicts connectedness of X.

Given f, let A consist of all y in $\mathbf{R}$ with $y < r$, and let B consist of all y with $y > r$. Then A and B are open in $\mathbf{R}$; if the set $f(X)$ does not contain r, then $f(X)$ is the union of the disjoint sets $f(X) \cap A$ and $f(X) \cap B$, each of which is open in $f(X)$. This contradicts connectedness of $f(X)$. □

If $\mathbf{a}$ and $\mathbf{b}$ are points of $\mathbf{R}^n$, then the **line segment** joining $\mathbf{a}$ and $\mathbf{b}$ is defined to be the set of all points $\mathbf{x}$ of the form $\mathbf{x} = \mathbf{a} + t(\mathbf{b} - \mathbf{a})$, where $0 \le t \le 1$. Any line segment is connected, for it is the image of the interval $[0,1]$ under the continuous map $t \to \mathbf{a} + t(\mathbf{b} - \mathbf{a})$.

A subset A of $\mathbf{R}^n$ is said to be **convex** if for every pair $\mathbf{a},\mathbf{b}$ of points of A, the line segment joining $\mathbf{a}$ and $\mathbf{b}$ is contained in A. Any convex subset A of $\mathbf{R}^n$ is automatically connected: For if A is the union of the disjoint sets U and V, each of which is open in A, we need merely choose $\mathbf{a}$ in U and $\mathbf{b}$ in V, and note that if L is the line segment joining $\mathbf{a}$ and $\mathbf{b}$, then the sets $U \cap L$ and $V \cap L$ are disjoint, non-empty, and open in L.

It follows that in $\mathbf{R}^n$ all open balls and open cubes and rectangles are connected. (See the exercises.)

EXERCISES

1. Let $\mathbf{R}_+$ denote the set of positive real numbers.
   (a) Show that the continuous function $f : \mathbf{R}_+ \to \mathbf{R}$ given by $f(x) = 1/(1+x)$ is bounded but has neither a maximum value nor a minimum value.
   (b) Show that the continuous function $g : \mathbf{R}_+ \to \mathbf{R}$ given by $g(x) = \sin(1/x)$ is bounded but does not satisfy the condition of uniform continuity on $\mathbf{R}_+$.
2. Let X denote the subset $(-1,1) \times 0$ of $\mathbf{R}^2$, and let U be the open ball $B(0;1)$ in $\mathbf{R}^2$, which contains X. Show there is no $\epsilon > 0$ such that the ϵ-neighborhood of X in $\mathbf{R}^n$ is contained in U.
3. Let $\mathbf{R}^\infty$ be the set of all "infinite-tuples" $\mathbf{x} = (x_1, x_2, \ldots)$ of real numbers that end in an infinite string of 0's. (See the exercises of § 1.) Define an inner product on $\mathbf{R}^\infty$ by the rule $\langle \mathbf{x}, \mathbf{y} \rangle = \Sigma x_i y_i$. (This is a finite sum, since all but finitely many terms vanish.) Let $\|\mathbf{x} - \mathbf{y}\|$ be the corresponding metric on $\mathbf{R}^\infty$. Define

$$\mathbf{e}_i = (0, \ldots, 0, 1, 0, \ldots, 0, \ldots),$$

   where 1 appears in the ith place. Then the $\mathbf{e}_i$ form a basis for $\mathbf{R}^\infty$. Let X be the set of all the points $\mathbf{e}_i$. Show that X is closed, bounded, and non-compact.

4. (a) Show that open balls and open cubes in $\mathbf{R}^n$ are convex.
   (b) Show that (open and closed) rectangles in $\mathbf{R}^n$ are convex.

2

Differentiation

In this chapter, we consider functions mapping $\mathbf{R}^m$ into $\mathbf{R}^n$, and we define what we mean by the derivative of such a function. Much of our discussion will simply generalize facts that are already familiar to you from calculus.

The two major results of this chapter are the *inverse function theorem,* which gives conditions under which a differentiable function from $\mathbf{R}^n$ to $\mathbf{R}^n$ has a differentiable inverse, and the *implicit function theorem,* which provides the theoretical underpinning for the technique of implicit differentiation as studied in calculus.

Recall that we write the elements of $\mathbf{R}^m$ and $\mathbf{R}^n$ as column matrices unless specifically stated otherwise.

§5. THE DERIVATIVE

First, let us recall how the derivative of a real-valued function of a real variable is defined.

Let A be a subset of $\mathbf{R}$; let $\phi : A \to \mathbf{R}$. Suppose A contains a neighborhood of the point a. We define the **derivative** of ϕ at a by the equation

$$\phi'(a) = \lim_{t \to 0} \frac{\phi(a+t) - \phi(a)}{t},$$

provided the limit exists. In this case, we say that ϕ is **differentiable** at a. The following facts are an immediate consequence:

(1) Differentiable functions are continuous.

(2) Composites of differentiable functions are differentiable.

We seek now to define the derivative of a function f mapping a subset of $\mathbf{R}^m$ into $\mathbf{R}^n$. We cannot simply replace a and t in the definition just given by points of $\mathbf{R}^m$, for we cannot divide a point of $\mathbf{R}^n$ by a point of $\mathbf{R}^m$ if $m > 1$! Here is a first attempt at a definition:

Definition. Let $A \subset \mathbf{R}^m$; let $f : A \to \mathbf{R}^n$. Suppose A contains a neighborhood of $\mathbf{a}$. Given $\mathbf{u} \in \mathbf{R}^m$ with $\mathbf{u} \neq \mathbf{0}$, define

$$f'(\mathbf{a};\mathbf{u}) = \lim_{t\to 0} \frac{f(\mathbf{a}+t\mathbf{u}) - f(\mathbf{a})}{t},$$

provided the limit exists. This limit depends both on $\mathbf{a}$ and on $\mathbf{u}$; it is called the **directional derivative** of f at $\mathbf{a}$ with respect to the vector $\mathbf{u}$. (In calculus, one usually requires $\mathbf{u}$ to be a unit vector, but that is not necessary.)

EXAMPLE 1. Let $f : \mathbf{R}^2 \to \mathbf{R}$ be given by the equation

$$f(x_1, x_2) = x_1 x_2.$$

The directional derivative of f at $\mathbf{a} = (a_1, a_2)$ with respect to the vector $\mathbf{u} = (1, 0)$ is

$$f'(\mathbf{a};\mathbf{u}) = \lim_{t\to 0} \frac{(a_1 + t)a_2 - a_1 a_2}{t} = a_2.$$

With respect to the vector $\mathbf{v} = (1, 2)$, the directional derivative is

$$f'(\mathbf{a};\mathbf{v}) = \lim_{t\to 0} \frac{(a_1 + t)(a_2 + 2t) - a_1 a_2}{t} = a_2 + 2a_1.$$

It is tempting to believe that the "directional derivative" is the appropriate generalization of the notion of "derivative," and to say that f is differentiable at $\mathbf{a}$ if $f'(\mathbf{a};\mathbf{u})$ exists for every $\mathbf{u} \neq \mathbf{0}$. This would not, however, be a very useful definition of differentiability. It would not follow, for instance, that differentiability implies continuity. (See Example 3 following.) Nor would it follow that composites of differentiable functions are differentiable. (See the exercises of § 7.) So we seek something stronger.

In order to motivate our eventual definition, let us reformulate the definition of differentiability in the single-variable case as follows:

Let A be a subset of $\mathbf{R}$; let $\phi : A \to \mathbf{R}$. Suppose A contains a neighborhood of a. We say that ϕ is **differentiable** at a if there is a number λ such that

$$\frac{\phi(a+t) - \phi(a) - \lambda t}{t} \to 0 \quad \text{as} \quad t \to 0.$$

The number λ, which is unique, is called the **derivative** of ϕ at a, and denoted $\phi'(a)$.

This formulation of the definition makes explicit the fact that if ϕ is differentiable, then the linear function λt is a good approximation to the "increment function" $\phi(a+t) - \phi(a)$; we often call λt the "first-order approximation" or the "linear approximation" to the increment function.

Let us generalize this version of the definition. If $A \subset \mathbf{R}^m$ and if $f : A \to \mathbf{R}^n$, what might we mean by a "first-order" or "linear" approximation to the increment function $f(\mathbf{a}+\mathbf{h}) - f(\mathbf{a})$? The natural thing to do is to take a function that is linear in the sense of linear algebra. This idea leads to the following definition:

Definition. Let $A \subset \mathbf{R}^m$, let $f : A \to \mathbf{R}^n$. Suppose A contains a neighborhood of $\mathbf{a}$. We say that f is **differentiable** at $\mathbf{a}$ if there is an n by m matrix B such that

$$\frac{f(\mathbf{a}+\mathbf{h}) - f(\mathbf{a}) - B \cdot \mathbf{h}}{|\mathbf{h}|} \to 0 \quad \text{as} \quad \mathbf{h} \to \mathbf{0}.$$

The matrix B, which is unique, is called the **derivative** of f at $\mathbf{a}$; it is denoted $Df(\mathbf{a})$.

Note that the quotient of which we are taking the limit is defined for $\mathbf{h}$ in some deleted neighborhood of $\mathbf{0}$, since the domain of f contains a neighborhood of $\mathbf{a}$. Use of the sup norm in the denominator is not essential; one obtains an equivalent definition if one replaces $|\mathbf{h}|$ by $\|\mathbf{h}\|$.

It is easy to see that B is unique. Suppose C is another matrix satisfying this condition. Subtracting, we have

$$\frac{(C - B) \cdot \mathbf{h}}{|\mathbf{h}|} \to 0$$

as $\mathbf{h} \to \mathbf{0}$. Let $\mathbf{u}$ be a fixed non-zero vector; set $\mathbf{h} = t\mathbf{u}$; let $t \to 0$. It follows that $(C - B) \cdot \mathbf{u} = \mathbf{0}$. Since $\mathbf{u}$ is arbitrary, $C = B$.

EXAMPLE 2. Let $f : \mathbf{R}^m \to \mathbf{R}^n$ be defined by the equation

$$f(\mathbf{x}) = B \cdot \mathbf{x} + \mathbf{b},$$

where B is an n by m matrix, and $\mathbf{b} \in \mathbf{R}^n$. Then f is differentiable and $Df(\mathbf{x}) = B$. Indeed, since

$$f(\mathbf{a}+\mathbf{h}) - f(\mathbf{a}) = B \cdot \mathbf{h},$$

the quotient used in defining the derivative vanishes identically.

We now show that this definition is stronger than the tentative one we gave earlier, and that it is indeed a "suitable" definition of differentiability. Specifically, we verify the following facts, in this section and those following:

(1) Differentiable functions are continuous.

(2) Composites of differentiable functions are differentiable.

(3) Differentiability of f at $\mathbf{a}$ implies the existence of all the directional derivatives of f at $\mathbf{a}$.

We also show how to compute the derivative when it exists.

Theorem 5.1. *Let $A \subset \mathbf{R}^m$; let $f : A \to \mathbf{R}^n$. If f is differentiable at $\mathbf{a}$, then all the directional derivatives of f at $\mathbf{a}$ exist, and*

$$f'(\mathbf{a};\mathbf{u}) = Df(\mathbf{a}) \cdot \mathbf{u}.$$

Proof. Let $B = Df(\mathbf{a})$. Set $\mathbf{h} = t\mathbf{u}$ in the definition of differentiability, where $t \neq 0$. Then by hypothesis,

$$(*) \qquad \frac{f(\mathbf{a}+t\mathbf{u}) - f(\mathbf{a}) - B \cdot t\mathbf{u}}{|t\mathbf{u}|} \to 0$$

as $t \to 0$. If t approaches 0 through positive values, we multiply $(*)$ by $|\mathbf{u}|$ to conclude that

$$\frac{f(\mathbf{a}+t\mathbf{u}) - f(\mathbf{a})}{t} - B \cdot \mathbf{u} \to 0$$

as $t \to 0$, as desired. If t approaches 0 through negative values, we multiply $(*)$ by $-|\mathbf{u}|$ to reach the same conclusion. Thus $f'(\mathbf{a};\mathbf{u}) = B \cdot \mathbf{u}$. □

EXAMPLE 3. Define $f : \mathbf{R}^2 \to \mathbf{R}$ by setting $f(\mathbf{0}) = 0$ and

$$f(x,y) = x^2y/(x^4 + y^2) \quad \text{if} \quad (x,y) \neq \mathbf{0}.$$

We show all directional derivatives of f exist at $\mathbf{0}$, but that f is not differentiable at $\mathbf{0}$. Let $\mathbf{u} \neq \mathbf{0}$. Then

$$\frac{f(\mathbf{0}+t\mathbf{u}) - f(\mathbf{0})}{t} = \frac{(th)^2(tk)}{(th)^4 + (tk)^2}\,\frac{1}{t} \quad \text{if} \quad \mathbf{u} = \begin{bmatrix} h \\ k \end{bmatrix}$$
$$= \frac{h^2k}{t^2h^4 + k^2},$$

so that

$$f'(\mathbf{0};\mathbf{u}) = \begin{cases} h^2/k & \text{if } k \neq 0, \\ 0 & \text{if } k = 0. \end{cases}$$

Thus $f'(\mathbf{0};\mathbf{u})$ exists for all $\mathbf{u} \neq \mathbf{0}$. However, the function f is not differentiable at $\mathbf{0}$. For if $g : \mathbb{R}^2 \to \mathbb{R}$ is a function that is differentiable at $\mathbf{0}$, then $Dg(\mathbf{0})$ is a 1 by 2 matrix of the form $[a\ b]$, and

$$g'(\mathbf{0};\mathbf{u}) = ah + bk,$$

which is a linear function of $\mathbf{u}$. But $f'(\mathbf{0};\mathbf{u})$ is *not* a linear function of $\mathbf{u}$.

The function f is particularly interesting. It is differentiable (and hence continuous) on each straight line through the origin. (In fact, on the straight line $y = mx$, it has the value $mx/(m^2 + x^2)$.) But f is not differentiable at the origin; in fact, f is not even *continuous* at the origin! For f has value 0 at the origin, while arbitrarily near the origin are points of the form (t, t^2), at which f has value 1/2. See Figure 5.1.

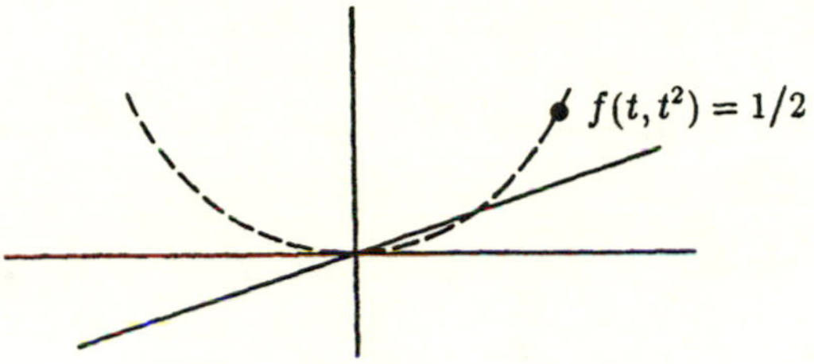

Figure 5.1

Theorem 5.2. *Let $A \subset \mathbb{R}^m$; let $f : A \to \mathbb{R}^n$. If f is differentiable at* $\mathbf{a}$, *then f is continuous at* $\mathbf{a}$.

Proof. Let $B = Df(\mathbf{a})$. For $\mathbf{h}$ near $\mathbf{0}$ but different from $\mathbf{0}$, write

$$f(\mathbf{a}+\mathbf{h}) - f(\mathbf{a}) = |\mathbf{h}| \left[\frac{f(\mathbf{a}+\mathbf{h}) - f(\mathbf{a}) - B \cdot \mathbf{h}}{|\mathbf{h}|} \right] + B \cdot \mathbf{h}.$$

By hypothesis, the expression in brackets approaches $\mathbf{0}$ as $\mathbf{h}$ approaches $\mathbf{0}$. Then, by our basic theorems on limits,

$$\lim_{\mathbf{h}\to\mathbf{0}} [f(\mathbf{a}+\mathbf{h}) - f(\mathbf{a})] = \mathbf{0}.$$

Thus f is continuous at $\mathbf{a}$. □

We shall deal with composites of differentiable functions in § 7.

Now we show how to calculate $Df(\mathbf{a})$, provided it exists. We first introduce the notion of the "partial derivatives" of a real-valued function.

Definition. Let $A \subset \mathbf{R}^m$; let $f : A \to \mathbf{R}$. We define the jth **partial derivative** of f at $\mathbf{a}$ to be the directional derivative of f at $\mathbf{a}$ with respect to the vector $\mathbf{e}_j$, provided this derivative exists; and we denote it by $D_j f(\mathbf{a})$. That is,

$$D_j f(\mathbf{a}) = \lim_{t \to 0} \big(f(\mathbf{a} + t\mathbf{e}_j) - f(\mathbf{a})\big)/t.$$

Partial derivatives are usually easy to calculate. Indeed, if we set

$$\phi(t) = f(a_1, \ldots, a_{j-1}, t, a_{j+1}, \ldots, a_m),$$

then the jth partial derivative of f at $\mathbf{a}$ equals, by definition, simply the ordinary derivative of the function ϕ at the point $t = a_j$. Thus the partial derivative $D_j f$ can be calculated by treating $x_1, \ldots, x_{j-1}, x_{j+1}, \ldots, x_m$ as constants, and differentiating the resulting function with respect to x_j, using the familiar differentiation rules for functions of a single variable.

We begin by calculating the derivative of f in the case where f is a real-valued function.

Theorem 5.3. *Let $A \subset \mathbf{R}^m$; let $f : A \to \mathbf{R}$. If f is differentiable at $\mathbf{a}$, then*

$$Df(\mathbf{a}) = [D_1 f(\mathbf{a}) \quad D_2 f(\mathbf{a}) \quad \cdots \quad D_m f(\mathbf{a})].$$

That is, if $Df(\mathbf{a})$ exists, it is the row matrix whose entries are the partial derivatives of f at $\mathbf{a}$.

Proof. By hypothesis, $Df(\mathbf{a})$ exists and is a matrix of size 1 by m. Let

$$Df(\mathbf{a}) = [\lambda_1 \ \lambda_2 \ \cdots \ \lambda_m].$$

It follows (using Theorem 5.1) that

$$D_j f(\mathbf{a}) = f'(\mathbf{a}; \mathbf{e}_j) = Df(\mathbf{a}) \cdot \mathbf{e}_j = \lambda_j. \quad \square$$

We generalize this theorem as follows:

Theorem 5.4. *Let $A \subset \mathbf{R}^m$; let $f : A \to \mathbf{R}^n$. Suppose A contains a neighborhood of $\mathbf{a}$. Let $f_i : A \to \mathbf{R}$ be the ith component function of f, so that*

$$f(\mathbf{x}) = \begin{bmatrix} f_1(\mathbf{x}) \\ \vdots \\ f_n(\mathbf{x}) \end{bmatrix}.$$

(a) *The function f is differentiable at* $\mathbf{a}$ *if and only if each component function f_i is differentiable at* $\mathbf{a}$.

(b) *If f is differentiable at* $\mathbf{a}$, *then its derivative is the n by m matrix whose i^{th} row is the derivative of the function f_i.*

This theorem tells us that

$$Df(\mathbf{a}) = \begin{bmatrix} Df_1(\mathbf{a}) \\ \vdots \\ Df_n(\mathbf{a}) \end{bmatrix},$$

so that $Df(\mathbf{a})$ is the matrix whose entry in row i and column j is $D_j f_i(\mathbf{a})$.

Proof. Let B be an arbitrary n by m matrix. Consider the function

$$F(\mathbf{h}) = \frac{f(\mathbf{a}+\mathbf{h}) - f(\mathbf{a}) - B \cdot \mathbf{h}}{|\mathbf{h}|},$$

which is defined for $0 < |\mathbf{h}| < \epsilon$ (for some ϵ). Now $F(\mathbf{h})$ is a column matrix of size n by 1. Its ith entry satisfies the equation

$$F_i(\mathbf{h}) = \frac{f_i(\mathbf{a}+\mathbf{h}) - f_i(\mathbf{a}) - (\text{row } i \text{ of } B) \cdot \mathbf{h}}{|\mathbf{h}|}.$$

Let $\mathbf{h}$ approach $\mathbf{0}$. Then the matrix $F(\mathbf{h})$ approaches $\mathbf{0}$ if and only if each of its entries approaches 0. Hence if B is a matrix for which $F(\mathbf{h}) \to \mathbf{0}$, then the ith row of B is a matrix for which $F_i(\mathbf{h}) \to 0$. And conversely. The theorem follows. □

Let $A \subset \mathbf{R}^m$ and $f : A \to \mathbf{R}^n$. If the partial derivatives of the component functions f_i of f exist at $\mathbf{a}$, then one can form the matrix that has $D_j f_i(\mathbf{a})$ as its entry in row i and column j. This matrix is called the **Jacobian matrix** of f. If f is differentiable at $\mathbf{a}$, this matrix equals $Df(\mathbf{a})$. However, it is possible for the partial derivatives, and hence the Jacobian matrix, to exist, *without* it following that f is differentiable at $\mathbf{a}$. (See Example 3 preceding.)

This fact leaves us in something of a quandary. We have no convenient way at present for determining whether or not a function is differentiable (other than going back to the definition). We know that such familiar functions as

$$\sin(xy) \quad \text{and} \quad xy^2 + ze^{xy}$$

have *partial* derivatives, for that fact is a consequence of familiar theorems from single-variable analysis. But we do not know they are differentiable.

We shall deal with this problem in the next section.

REMARK. If $m = 1$ or $n = 1$, our definition of the derivative is simply a reformulation, in matrix notation, of concepts familiar from calculus. For instance, if $f : \mathbb{R}^1 \to \mathbb{R}^3$ is a differentiable function, its derivative is the column matrix

$$Df(t) = \begin{bmatrix} f_1'(t) \\ f_2'(t) \\ f_3'(t) \end{bmatrix}.$$

In calculus, f is often interpreted as a *parametrized-curve*, and the vector

$$\vec{v} = f_1'(t)\mathbf{e}_1 + f_2'(t)\mathbf{e}_2 + f_3'(t)\mathbf{e}_3$$

is called the *velocity vector* of the curve. (Of course, in calculus one is apt to use $\vec{i}, \vec{j}$, and $\vec{k}$ for the unit basis vectors in $\mathbb{R}^3$ rather than $\mathbf{e}_1, \mathbf{e}_2$, and $\mathbf{e}_3$.)

For another example, consider a differentiable function $g : \mathbb{R}^3 \to \mathbb{R}^1$. Its derivative is the row matrix

$$Dg(\mathbf{x}) = [D_1g(\mathbf{x}) \quad D_2g(\mathbf{x}) \quad D_3g(\mathbf{x})],$$

and the directional derivative equals the matrix product $Dg(\mathbf{x})\cdot\mathbf{u}$. In calculus, the function g is often interpreted as a *scalar field*, and the vector field

$$\text{grad } g = (D_1g)\mathbf{e}_1 + (D_2g)\mathbf{e}_2 + (D_3g)\mathbf{e}_3$$

is called the *gradient* of g. (It is often denoted by the symbol $\vec{\nabla}g$.) The directional derivative of g with respect to $\mathbf{u}$ is written in calculus as the dot product of the vectors grad g and $\mathbf{u}$.

Note that vector notation is adequate for dealing with the derivative of f when either the domain or the range of f has dimension 1. For a general function $f : \mathbb{R}^m \to \mathbb{R}^n$, matrix notation is needed.

EXERCISES

1. Let $A \subset \mathbb{R}^m$; let $f : A \to \mathbb{R}^n$. Show that if $f'(\mathbf{a};\mathbf{u})$ exists, then $f'(\mathbf{a};c\mathbf{u})$ exists and equals $cf'(\mathbf{a};\mathbf{u})$.
2. Let $f : \mathbb{R}^2 \to \mathbb{R}$ be defined by setting $f(\mathbf{0}) = 0$ and

$$f(x,y) = xy/(x^2+y^2) \quad \text{if} \quad (x,y) \neq \mathbf{0}.$$

   (a) For which vectors $\mathbf{u} \neq \mathbf{0}$ does $f'(\mathbf{0};\mathbf{u})$ exist? Evaluate it when it exists.
   (b) Do D_1f and D_2f exist at $\mathbf{0}$?
   (c) Is f differentiable at $\mathbf{0}$?
   (d) Is f continuous at $\mathbf{0}$?

3. Repeat Exercise 2 for the function f defined by setting $f(\mathbf{0}) = 0$ and

$$f(x,y) = x^2y^2/(x^2y^2 + (y-x)^2) \quad \text{if} \quad (x,y) \neq \mathbf{0}.$$

4. Repeat Exercise 2 for the function f defined by setting $f(\mathbf{0}) = 0$ and

$$f(x,y) = x^3/(x^2 + y^2) \quad \text{if} \quad (x,y) \neq \mathbf{0}.$$

5. Repeat Exercise 2 for the function

$$f(x,y) = |x| + |y|.$$

6. Repeat Exercise 2 for the function

$$f(x,y) = |xy|^{1/2}.$$

7. Repeat Exercise 2 for the function f defined by setting $f(\mathbf{0}) = 0$ and

$$f(x,y) = x|y|/(x^2 + y^2)^{1/2} \quad \text{if} \quad (x,y) \neq \mathbf{0}.$$

§6. CONTINUOUSLY DIFFERENTIABLE FUNCTIONS

In this section, we obtain a useful criterion for differentiability. We know that mere *existence* of the partial derivatives does not imply differentiability. If, however, we impose the (comparatively mild) additional condition that these partial derivatives be *continuous*, then differentiability is assured.

We begin by recalling the mean-value theorem of single-variable analysis:

Theorem 6.1 (Mean-value theorem). *If $\phi : [a,b] \to \mathbf{R}$ is continuous at each point of the closed interval $[a,b]$, and differentiable at each point of the open interval (a,b), then there exists a point c of (a,b) such that*

$$\phi(b) - \phi(a) = \phi'(c)\,(b-a). \quad \square$$

In practice, we most often apply this theorem when ϕ is differentiable on an open interval containing $[a,b]$. In this case, of course, ϕ is continuous on $[a,b]$.

Theorem 6.2. *Let A be open in $\mathbf{R}^m$. Suppose that the partial derivatives $D_j f_i(\mathbf{x})$ of the component functions of f exist at each point $\mathbf{x}$ of A and are continuous on A. Then f is differentiable at each point of A.*

A function satisfying the hypotheses of this theorem is often said to be **continuously differentiable, or of class C^1,** on A.

Proof. In view of Theorem 5.4, it suffices to prove that each component function of f is differentiable. Therefore we may restrict ourselves to the case of a real-valued function $f : A \to \mathbf{R}$.

Let $\mathbf{a}$ be a point of A. We are given that, for some ϵ, the partial derivatives $D_j f(\mathbf{x})$ exist and are continuous for $|\mathbf{x} - \mathbf{a}| < \epsilon$. We wish to show that f is differentiable at $\mathbf{a}$.

Step 1. Let $\mathbf{h}$ be a point of $\mathbf{R}^m$ with $0 < |\mathbf{h}| < \epsilon$; let $h_1, \ldots, h_m$ be the components of $\mathbf{h}$. Consider the following sequence of points of $\mathbf{R}^m$:

$$\begin{aligned}
\mathbf{p}_0 &= \mathbf{a}, \\
\mathbf{p}_1 &= \mathbf{a} + h_1\mathbf{e}_1, \\
\mathbf{p}_2 &= \mathbf{a} + h_1\mathbf{e}_1 + h_2\mathbf{e}_2, \\
&\ldots \\
\mathbf{p}_m &= \mathbf{a} + h_1\mathbf{e}_1 + \cdots + h_m\mathbf{e}_m = \mathbf{a} + \mathbf{h}.
\end{aligned}$$

The points $\mathbf{p}_i$ all belong to the (closed) cube C of radius $|\mathbf{h}|$ centered at $\mathbf{a}$. Figure 6.1 illustrates the case where $m = 3$ and all h_i are positive.

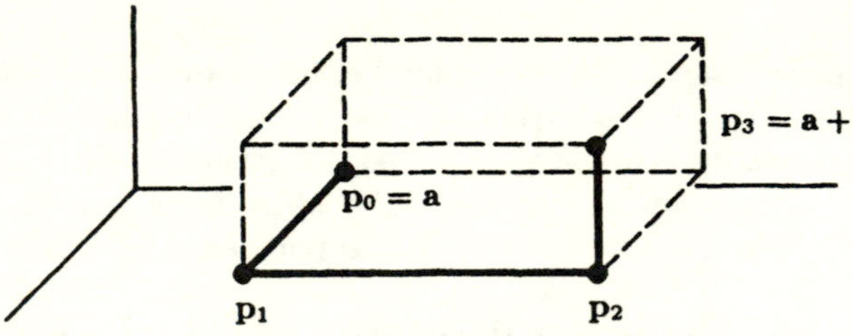

Figure 6.1

Since we are concerned with the differentiability of f, we shall need to deal with the difference $f(\mathbf{a} + \mathbf{h}) - f(\mathbf{a})$. We begin by writing it in the form

$$(*) \qquad f(\mathbf{a} + \mathbf{h}) - f(\mathbf{a}) = \sum_{j=1}^{m} [f(\mathbf{p}_j) - f(\mathbf{p}_{j-1})].$$

Consider the general term of this summation. Let j be fixed, and define

$$\phi(t) = f(\mathbf{p}_{j-1} + t\mathbf{e}_j).$$

Assume $h_j \neq 0$ for the moment. As t ranges over the closed interval I with end points 0 and h_j, the point $\mathbf{p}_{j-1} + t\mathbf{e}_j$ ranges over the line segment from $\mathbf{p}_{j-1}$ to $\mathbf{p}_j$; this line segment lies in C, and hence in A. Thus ϕ is defined for t in an open interval about I.

As t varies, only the jth component of the point $\mathbf{p}_{j-1} + t\mathbf{e}_j$ varies. Hence because $D_j f$ exists at each point of A, the function ϕ is differentiable on an open interval containing I. Applying the mean-value theorem to ϕ, we conclude that

$$\phi(h_j) - \phi(0) = \phi'(c_j)h_j$$

for some point c_j between 0 and h_j. (This argument applies whether h_j is positive or negative.) We can rewrite this equation in the form

$$(**) \qquad f(\mathbf{p}_j) - f(\mathbf{p}_{j-1}) = D_j f(\mathbf{q}_j)h_j,$$

where $\mathbf{q}_j$ is the point $\mathbf{p}_{j-1} + c_j\mathbf{e}_j$ of the line segment from $\mathbf{p}_{j-1}$ to $\mathbf{p}_j$, which lies in C.

We derived (**) under the assumption that $h_j \neq 0$. If $h_j = 0$, then (**) holds automatically, for any point $\mathbf{q}_j$ of C.

Using (**), we rewrite (*) in the form

$$(***) \qquad f(\mathbf{a}+\mathbf{h}) - f(\mathbf{a}) = \sum_{j=1}^{m} D_j f(\mathbf{q}_j)h_j,$$

where each point $\mathbf{q}_j$ lies in the cube C of radius $|\mathbf{h}|$ centered at $\mathbf{a}$.

Step 2. We prove the theorem. Let B be the matrix

$$B = [D_1 f(\mathbf{a}) \ \cdots \ D_m f(\mathbf{a})].$$

Then

$$B \cdot \mathbf{h} = \sum_{j=1}^{m} D_j f(\mathbf{a})h_j.$$

Using (***), we have

$$\frac{f(\mathbf{a}+\mathbf{h}) - f(\mathbf{a}) - B \cdot \mathbf{h}}{|\mathbf{h}|} = \sum_{j=1}^{m} \frac{[D_j f(\mathbf{q}_j) - D_j f(\mathbf{a})]h_j}{|\mathbf{h}|};$$

then we let $\mathbf{h} \to \mathbf{0}$. Since $\mathbf{q}_j$ lies in the cube C of radius $|\mathbf{h}|$ centered at $\mathbf{a}$, we have $\mathbf{q}_j \to \mathbf{a}$. Since the partials of f are continuous at $\mathbf{a}$, the factors in brackets all go to zero. The factors $h_j/|\mathbf{h}|$ are of course bounded in absolute value by 1. Hence the entire expression goes to zero, as desired. □

One effect of this theorem is to reassure us that the functions familiar to us from calculus are in fact differentiable. We know how to compute the partial derivatives of such functions as $\sin(xy)$ and $xy^2 + ze^{xy}$, and we know that these partials are continuous. Therefore these functions are differentiable.

In practice, we usually deal only with functions that are of class C^1. While it is interesting to know there are functions that are differentiable but not of class C^1, such functions occur rarely enough that we need not be concerned with them.

Suppose f is a function mapping an open set A of $\mathbf{R}^m$ into $\mathbf{R}^n$, and suppose the partial derivatives $D_j f_i$ of the component functions of f exist on A. These then are functions from A to $\mathbf{R}$, and we may consider *their* partial derivatives, which have the form $D_k(D_j f_i)$ and are called the **second-order partial derivatives** of f. Similarly, one defines the third-order partial derivatives of the functions f_i, or more generally the partial derivatives of order r for arbitrary r.

If the partial derivatives of the functions f_i of order less than or equal to r are continuous on A, we say f **is of class C^r** on A. Then the function f is of class C^r on A if and only if each function $D_j f_i$ is of class C^{r-1} on A. We say f **is of class C^∞** on A if the partials of the functions f_i of *all* orders are continuous on A.

As you may recall, for most functions the "mixed" partial derivatives

$$D_k D_j f_i \quad \text{and} \quad D_j D_k f_i$$

are equal. This result in fact holds under the hypothesis that the function f is of class C^2, as we now show.

Theorem 6.3. *Let A be open in $\mathbf{R}^m$; let $f : A \to \mathbf{R}$ be a function of class C^2. Then for each $\mathbf{a} \in A$,*

$$D_k D_j f(\mathbf{a}) = D_j D_k f(\mathbf{a}).$$

Proof. Since one calculates the partial derivatives in question by letting all variables other than x_k and x_j remain constant, it suffices to consider the case where f is a function merely of two variables. So we assume that A is open in $\mathbf{R}^2$, and that $f : A \to \mathbf{R}$ is of class C^2.

Step 1. We first prove a certain "second-order" mean-value theorem for f. Let

$$Q = [a, a+h] \times [b, b+k]$$

be a rectangle contained in A. Define

$$\lambda(h,k) = f(a,b) - f(a+h,b) - f(a,b+k) + f(a+h,b+k).$$

Then λ is the sum, with appropriate signs, of the values of f at the four vertices of Q. See Figure 6.2. We show that there are points $\mathbf{p}$ and $\mathbf{q}$ of Q such that

$$\lambda(h,k) = D_2D_1f(\mathbf{p}) \cdot hk, \quad \text{and}$$
$$\lambda(h,k) = D_1D_2f(\mathbf{q}) \cdot hk.$$

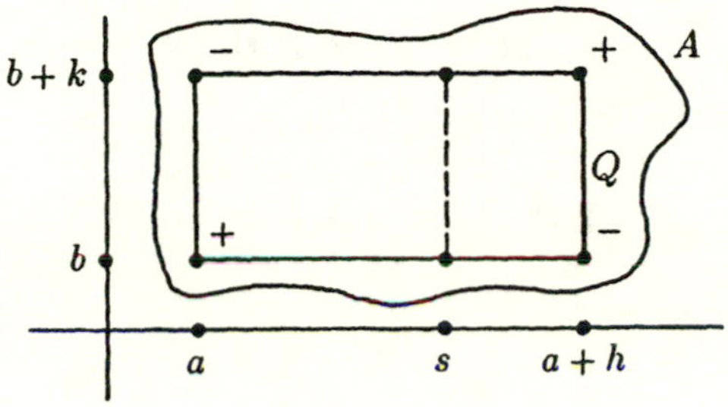

Figure 6.2

By symmetry, it suffices to prove the first of these equations. To begin, we define

$$\phi(s) = f(s,b+k) - f(s,b).$$

Then $\phi(a+h) - \phi(a) = \lambda(h,k)$, as you can check. Because D_1f exists in A, the function ϕ is differentiable in an open interval containing $[a,a+h]$. The mean-value theorem implies that

$$\phi(a+h) - \phi(a) = \phi'(s_0) \cdot h$$

for some s_0 between a and $a+h$. This equation can be rewritten in the form

$$(*) \qquad \lambda(h,k) = [D_1f(s_0,b+k) - D_1f(s_0,b)] \cdot h.$$

Now s_0 is fixed, and we consider the function $D_1f(s_0,t)$. Because D_2D_1f exists in A, this function is differentiable for t in an open interval about $[b,b+k]$. We apply the mean-value theorem once more to conclude that

$$(**) \qquad D_1f(s_0,b+k) - D_1f(s_0,b) = D_2D_1f(s_0,t_0) \cdot k$$

for some t_0 between b and $b+k$. Combining (*) and (**) gives our desired result.

Step 2. We prove the theorem. Given the point $\mathbf{a} = (a,b)$ of A and given $t > 0$, let Q_t be the rectangle

$$Q_t = [a, a+t] \times [b, b+t].$$

If t is sufficiently small, Q_t is contained in A; then Step 1 implies that

$$\lambda(t,t) = D_2D_1f(\mathbf{p}_t) \cdot t^2$$

for some point $\mathbf{p}_t$ in Q_t. If we let $t \to 0$, then $\mathbf{p}_t \to \mathbf{a}$. *Because* D_2D_1f is continuous, it follows that

$$\lambda(t,t)/t^2 \to D_2D_1f(\mathbf{a}) \quad \text{as} \quad t \to 0.$$

A similar argument, using the other equation from Step 1, implies that

$$\lambda(t,t)/t^2 \to D_1D_2f(\mathbf{a}) \quad \text{as} \quad t \to 0.$$

The theorem follows. □

EXERCISES

1. Show that the function $f(x,y) = |xy|$ is differentiable at $\mathbf{0}$, but is not of class C^1 in any neighborhood of $\mathbf{0}$.
2. Define $f : \mathbf{R} \to \mathbf{R}$ by setting $f(0) = 0$, and

$$f(t) = t^2 \sin(1/t) \quad \text{if} \quad t \neq 0.$$

   (a) Show f is differentiable at 0, and calculate $f'(0)$.
   (b) Calculate $f'(t)$ if $t \neq 0$.
   (c) Show f' is not continuous at 0.
   (d) Conclude that f is differentiable on $\mathbf{R}$ but not of class C^1 on $\mathbf{R}$.
3. Show that the proof of Theorem 6.2 goes through if we assume merely that the partials D_jf exist in a neighborhood of $\mathbf{a}$ and are continuous at $\mathbf{a}$.
4. Show that if $A \subset \mathbf{R}^m$ and $f : A \to \mathbf{R}$, and if the partials D_jf exist and are bounded in a neighborhood of $\mathbf{a}$, then f is continuous at $\mathbf{a}$.
5. Let $f : \mathbf{R}^2 \to \mathbf{R}^2$ be defined by the equation

$$f(r,\theta) = (r\cos\theta, r\sin\theta).$$

It is called the polar coordinate transformation.

(a) Calculate Df and $\det Df$.

(b) Sketch the image under f of the set $S = [1,2] \times [0,\pi]$. [*Hint:* Find the images under f of the line segments that bound S.]

6. Repeat Exercise 5 for the function $f : \mathbf{R}^2 \to \mathbf{R}^2$ given by

$$f(x,y) = (x^2 - y^2, 2xy).$$

Take S to be the set

$$S = \{(x,y) \mid x^2 + y^2 \leq a^2 \quad \text{and} \quad x \geq 0 \quad \text{and} \quad y \geq 0\}.$$

[*Hint:* Parametrize part of the boundary of S by setting $x = a\cos t$ and $y = a \sin t$; find the image of this curve. Proceed similarly for the rest of the boundary of S.]

We remark that if one identifies the complex numbers $\mathbf{C}$ with $\mathbf{R}^2$ in the usual way, then f is just the function $f(z) = z^2$.

7. Repeat Exercise 5 for the function $f : \mathbf{R}^2 \to \mathbf{R}^2$ given by

$$f(x,y) = (e^x \cos y,\ e^x \sin y).$$

Take S to be the set $S = [0,1] \times [0,\pi]$.

We remark that if one identifies $\mathbf{C}$ with $\mathbf{R}^2$ as usual, then f is the function $f(z) = e^z$.

8. Repeat Exercise 5 for the function $f : \mathbf{R}^3 \to \mathbf{R}^3$ given by

$$f(\rho,\phi,\theta) = (\rho\cos\theta\sin\phi,\ \rho\sin\theta\sin\phi,\ \rho\cos\phi).$$

It is called the **spherical coordinate transformation.** Take S to be the set

$$S = [1,2] \times [0,\pi/2] \times [0,\pi/2].$$

9. Let $g : \mathbf{R} \to \mathbf{R}$ be a function of class C^2. Show that

$$\lim_{h \to 0} \frac{g(a+h) - 2g(a) + g(a-h)}{h^2} = g''(a).$$

[*Hint:* Consider Step 1 of Theorem 6.3 in the case $f(x,y) = g(x+y)$.]

*10. Define $f : \mathbf{R}^2 \to \mathbf{R}$ by setting $f(\mathbf{0}) = 0$, and

$$f(x,y) = xy(x^2 - y^2)/(x^2 + y^2) \quad \text{if} \quad (x,y) \neq \mathbf{0}.$$

(a) Show $D_1 f$ and $D_2 f$ exist at $\mathbf{0}$.

(b) Calculate $D_1 f$ and $D_2 f$ at $(x,y) \neq \mathbf{0}$.

(c) Show f is of class C^1 on $\mathbf{R}^2$. [*Hint:* Show $D_1 f(x,y)$ equals the product of y and a bounded function, and $D_2 f(x,y)$ equals the product of x and a bounded function.]

(d) Show that $D_2 D_1 f$ and $D_1 D_2 f$ exist at $\mathbf{0}$, but are not equal there.

§7. THE CHAIN RULE

In this section we show that the composite of two differentiable functions is differentiable, and we derive a formula for its derivative. This formula is commonly called the "chain rule."

Theorem 7.1. *Let $A \subset \mathbb{R}^m$; let $B \subset \mathbb{R}^n$. Let*

$$f : A \rightarrow \mathbb{R}^n \quad \textit{and} \quad g : B \rightarrow \mathbb{R}^p,$$

with $f(A) \subset B$. Suppose $f(\mathbf{a}) = \mathbf{b}$. If f is differentiable at $\mathbf{a}$, and if g is differentiable at $\mathbf{b}$, then the composite function $g \circ f$ is differentiable at $\mathbf{a}$. Furthermore,

$$D(g \circ f)(\mathbf{a}) = Dg(\mathbf{b}) \cdot Df(\mathbf{a}),$$

where the indicated product is matrix multiplication.

Although this version of the chain rule may look a bit strange, it is really just the familiar chain rule of calculus in a new guise. You can convince yourself of this fact by writing the formula out in terms of partial derivatives. We shall return to this matter later.

Proof. For convenience, let $\mathbf{x}$ denote the general point of $\mathbb{R}^m$, and let $\mathbf{y}$ denote the general point of $\mathbb{R}^n$.

By hypothesis, g is defined in a neighborhood of $\mathbf{b}$; choose ϵ so that $g(\mathbf{y})$ is defined for $|\mathbf{y} - \mathbf{b}| < \epsilon$. Similarly, since f is defined in a neighborhood of $\mathbf{a}$ and is continuous at $\mathbf{a}$, we can choose δ so that $f(\mathbf{x})$ is defined and satisfies the condition $|f(\mathbf{x}) - \mathbf{b}| < \epsilon$, for $|\mathbf{x} - \mathbf{a}| < \delta$. Then the composite function $(g \circ f)(\mathbf{x}) = g(f(\mathbf{x}))$ is defined for $|\mathbf{x} - \mathbf{a}| < \delta$. See Figure 7.1.

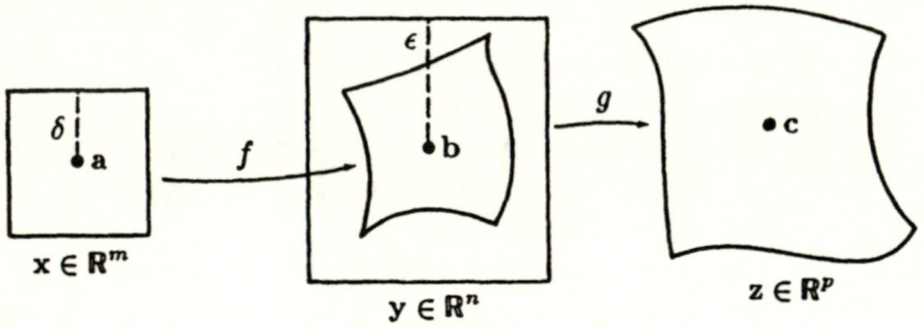

Figure 7.1

Step 1. Throughout, let $\Delta(\mathbf{h})$ denote the function

$$\Delta(\mathbf{h}) = f(\mathbf{a}+\mathbf{h}) - f(\mathbf{a}),$$

which is defined for $|\mathbf{h}| < \delta$. First, we show that the quotient $|\Delta(\mathbf{h})|/|\mathbf{h}|$ is bounded for $\mathbf{h}$ in some deleted neighborhood of $\mathbf{0}$.

For this purpose, let us introduce the function $F(\mathbf{h})$ defined by setting $F(\mathbf{0}) = \mathbf{0}$ and

$$F(\mathbf{h}) = \frac{[\Delta(\mathbf{h}) - Df(\mathbf{a})\cdot\mathbf{h}]}{|\mathbf{h}|} \quad \text{for} \quad 0 < |\mathbf{h}| < \delta.$$

Because f is differentiable at $\mathbf{a}$, the function F is continuous at $\mathbf{0}$. Furthermore, one has the equation

$$(*) \qquad \Delta(\mathbf{h}) = Df(\mathbf{a})\cdot\mathbf{h} + |\mathbf{h}|F(\mathbf{h})$$

for $0 < |\mathbf{h}| < \delta$, and also for $\mathbf{h} = \mathbf{0}$ (trivially). The triangle inequality implies that

$$|\Delta(\mathbf{h})| \leq m|Df(\mathbf{a})|\,|\mathbf{h}| \;+\; |\mathbf{h}|\,|F(\mathbf{h})|.$$

Now $|F(\mathbf{h})|$ is bounded for $\mathbf{h}$ in a neighborhood of $\mathbf{0}$; in fact, it approaches $\mathbf{0}$ as $\mathbf{h}$ approaches $\mathbf{0}$. Therefore $|\Delta(\mathbf{h})|\,/\,|\mathbf{h}|$ is bounded on a deleted neighborhood of $\mathbf{0}$.

Step 2. We repeat the construction of Step 1 for the function g. We define a function $G(\mathbf{k})$ by setting $G(\mathbf{0}) = \mathbf{0}$ and

$$G(\mathbf{k}) = \frac{g(\mathbf{b}+\mathbf{k}) - g(\mathbf{b}) - Dg(\mathbf{b})\cdot\mathbf{k}}{|\mathbf{k}|} \quad \text{for} \quad 0 < |\mathbf{k}| < \epsilon.$$

Because g is differentiable at $\mathbf{b}$, the function G is continuous at $\mathbf{0}$. Furthermore, for $|\mathbf{k}| < \epsilon$, G satisfies the equation

$$(**) \qquad g(\mathbf{b}+\mathbf{k}) - g(\mathbf{b}) = Dg(\mathbf{b})\cdot\mathbf{k} + |\mathbf{k}|G(\mathbf{k}).$$

Step 3. We prove the theorem. Let $\mathbf{h}$ be any point of $\mathbf{R}^m$ with $|\mathbf{h}| < \delta$. Then $|\Delta(\mathbf{h})| < \epsilon$, so we may substitute $\Delta(\mathbf{h})$ for $\mathbf{k}$ in formula $(**)$. After this substitution, $\mathbf{b}+\mathbf{k}$ becomes

$$\mathbf{b} + \Delta(\mathbf{h}) = f(\mathbf{a}) + \Delta(\mathbf{h}) = f(\mathbf{a}+\mathbf{h}),$$

so formula $(**)$ takes the form

$$g(f(\mathbf{a}+\mathbf{h})) - g(f(\mathbf{a})) = Dg(\mathbf{b})\cdot\Delta(\mathbf{h}) + |\Delta(\mathbf{h})|G(\Delta(\mathbf{h})).$$

Now we use (∗) to rewrite this equation in the form

$$\frac{1}{|\mathbf{h}|}[g(f(\mathbf{a}+\mathbf{h}))-g(f(\mathbf{a}))-Dg(\mathbf{b})\cdot Df(\mathbf{a})\cdot\mathbf{h}]$$
$$=Dg(\mathbf{b})\cdot F(\mathbf{h})+\frac{1}{|\mathbf{h}|}|\Delta(\mathbf{h})|G(\Delta(\mathbf{h})).$$

This equation holds for $0<|\mathbf{h}|<\delta$. In order to show that $g\circ f$ is differentiable at $\mathbf{a}$ with derivative $Dg(\mathbf{b})\cdot Df(\mathbf{a})$, it suffices to show that the right side of this equation goes to zero as $\mathbf{h}$ approaches $\mathbf{0}$.

The matrix $Dg(\mathbf{b})$ is constant, while $F(\mathbf{h})\to\mathbf{0}$ as $\mathbf{h}\to\mathbf{0}$ (because F is continuous at $\mathbf{0}$ and vanishes there). The factor $G(\Delta(\mathbf{h}))$ also approaches zero as $\mathbf{h}\to\mathbf{0}$; for it is the composite of two functions G and Δ, both of which are continuous at $\mathbf{0}$ and vanish there. Finally, $|\Delta(\mathbf{h})|/|\mathbf{h}|$ is bounded in a deleted neighborhood of $\mathbf{0}$, by Step 1. The theorem follows. □

Here is an immediate consequence:

Corollary 7.2. *Let A be open in $\mathbb{R}^m$; let B be open in $\mathbb{R}^n$. Let*

$$f:A\to\mathbb{R}^n \quad \text{and} \quad g:B\to\mathbb{R}^p,$$

with $f(A)\subset B$. If f and g are of class C^r, so is the composite function $g\circ f$.

Proof. The chain rule gives us the formula

$$D(g\circ f)(\mathbf{x})=Dg(f(\mathbf{x}))\cdot Df(\mathbf{x}),$$

which holds for $\mathbf{x}\in A$.

Suppose first that f and g are of class C^1. Then the entries of Dg are continuous real-valued functions defined on B; because f is continuous on A, the composite function $Dg(f(\mathbf{x}))$ is also continuous on A. Similarly, the entries of the matrix $Df(\mathbf{x})$ are continuous on A. Because the entries of the matrix product are algebraic functions of the entries of the matrices involved, the entries of the product $Dg(f(\mathbf{x}))\cdot Df(\mathbf{x})$ are also continuous on A. Then $g\circ f$ is of class C^1 on A.

To prove the general case, we proceed by induction. Suppose the theorem is true for functions of class C^{r-1}. Let f and g be of class C^r. Then the entries of Dg are real-valued functions of class C^{r-1} on B. Now f is of class C^{r-1} on A (being in fact of class C^r); hence the induction hypothesis implies that the function $D_jg_i(f(\mathbf{x}))$, which is a composite of two functions of class C^{r-1}, is of class C^{r-1}. Since the entries of the matrix $Df(\mathbf{x})$ are also of class C^{r-1} on A by hypothesis, the entries of the product $Dg(f(\mathbf{x}))\cdot Df(\mathbf{x})$ are of class C^{r-1} on A. Hence $g\circ f$ is of class C^r on A, as desired.

The theorem follows for r finite. If now f and g are of class C^∞, then they are of class C^r for every r, whence $g \circ f$ is also of class C^r for every r. □

As another application of the chain rule, we generalize the mean-value theorem of single-variable analysis to real-valued functions defined in $\mathbf{R}^m$. We will use this theorem in the next section.

Theorem 7.3 (Mean-value theorem). *Let A be open in $\mathbf{R}^m$; let $f : A \to \mathbf{R}$ be differentiable on A. If A contains the line segment with end points $\mathbf{a}$ and $\mathbf{a}+\mathbf{h}$, then there is a point $\mathbf{c} = \mathbf{a} + t_0\mathbf{h}$ with $0 < t_0 < 1$ of this line segment such that*

$$f(\mathbf{a}+\mathbf{h}) - f(\mathbf{a}) = Df(\mathbf{c}) \cdot \mathbf{h}.$$

Proof. Set $\phi(t) = f(\mathbf{a} + t\mathbf{h})$; then ϕ is defined for t in an open interval about $[0,1]$. Being the composite of differentiable functions, ϕ is differentiable; its derivative is given by the formula

$$\phi'(t) = Df(\mathbf{a} + t\mathbf{h}) \cdot \mathbf{h}.$$

The ordinary mean-value theorem implies that

$$\phi(1) - \phi(0) = \phi'(t_0) \cdot 1$$

for some t_0 with $0 < t_0 < 1$. This equation can be rewritten in the form

$$f(\mathbf{a}+\mathbf{h}) - f(\mathbf{a}) = Df(\mathbf{a} + t_0\mathbf{h}) \cdot \mathbf{h}. \quad \square$$

As yet another application of the chain rule, we consider the problem of differentiating an inverse function.

Recall the situation that occurs in single-variable analysis. Suppose $\phi(x)$ is differentiable on an open interval, with $\phi'(x) > 0$ on that interval. Then ϕ is strictly increasing and has an inverse function ψ, which is defined by letting $\psi(y)$ be that unique number x such that $\phi(x) = y$. The function ψ is in fact differentiable, and its derivative satisfies the equation

$$\psi'(y) = 1/\phi'(x),$$

where $y = \phi(x)$.

There is a similar formula for differentiating the inverse of a function f of several variables. In the present section, we do not consider the question whether the function f *has* an inverse, or whether that inverse is differentiable. We consider only the problem of finding the derivative of the inverse function.

Theorem 7.4. *Let A be open in $\mathbb{R}^n$; let $f : A \to \mathbb{R}^n$; let $f(\mathbf{a}) = \mathbf{b}$. Suppose that g maps a neighborhood of $\mathbf{b}$ into $\mathbb{R}^n$, that $g(\mathbf{b}) = \mathbf{a}$, and*

$$g(f(\mathbf{x})) = \mathbf{x}$$

for all $\mathbf{x}$ in a neighborhood of $\mathbf{a}$. If f is differentiable at $\mathbf{a}$ and if g is differentiable at $\mathbf{b}$, then

$$Dg(\mathbf{b}) = [Df(\mathbf{a})]^{-1}.$$

Proof. Let $i : \mathbb{R}^n \to \mathbb{R}^n$ be the identity function; its derivative is the identity matrix I_n. We are given that

$$g(f(\mathbf{x})) = i(\mathbf{x})$$

for all $\mathbf{x}$ in a neighborhood of $\mathbf{a}$. The chain rule implies that

$$Dg(\mathbf{b}) \cdot Df(\mathbf{a}) = I_n.$$

Thus $Dg(\mathbf{b})$ is the inverse matrix to $Df(\mathbf{a})$ (see Theorem 2.5). □

The preceding theorem implies that if a differentiable function f is to have a differentiable inverse, it is *necessary* that the matrix Df be non-singular. It is a somewhat surprising fact that this condition is also *sufficient* for a function f of class C^1 to have an inverse, at least locally. We shall prove this fact in the next section.

REMARK. Let us make a comment on notation. The usefulness of well-chosen notation can hardly be overemphasized. Arguments that are obscure, and formulas that are complicated, sometimes become beautifully simple once the proper notation is chosen. Our use of matrix notation for the derivative is a case in point. The formulas for the derivatives of a composite function and an inverse function could hardly be simpler.

Nevertheless, a word may be in order for those who remember the notation used in calculus for partial derivatives, and the version of the chain rule proved there.

In advanced mathematics, it is usual to use either the functional notation ϕ' or the operator notation $D\phi$ for the derivative of a real-valued function of a real variable. ($D\phi$ denotes a 1 by 1 matrix in this case!) In calculus, however, another notation is common. One often denotes the derivative $\phi'(x)$ by the symbol $d\phi/dx$, or, introducing the "variable" y by setting $y = \phi(x)$, by the symbol dy/dx. This notation was introduced by Leibnitz, one of the originators of calculus. It comes from the time when the focus of every physical and mathematical problem was on the *variables* involved, and when *functions* as such were hardly even thought about.

The Leibnitz notation has some familiar virtues. For one thing, it makes the chain rule easy to remember. Given functions $\phi : \mathbf{R} \to \mathbf{R}$ and $\psi : \mathbf{R} \to \mathbf{R}$, the derivative of the composite function $\psi \circ \phi$ is given by the formula

$$D(\psi \circ \phi)(x) = D\psi\big(\phi(x)\big) \cdot D\phi(x).$$

If we introduce variables by setting $y = \phi(x)$ and $z = \psi(y)$, then the derivative of the composite function $z = \psi\big(\phi(x)\big)$ can be expressed in the Leibnitz notation by the formula

$$\frac{dz}{dx} = \frac{dz}{dy} \cdot \frac{dy}{dx}.$$

The latter formula is easy to remember because it looks like the formula for multiplying fractions! However, this notation has its ambiguities. The letter "z," when it appears on the left side of this equation, denotes one function (a function of x); and when it appears on the right side, it denotes a different function (a function of y). This can lead to difficulties when it comes to computing higher derivatives unless one is very careful.

The formula for the derivative of an inverse function is also easy to remember. If $y = \phi(x)$ has the inverse function $x = \psi(y)$, then the derivative of ψ is expressed in Leibnitz notation by the equation

$$dx/dy = \frac{1}{dy/dx},$$

which looks like the formula for the reciprocal of a fraction!

The Leibnitz notation can easily be extended to functions of several variables. If $A \subset \mathbf{R}^m$ and $f : A \to \mathbf{R}$, we often set

$$y = f(\mathbf{x}) = f(x_1, \dots, x_m),$$

and denote the partial derivative $D_i f$ by one of the symbols

$$\frac{\partial f}{\partial x_i} \quad \text{or} \quad \frac{\partial y}{\partial x_i}.$$

The Leibnitz notation is not nearly as convenient in this situation. Consider the chain rule, for example. If

$$f : \mathbf{R}^m \to \mathbf{R}^n \quad \text{and} \quad g : \mathbf{R}^n \to \mathbf{R},$$

then the composite function $F = g \circ f$ maps $\mathbf{R}^m$ into $\mathbf{R}$, and its derivative is given by the formula

$$(*) \qquad DF(\mathbf{x}) = Dg\big(f(\mathbf{x})\big) \cdot Df(\mathbf{x}),$$

which can be written out in the form

$$[D_1F(\mathbf{x}) \quad \cdots \quad D_mF(\mathbf{x})]$$
$$= [D_1g(f(\mathbf{x})) \quad \cdots \quad D_ng(f(\mathbf{x}))] \begin{bmatrix} D_1f_1(\mathbf{x}) & \cdots & D_mf_1(\mathbf{x}) \\ \cdots & & \cdots \\ D_1f_n(\mathbf{x}) & \cdots & D_mf_n(\mathbf{x}) \end{bmatrix}.$$

The formula for the jth partial derivative of F is thus given by the equation

$$D_jF(\mathbf{x}) = \sum_{k=1}^{n} D_kg(f(\mathbf{x}))\, D_jf_k(\mathbf{x}).$$

If we shift to "variable" notation by setting $\mathbf{y} = f(\mathbf{x})$ and $z = g(\mathbf{y})$, this equation becomes

$$\frac{\partial z}{\partial x_j} = \sum_{k=1}^{n} \frac{\partial z}{\partial y_k} \frac{\partial y_k}{\partial x_j};$$

this is probably the version of the chain rule you learned in calculus. Only familiarity would suggest that it is easier to remember than (*)! Certainly one cannot obtain the formula for $\partial z/\partial x_j$ by a simple-minded multiplication of fractions, as in the single-variable case.

The formula for the derivative of an inverse function is even more troublesome. Suppose $f : \mathbf{R}^2 \to \mathbf{R}^2$ is differentiable and has a differentiable inverse function g. The derivative of g is given by the formula

$$Dg(\mathbf{y}) = [Df(\mathbf{x})]^{-1}.$$

where $\mathbf{y} = f(\mathbf{x})$. In Leibnitz notation, this formula takes the form

$$\begin{bmatrix} \partial x_1/\partial y_1 & \partial x_1/\partial y_2 \\ \partial x_2/\partial y_1 & \partial x_2/\partial y_2 \end{bmatrix} = \begin{bmatrix} \partial y_1/\partial x_1 & \partial y_1/\partial x_2 \\ \partial y_2/\partial x_1 & \partial y_2/\partial x_2 \end{bmatrix}^{-1}.$$

Recalling the formula for the inverse of a matrix, we see that the partial derivative $\partial x_i/\partial y_j$ is about as far from being the reciprocal of the partial derivative $\partial y_j/\partial x_i$ as one could imagine!

EXERCISES

1. Let $f : \mathbf{R}^3 \to \mathbf{R}^2$ satisfy the conditions $f(\mathbf{0}) = (1,2)$ and
$$Df(\mathbf{0}) = \begin{bmatrix} 1 & 2 & 3 \\ 0 & 0 & 1 \end{bmatrix}.$$
Let $g : \mathbf{R}^2 \to \mathbf{R}^2$ be defined by the equation
$$g(x,y) = (x + 2y + 1, 3xy).$$
Find $D(g \circ f)(\mathbf{0})$.

2. Let $f : \mathbf{R}^2 \to \mathbf{R}^3$ and $g : \mathbf{R}^3 \to \mathbf{R}^2$ be given by the equations
$$f(\mathbf{x}) = (e^{2x_1+x_2}, 3x_2 - \cos x_1, x_1^2 + x_2 + 2),$$
$$g(\mathbf{y}) = (3y_1 + 2y_2 + y_3^2, y_1^2 - y_3 + 1).$$
   (a) If $F(\mathbf{x}) = g(f(\mathbf{x}))$, find $DF(\mathbf{0})$. [*Hint:* Don't compute F explicitly.]
   (b) If $G(\mathbf{y}) = f(g(\mathbf{y}))$, find $DG(\mathbf{0})$.

3. Let $f : \mathbf{R}^3 \to \mathbf{R}$ and $g : \mathbf{R}^2 \to \mathbf{R}$ be differentiable. Let $F : \mathbf{R}^2 \to \mathbf{R}$ be defined by the equation
$$F(x,y) = f(x,y,g(x,y)).$$
   (a) Find DF in terms of the partials of f and g.
   (b) If $F(x,y) = 0$ for all (x,y), find $D_1 g$ and $D_2 g$ in terms of the partials of f.

4. Let $g : \mathbf{R}^2 \to \mathbf{R}^2$ be defined by the equation $g(x,y) = (x, y + x^2)$. Let $f : \mathbf{R}^2 \to \mathbf{R}$ be the function defined in Example 3 of § 5. Let $h = f \circ g$. Show that the directional derivatives of f and g exist everywhere, but that there is a $\mathbf{u} \neq \mathbf{0}$ for which $h'(\mathbf{0};\mathbf{u})$ does not exist.

§8. THE INVERSE FUNCTION THEOREM

Let A be open in $\mathbf{R}^n$; let $f : A \to \mathbf{R}^n$ be of class C^1. We know that for f to have a differentiable inverse, it is necessary that the derivative $Df(\mathbf{x})$ of f be non-singular. We now prove that this condition is also sufficient for f to have a differentiable inverse, at least locally. This result is called the *inverse function theorem.*

We begin by showing that non-singularity of Df implies that f is locally one-to-one.

Lemma 8.1. *Let A be open in $\mathbb{R}^n$; let $f : A \to \mathbb{R}^n$ be of class C^1. If $Df(\mathbf{a})$ is non-singular, then there exists an $\alpha > 0$ such that the inequality*

$$|f(\mathbf{x}_0) - f(\mathbf{x}_1)| \geq \alpha|\mathbf{x}_0 - \mathbf{x}_1|$$

holds for all $\mathbf{x}_0, \mathbf{x}_1$ in some open cube $C(\mathbf{a};\epsilon)$ centered at $\mathbf{a}$. It follows that f is one-to-one on this open cube.

Proof. Let $E = Df(\mathbf{a})$; then E is non-singular. We first consider the linear transformation that maps $\mathbf{x}$ to $E \cdot \mathbf{x}$. We compute

$$\begin{aligned}|\mathbf{x}_0 - \mathbf{x}_1| &= |E^{-1} \cdot (E \cdot \mathbf{x}_0 - E \cdot \mathbf{x}_1)| \\ &\leq n|E^{-1}| \cdot |E \cdot \mathbf{x}_0 - E \cdot \mathbf{x}_1|.\end{aligned}$$

If we set $2\alpha = 1/n|E^{-1}|$, then for all $\mathbf{x}_0, \mathbf{x}_1$ in $\mathbb{R}^n$,

$$|E \cdot \mathbf{x}_0 - E \cdot \mathbf{x}_1| \geq 2\alpha|\mathbf{x}_0 - \mathbf{x}_1|.$$

Now we prove the lemma. Consider the function

$$H(\mathbf{x}) = f(\mathbf{x}) - E \cdot \mathbf{x}.$$

Then $DH(\mathbf{x}) = Df(\mathbf{x}) - E$, so that $DH(\mathbf{a}) = \mathbf{0}$. Because H is of class C^1, we can choose $\epsilon > 0$ so that $|DH(\mathbf{x})| < \alpha/n$ for $\mathbf{x}$ in the open cube $C = C(\mathbf{a};\epsilon)$. The mean-value theorem, applied to the ith component function of H, tells us that, given $\mathbf{x}_0, \mathbf{x}_1 \in C$, there is a $\mathbf{c} \in C$ such that

$$|H_i(\mathbf{x}_0) - H_i(\mathbf{x}_1)| = |DH_i(\mathbf{c}) \cdot (\mathbf{x}_0 - \mathbf{x}_1)| \leq n(\alpha/n)|\mathbf{x}_0 - \mathbf{x}_1|.$$

Then for $\mathbf{x}_0, \mathbf{x}_1 \in C$, we have

$$\begin{aligned}\alpha|\mathbf{x}_0 - \mathbf{x}_1| &\geq |H(\mathbf{x}_0) - H(\mathbf{x}_1)| \\ &= |f(\mathbf{x}_0) - E \cdot \mathbf{x}_0 - f(\mathbf{x}_1) + E \cdot \mathbf{x}_1| \\ &\geq |E \cdot \mathbf{x}_1 - E \cdot \mathbf{x}_0| - |f(\mathbf{x}_1) - f(\mathbf{x}_0)| \\ &\geq 2\alpha|\mathbf{x}_1 - \mathbf{x}_0| - |f(\mathbf{x}_1) - f(\mathbf{x}_0)|.\end{aligned}$$

The lemma follows. □

Now we show that non-singularity of Df, in the case where f is one-to-one, implies that the inverse function is differentiable.

Theorem 8.2. *Let A be open in $\mathbf{R}^n$; let $f : A \to \mathbf{R}^n$ be of class C^r; let $B = f(A)$. If f is one-to-one on A and if $Df(\mathbf{x})$ is non-singular for $\mathbf{x} \in A$, then the set B is open in $\mathbf{R}^n$ and the inverse function $g : B \to A$ is of class C^r.*

Proof. *Step 1.* We prove the following elementary result: If $\phi : A \to \mathbf{R}$ is differentiable and if ϕ has a local minimum at $\mathbf{x}_0 \in A$, then $D\phi(\mathbf{x}_0) = \mathbf{0}$.

To say that ϕ has a local minimum at $\mathbf{x}_0$ means that $\phi(\mathbf{x}) \geq \phi(\mathbf{x}_0)$ for all $\mathbf{x}$ in a neighborhood of $\mathbf{x}_0$. Then given $\mathbf{u} \neq \mathbf{0}$,

$$\phi(\mathbf{x}_0 + t\mathbf{u}) - \phi(\mathbf{x}_0) \geq 0$$

for all sufficiently small values of t. Therefore

$$\phi'(\mathbf{x}_0;\mathbf{u}) = \lim_{t\to 0} [\phi(\mathbf{x}_0 + t\mathbf{u}) - \phi(\mathbf{x}_0)]/t$$

is non-negative if t approaches 0 through positive values, and is non-positive if t approaches 0 through negative values. It follows that $\phi'(\mathbf{x}_0;\mathbf{u}) = 0$. In particular, $D_j\phi(\mathbf{x}_0) = 0$ for all j, so that $D\phi(\mathbf{x}_0) = \mathbf{0}$.

Step 2. We show that the set B is open in $\mathbf{R}^n$. Given $\mathbf{b} \in B$, we show B contains some open ball $B(\mathbf{b};\delta)$ about $\mathbf{b}$.

We begin by choosing a rectangle Q lying in A whose interior contains the point $\mathbf{a} = f^{-1}(\mathbf{b})$ of A. The set $\operatorname{Bd} Q$ is compact, being closed and bounded in $\mathbf{R}^n$. Then the set $f(\operatorname{Bd} Q)$ is also compact, and thus is closed and bounded in $\mathbf{R}^n$. Because f is one-to-one, $f(\operatorname{Bd} Q)$ is disjoint from $\mathbf{b}$; because $f(\operatorname{Bd} Q)$ is closed, we can choose $\delta > 0$ so that the ball $B(\mathbf{b};2\delta)$ is disjoint from $f(\operatorname{Bd} Q)$. Given $\mathbf{c} \in B(\mathbf{b};\delta)$ we show that $\mathbf{c} = f(\mathbf{x})$ for some $\mathbf{x} \in Q$; it then follows that the set $f(A) = B$ contains each point of $B(\mathbf{b};\delta)$, as desired. See Figure 8.1.

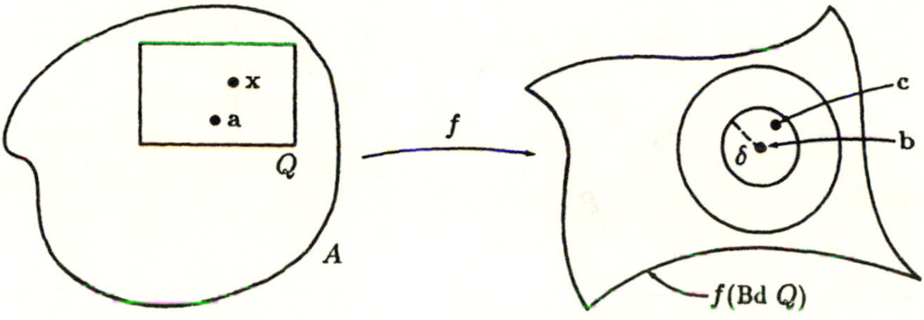

Figure 8.1

Given $\mathbf{c} \in B(\mathbf{b};\delta)$, consider the real-valued function

$$\phi(\mathbf{x}) = \|f(\mathbf{x}) - \mathbf{c}\|^2,$$

which is of class C^r. Because Q is compact, this function has a minimum value on Q; suppose that this minimum value occurs at the point $\mathbf{x}$ of Q. We show that $f(\mathbf{x}) = \mathbf{c}$.

Now the value of ϕ at the point $\mathbf{a}$ is

$$\phi(\mathbf{a}) = \|f(\mathbf{a}) - \mathbf{c}\|^2 = \|\mathbf{b} - \mathbf{c}\|^2 < \delta^2.$$

Hence the minimum value of ϕ on Q must be less than δ^2. It follows that this minimum value cannot occur on $\operatorname{Bd} Q$, for if $\mathbf{x} \in \operatorname{Bd} Q$, the point $f(\mathbf{x})$ lies outside the ball $B(\mathbf{b};2\delta)$, so that $\|f(\mathbf{x}) - \mathbf{c}\| \geq \delta$. Thus the minimum value of ϕ occurs at a point $\mathbf{x}$ of $\operatorname{Int} Q$.

Because $\mathbf{x}$ is interior to Q, it follows that ϕ has a *local* minimum at $\mathbf{x}$; then by Step 1, the derivative of ϕ vanishes at $\mathbf{x}$. Since

$$\phi(\mathbf{x}) = \sum_{k=1}^{n} (f_k(\mathbf{x}) - c_k)^2,$$

$$D_j\phi(\mathbf{x}) = \sum_{k=1}^{n} 2(f_k(\mathbf{x}) - c_k)D_j f_k(\mathbf{x}).$$

The equation $D\phi(\mathbf{x}) = \mathbf{0}$ can be written in matrix form as

$$2[(f_1(\mathbf{x}) - c_1) \quad \cdots \quad (f_n(\mathbf{x}) - c_n)] \cdot Df(\mathbf{x}) = \mathbf{0}.$$

Now $Df(\mathbf{x})$ is non-singular, by hypothesis. Multiplying both sides of this equation on the right by the inverse of $Df(\mathbf{x})$, we see that $f(\mathbf{x}) - \mathbf{c} = \mathbf{0}$, as desired.

Step 3. The function $f : A \to B$ is one-to-one by hypothesis; let $g : B \to A$ be the inverse function. We show g is continuous.

Continuity of g is equivalent to the statement that for each open set U of A, the set $V = g^{-1}(U)$ is open in B. But $V = f(U)$; and Step 2, applied to the set U, which is open in A and hence open in $\mathbf{R}^n$, tells us that V is open in $\mathbf{R}^n$ and hence open in B. See Figure 8.2.

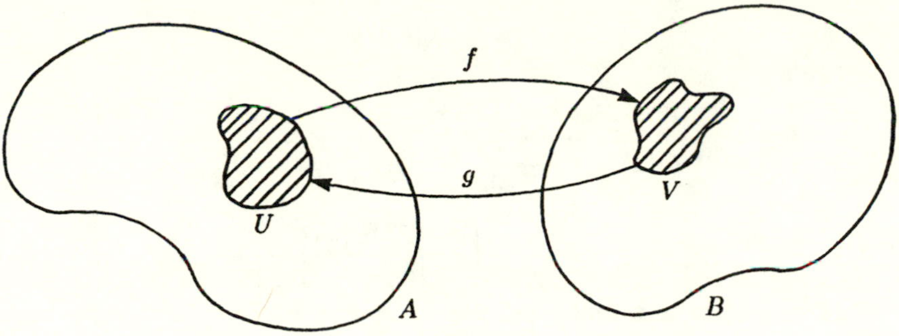

Figure 8.2

It is an interesting fact that the results of Steps 2 and 3 hold without assuming that $Df(\mathbf{x})$ is non-singular, or even that f is differentiable. If A is open in $\mathbf{R}^n$ and $f : A \to \mathbf{R}^n$ is continuous and one-to-one, then it is true that $f(A)$ is open in $\mathbf{R}^n$ and the inverse function g is continuous. This result is known as the *Brouwer theorem on invariance of domain.* Its proof requires the tools of algebraic topology and is quite difficult. We have proved the differentiable version of this theorem.

Step 4. Given $\mathbf{b} \in B$, we show that g is differentiable at $\mathbf{b}$.

Let $\mathbf{a}$ be the point $g(\mathbf{b})$, and let $E = Df(\mathbf{a})$. We show that the function

$$G(\mathbf{k}) = \frac{[g(\mathbf{b}+\mathbf{k}) - g(\mathbf{b}) - E^{-1}\cdot\mathbf{k}]}{|\mathbf{k}|},$$

which is defined for $\mathbf{k}$ in a deleted neighborhood of $\mathbf{0}$, approaches $\mathbf{0}$ as $\mathbf{k}$ approaches $\mathbf{0}$. Then g is differentiable at $\mathbf{b}$ with derivative E^{-1}.

Let us define

$$\Delta(\mathbf{k}) = g(\mathbf{b}+\mathbf{k}) - g(\mathbf{b})$$

for $\mathbf{k}$ near $\mathbf{0}$. We first show that there is an $\epsilon > 0$ such that $|\Delta(\mathbf{k})|/|\mathbf{k}|$ is bounded for $0 < |\mathbf{k}| < \epsilon$. (This would follow from differentiability of g, but that is what we are trying to prove!) By the preceding lemma, there is a neighborhood C of $\mathbf{a}$ and an $\alpha > 0$ such that

$$|f(\mathbf{x}_0) - f(\mathbf{x}_1)| \geq \alpha|\mathbf{x}_0 - \mathbf{x}_1|$$

for $\mathbf{x}_0, \mathbf{x}_1 \in C$. Now $f(C)$ is a neighborhood of $\mathbf{b}$, by Step 2; choose ϵ so that $\mathbf{b}+\mathbf{k}$ is in $f(C)$ whenever $|\mathbf{k}| < \epsilon$. Then for $|\mathbf{k}| < \epsilon$, we can set $\mathbf{x}_0 = g(\mathbf{b}+\mathbf{k})$ and $\mathbf{x}_1 = g(\mathbf{b})$ and rewrite the preceding inequality in the form

$$|(\mathbf{b}+\mathbf{k}) - \mathbf{b}| \geq \alpha|g(\mathbf{b}+\mathbf{k}) - g(\mathbf{b})|,$$

which implies that

$$1/\alpha \geq |\Delta(\mathbf{k})|/|\mathbf{k}|,$$

as desired.

Now we show that $G(\mathbf{k}) \to 0$ as $\mathbf{k} \to 0$. Let $0 < |\mathbf{k}| < \epsilon$. We have

$$G(\mathbf{k}) = \frac{\Delta(\mathbf{k}) - E^{-1} \cdot \mathbf{k}}{|\mathbf{k}|} \qquad \text{by definition,}$$

$$= -E^{-1} \cdot \left[\frac{\mathbf{k} - E \cdot \Delta(\mathbf{k})}{|\Delta(\mathbf{k})|}\right] \frac{|\Delta(\mathbf{k})|}{|\mathbf{k}|}.$$

(Here we use the fact that $\Delta(\mathbf{k}) \neq \mathbf{0}$ for $\mathbf{k} \neq \mathbf{0}$, which follows from the fact that g is one-to-one.) Now E^{-1} is constant, and $|\Delta(\mathbf{k})|/|\mathbf{k}|$ is bounded. It remains to show that the expression in brackets goes to zero. We have

$$\mathbf{b} + \mathbf{k} = f(g(\mathbf{b} + \mathbf{k})) = f(g(\mathbf{b}) + \Delta(\mathbf{k})) = f(\mathbf{a} + \Delta(\mathbf{k})).$$

Thus the expression in brackets equals

$$\frac{f(\mathbf{a} + \Delta(\mathbf{k})) - f(\mathbf{a}) - E \cdot \Delta(\mathbf{k})}{|\Delta(\mathbf{k})|}.$$

Let $\mathbf{k} \to \mathbf{0}$. Then $\Delta(\mathbf{k}) \to \mathbf{0}$ as well, *because* g is continuous. Since f is differentiable at $\mathbf{a}$ with derivative E, this expression goes to zero, as desired.

Step 5. Finally, we show the inverse function g is of class C^r.

Because g is differentiable, Theorem 7.4 applies to show that its derivative is given by the formula

$$Dg(\mathbf{y}) = [Df(g(\mathbf{y}))]^{-1},$$

for $\mathbf{y} \in B$. The function Dg thus equals the composite of three functions:

$$B \xrightarrow{g} A \xrightarrow{Df} GL(n) \xrightarrow{I} GL(n),$$

where $GL(n)$ is the set of non-singular n by n matrices, and I is the function that maps each non-singular matrix to its inverse. Now the function I is given by a specific formula involving determinants. In fact, the entries of $I(C)$ are *rational functions* of the entries of C; as such, they are C^∞ functions of the entries of C.

We proceed by induction on r. Suppose f is of class C^1. Then Df is continuous. Because g and I are also continuous (indeed, g is differentiable and I is of class C^∞), the composite function, which equals Dg, is also continuous. Hence g is of class C^1.

Suppose the theorem holds for functions of class C^{r-1}. Let f be of class C^r. Then in particular f is of class C^{r-1}, so that (by the induction hypothesis), the inverse function g is of class C^{r-1}. Furthermore, the function Df is of class C^{r-1}. We invoke Corollary 7.2 to conclude that the composite function, which equals Dg, is of class C^{r-1}. Then g is of class C^r. □

Finally, we prove the inverse function theorem.

Theorem 8.3 (The inverse function theorem). *Let A be open in $\mathbf{R}^n$; let $f : A \to \mathbf{R}^n$ be of class C^r. If $Df(\mathbf{x})$ is non-singular at the point $\mathbf{a}$ of A, there is a neighborhood U of the point $\mathbf{a}$ such that f carries U in a one-to-one fashion onto an open set V of $\mathbf{R}^n$ and the inverse function is of class C^r.*

Proof. By Lemma 8.1, there is a neighborhood U_0 of $\mathbf{a}$ on which f is one-to-one. Because $\det Df(\mathbf{x})$ is a continuous function of $\mathbf{x}$, and $\det Df(\mathbf{a}) \neq 0$, there is a neighborhood U_1 of $\mathbf{a}$ such that $\det Df(\mathbf{x}) \neq 0$ on U_1. If U equals the intersection of U_0 and U_1, then the hypotheses of the preceding theorem are satisfied for $f : U \to \mathbf{R}^n$. The theorem follows. □

This theorem is the strongest one that can be proved in general. While the non-singularity of Df on A implies that f is locally one-to-one at each point of A, it does not imply that f is one-to-one on all of A. Consider the following example:

EXAMPLE 1. Let $f : \mathbf{R}^2 \to \mathbf{R}^2$ be defined by the equation

$$f(r,\theta) = (r\cos\theta, r\sin\theta).$$

Then

$$Df(r,\theta) = \begin{bmatrix} \cos\theta & -r\sin\theta \\ \sin\theta & r\cos\theta \end{bmatrix},$$

so that $\det Df(r,\theta) = r$.

Let A be the open set $(0,1) \times (0,b)$ in the (r,θ) plane. Then Df is non-singular at each point of A. However, f is one-to-one on A only if $b \leq 2\pi$. See Figures 8.3 and 8.4.

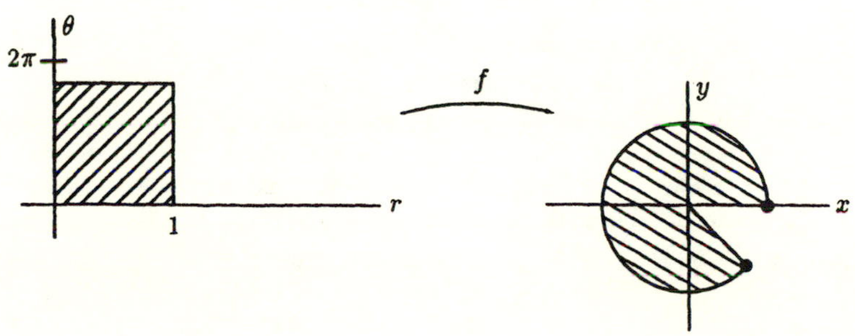

Figure 8.3

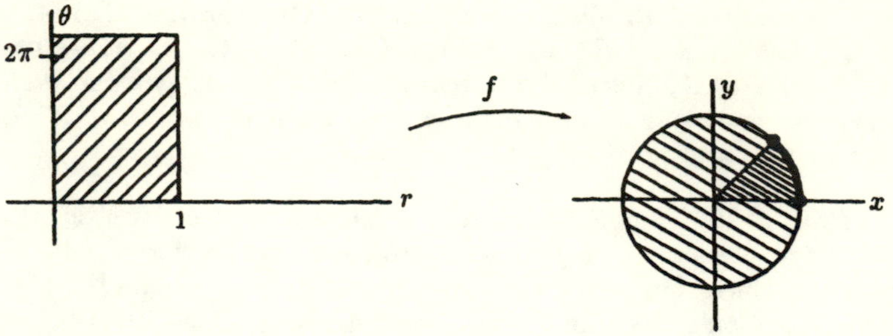

Figure 8.4

EXERCISES

1. Let $f : \mathbf{R}^2 \to \mathbf{R}^2$ be defined by the equation

$$f(x, y) = (x^2 - y^2, 2xy).$$

   (a) Show that f is one-to-one on the set A consisting of all (x, y) with $x > 0$. [*Hint:* If $f(x, y) = f(a, b)$, then $\|f(x, y)\| = \|f(a, b)\|$.]
   (b) What is the set $B = f(A)$?
   (c) If g is the inverse function, find $Dg(0, 1)$.

2. Let $f : \mathbf{R}^2 \to \mathbf{R}^2$ be defined by the equation

$$f(x, y) = (e^x \cos y, e^x \sin y).$$

   (a) Show that f is one-to-one on the set A consisting of all (x, y) with $0 < y < 2\pi$. [*Hint:* See the hint in the preceding exercise.]
   (b) What is the set $B = f(A)$?
   (c) If g is the inverse function, find $Dg(0, 1)$.

3. Let $f : \mathbf{R}^n \to \mathbf{R}^n$ be given by the equation $f(\mathbf{x}) = \|\mathbf{x}\|^2 \cdot \mathbf{x}$. Show that f is of class C^∞ and that f carries the unit ball $B(\mathbf{0}; 1)$ onto itself in a one-to-one fashion. Show, however, that the inverse function is not differentiable at $\mathbf{0}$.

4. Let $g : \mathbf{R}^2 \to \mathbf{R}^2$ be given by the equation

$$g(x, y) = (2ye^{2x}, xe^y).$$

   Let $f : \mathbf{R}^2 \to \mathbf{R}^3$ be given by the equation

$$f(x, y) = (3x - y^2, 2x + y, xy + y^3).$$

(a) Show that there is a neighborhood of $(0,1)$ that g carries in a one-to-one fashion onto a neighborhood of $(2,0)$.

(b) Find $D(f \circ g^{-1})$ at $(2,0)$.

5. Let A be open in $\mathbf{R}^n$; let $f : A \to \mathbf{R}^n$ be of class C^r; assume $Df(\mathbf{x})$ is non-singular for $\mathbf{x} \in A$. Show that even if f is not one-to-one on A, the set $B = f(A)$ is open in $\mathbf{R}^n$.

*§9. THE IMPLICIT FUNCTION THEOREM

The topic of implicit differentiation is one that is probably familiar to you from calculus. Here is a typical problem:

> "Assume that the equation $x^3y + 2e^{xy} = 0$ determines y as a differentiable function of x. Find dy/dx."

One solves this calculus problem by "looking at y as a function of x," and differentiating with respect to x. One obtains the equation

$$3x^2y + x^3dy/dx + 2e^{xy}(y + x\,dy/dx) = 0,$$

which one solves for dy/dx. The derivative dy/dx is of course expressed in terms of x and the unknown function y.

The case of an arbitrary function f is handled similarly. Supposing that the equation $f(x,y) = 0$ determines y as a differentiable function of x, say $y = g(x)$, the equation $f\big(x,g(x)\big) = 0$ is an identity. One applies the chain rule to calculate

$$\partial f/\partial x + (\partial f/\partial y)g'(x) = 0,$$

so that

$$g'(x) = -\frac{\partial f/\partial x}{\partial f/\partial y},$$

where the partial derivatives are evaluated at the point $\big(x,g(x)\big)$. Note that the solution involves a hypothesis not given in the statement of the problem. In order to find $g'(x)$, it is necessary to assume that $\partial f/\partial y$ is non-zero at the point in question.

It in fact turns out that the non-vanishing of $\partial f/\partial y$ is also *sufficient* to justify the assumptions we made in solving the problem. That is, if the function $f(x,y)$ has the property that $\partial f/\partial y \neq 0$ at a point (a,b) that is a solution of the equation $f(x,y) = 0$, then this equation *does* determine y as a function of x, for x near a, and this function of x *is* differentiable.

This result is a special case of a theorem called the *implicit function theorem,* which we prove in this section.

The general case of the implicit function theorem involves a system of equations rather than a single equation. One seeks to solve this system for some of the unknowns in terms of the others. Specifically, suppose that $f : \mathbf{R}^{k+n} \to \mathbf{R}^n$ is a function of class C^1. Then the vector equation

$$f(x_1, \dots, x_{k+n}) = \mathbf{0}$$

is equivalent to a system of n scalar equations in $k+n$ unknowns. One would expect to be able to assign arbitrary values to k of the unknowns and to solve for the remaining unknowns in terms of these. One would also expect that the resulting functions would be differentiable, and that one could by implicit differentiation find their derivatives.

There are two separate problems here. The first is the problem of finding the derivatives of these implicitly defined functions, assuming they exist; the solution to this problem generalizes the computation of $g'(x)$ just given. The second involves showing that (under suitable conditions) the implicitly defined functions exist and are differentiable.

In order to state our results in a convenient form, we introduce a new notation for the matrix Df and its submatrices:

Definition. Let A be open in $\mathbf{R}^m$; let $f : A \to \mathbf{R}^n$ be differentiable. Let $f_1, \dots, f_n$ be the component functions of f. We sometimes use the notation

$$Df = \frac{\partial(f_1, \dots, f_n)}{\partial(x_1, \dots, x_m)}$$

for the derivative of f. On occasion we shorten this to the notation

$$Df = \partial f / \partial \mathbf{x}.$$

More generally, we shall use the notation

$$\frac{\partial(f_{i_1}, \dots, f_{i_k})}{\partial(x_{j_1}, \dots, x_{j_\ell})}$$

to denote the k by ℓ matrix that consists of the entries of Df lying in rows $i_1, \dots, i_k$ and columns $j_1, \dots, j_\ell$. The general entry of this matrix, in row p and column q, is the partial derivative $\partial f_{i_p} / \partial x_{j_q}$.

Now we deal with the problem of finding the derivative of an implicitly defined function, assuming it exists and is differentiable. For simplicity, we shall assume that we have solved a system of n equations in $k+n$ unknowns for the *last* n unknowns in terms of the *first* k unknowns.

Theorem 9.1. *Let A be open in $\mathbf{R}^{k+n}$; let $f : A \to \mathbf{R}^n$ be differentiable. Write f in the form $f(\mathbf{x}, \mathbf{y})$, for $\mathbf{x} \in \mathbf{R}^k$ and $\mathbf{y} \in \mathbf{R}^n$; then Df has the form*

$$Df = \begin{bmatrix} \partial f/\partial \mathbf{x} & \partial f/\partial \mathbf{y} \end{bmatrix}.$$

Suppose there is a differentiable function $g : B \to \mathbf{R}^n$ defined on an open set B in $\mathbf{R}^k$, such that

$$f(\mathbf{x}, g(\mathbf{x})) = 0$$

for all $\mathbf{x} \in B$. Then for $\mathbf{x} \in B$,

$$\frac{\partial f}{\partial \mathbf{x}}(\mathbf{x}, g(\mathbf{x})) + \frac{\partial f}{\partial \mathbf{y}}(\mathbf{x}, g(\mathbf{x})) \cdot Dg(\mathbf{x}) = 0.$$

This equation implies that if the n by n matrix $\partial f/\partial \mathbf{y}$ is non-singular at the point $(\mathbf{x}, g(\mathbf{x}))$, then

$$Dg(\mathbf{x}) = -\left[\frac{\partial f}{\partial \mathbf{y}}(\mathbf{x}, g(\mathbf{x}))\right]^{-1} \cdot \frac{\partial f}{\partial \mathbf{x}}(\mathbf{x}, g(\mathbf{x})).$$

Note that in the case $n = k = 1$, this is the same formula for the derivative that was derived earlier; the matrices involved are 1 by 1 matrices in that case.

Proof. Given g, let us define $h : B \to \mathbf{R}^{k+n}$ by the equation

$$h(\mathbf{x}) = (\mathbf{x}, g(\mathbf{x})).$$

The hypotheses of the theorem imply that the composite function

$$H(\mathbf{x}) = f(h(\mathbf{x})) = f(\mathbf{x}, g(\mathbf{x}))$$

is defined and equals zero for all $\mathbf{x} \in B$. The chain rule then implies that

$$\begin{aligned} 0 = DH(\mathbf{x}) &= Df(h(\mathbf{x})) \cdot Dh(\mathbf{x}) \\ &= \begin{bmatrix} \frac{\partial f}{\partial \mathbf{x}}(h(\mathbf{x})) & \frac{\partial f}{\partial \mathbf{y}}(h(\mathbf{x})) \end{bmatrix} \cdot \begin{bmatrix} I_k \\ Dg(\mathbf{x}) \end{bmatrix} \\ &= \frac{\partial f}{\partial \mathbf{x}}(h(\mathbf{x})) + \frac{\partial f}{\partial \mathbf{y}}(h(\mathbf{x})) \cdot Dg(\mathbf{x}), \end{aligned}$$

as desired. □

The preceding theorem tells us that in order to compute Dg, we must assume that the matrix $\partial f/\partial \mathbf{y}$ is non-singular. Now we prove that the non-singularity of $\partial f/\partial \mathbf{y}$ suffices to guarantee that the function g exists and is differentiable.

Theorem 9.2 (Implicit function theorem). *Let A be open in $\mathbf{R}^{k+n}$; let $f : A \to \mathbf{R}^n$ be of class C^r. Write f in the form $f(\mathbf{x},\mathbf{y})$, for $\mathbf{x} \in \mathbf{R}^k$ and $\mathbf{y} \in \mathbf{R}^n$. Suppose that $(\mathbf{a},\mathbf{b})$ is a point of A such that $f(\mathbf{a},\mathbf{b}) = 0$ and*

$$\det \frac{\partial f}{\partial \mathbf{y}}(\mathbf{a},\mathbf{b}) \neq 0.$$

Then there is a neighborhood B of $\mathbf{a}$ in $\mathbf{R}^k$ and a unique continuous function $g : B \to \mathbf{R}^n$ such that $g(\mathbf{a}) = \mathbf{b}$ and

$$f(\mathbf{x}, g(\mathbf{x})) = 0$$

for all $\mathbf{x} \in B$. The function g is in fact of class C^r.

Proof. We construct a function F to which we can apply the inverse function theorem. Define $F : A \to \mathbf{R}^{k+n}$ by the equation

$$F(\mathbf{x},\mathbf{y}) = (\mathbf{x}, f(\mathbf{x},\mathbf{y})).$$

Then F maps the open set A of $\mathbf{R}^{k+n}$ into $\mathbf{R}^k \times \mathbf{R}^n = \mathbf{R}^{k+n}$. Furthermore,

$$DF = \begin{bmatrix} I_k & 0 \\ \partial f/\partial \mathbf{x} & \partial f/\partial \mathbf{y} \end{bmatrix}.$$

Computing $\det DF$ by repeated application of Lemma 2.12, we have $\det DF = \det \partial f/\partial \mathbf{y}$. Thus DF is non-singular at the point $(\mathbf{a},\mathbf{b})$.

Now $F(\mathbf{a},\mathbf{b}) = (\mathbf{a},\mathbf{0})$. Applying the inverse function theorem to the map F, we conclude that there exists an open set $U \times V$ of $\mathbf{R}^{k+n}$ about $(\mathbf{a},\mathbf{b})$ (where U is open in $\mathbf{R}^k$ and V is open in $\mathbf{R}^n$) such that:

(1) F maps $U \times V$ in a one-to-one fashion onto an open set W in $\mathbf{R}^{k+n}$ containing $(\mathbf{a},\mathbf{0})$.

(2) The inverse function $G : W \to U \times V$ is of class C^r.

Note that because $F(\mathbf{x},\mathbf{y}) = (\mathbf{x}, f(\mathbf{x},\mathbf{y}))$, we have

$$(\mathbf{x},\mathbf{y}) = G(\mathbf{x}, f(\mathbf{x},\mathbf{y})).$$

Thus G preserves the first k coordinates, as F does. Then we can write G in the form

$$G(\mathbf{x},\mathbf{z}) = (\mathbf{x}, h(\mathbf{x},\mathbf{z}))$$

for $\mathbf{x} \in \mathbf{R}^k$ and $\mathbf{z} \in \mathbf{R}^n$; here h is a function of class C^r mapping W into $\mathbf{R}^n$.

Let B be a connected neighborhood of $\mathbf{a}$ in $\mathbf{R}^k$, chosen small enough that $B \times \mathbf{0}$ is contained in W. See Figure 9.1. We prove existence of the function $g : B \to \mathbf{R}^n$. If $\mathbf{x} \in B$, then $(\mathbf{x},\mathbf{0}) \in W$, so we have:

$$\begin{aligned} G(\mathbf{x},\mathbf{0}) &= (\mathbf{x}, h(\mathbf{x},\mathbf{0})), \\ (\mathbf{x},\mathbf{0}) &= F(\mathbf{x}, h(\mathbf{x},\mathbf{0})) = (\mathbf{x}, f(\mathbf{x}, h(\mathbf{x},\mathbf{0}))), \\ \mathbf{0} &= f(\mathbf{x}, h(\mathbf{x},\mathbf{0})). \end{aligned}$$

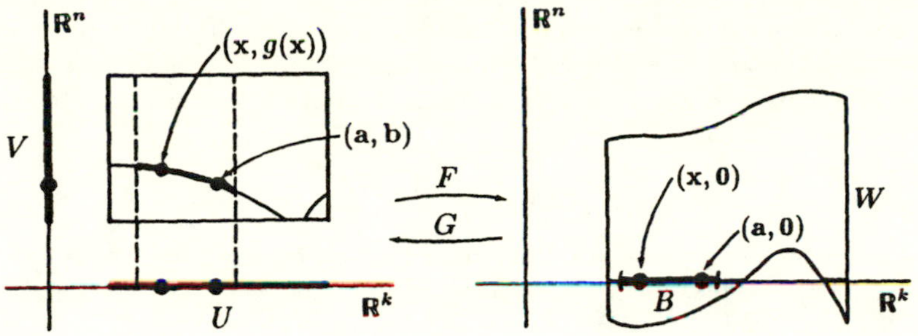

Figure 9.1

We set $g(\mathbf{x}) = h(\mathbf{x}, \mathbf{0})$ for $\mathbf{x} \in B$; then g satisfies the equation $f(\mathbf{x}, g(\mathbf{x})) = \mathbf{0}$, as desired. Furthermore,

$$(\mathbf{a}, \mathbf{b}) = G(\mathbf{a}, \mathbf{0}) = (\mathbf{a}, h(\mathbf{a}, \mathbf{0}));$$

then $\mathbf{b} = g(\mathbf{a})$, as desired.

Now we prove uniqueness of g. Let $g_0 : B \to \mathbf{R}^n$ be a continuous function satisfying the conditions in the conclusion of our theorem. Then in particular, g_0 agrees with g at the point $\mathbf{a}$. We show that if g_0 agrees with g at the point $\mathbf{a}_0 \in B$, then g_0 agrees with g in a neighborhood B_0 of $\mathbf{a}_0$. This is easy. The map g carries $\mathbf{a}_0$ into V. Since g_0 is continuous, there is a neighborhood B_0 of $\mathbf{a}_0$ contained in B such that g_0 also maps B_0 into V. The fact that $f(\mathbf{x}, g_0(\mathbf{x})) = \mathbf{0}$ for $\mathbf{x} \in B_0$ implies that

$$F(\mathbf{x}, g_0(\mathbf{x})) = (\mathbf{x}, \mathbf{0}), \quad \text{so}$$
$$(\mathbf{x}, g_0(\mathbf{x})) = G(\mathbf{x}, \mathbf{0}) = (\mathbf{x}, h(\mathbf{x}, \mathbf{0})).$$

Thus g_0 and g agree on B_0. It follows that g_0 and g agree on all of B: The set of points of B for which $|g(\mathbf{x}) - g_0(\mathbf{x})| = 0$ is open in B (as we just proved), and so is the set of points of B for which $|g(\mathbf{x}) - g_0(\mathbf{x})| > 0$ (by continuity of g and g_0). Since B is connected, the latter set must be empty. □

In our proof of the implicit function theorem, there was of course nothing special about solving for the *last* n coordinates; that choice was made simply for convenience. The same argument applies to the problem of solving for any n coordinates in terms of the others.

For example, suppose A is open in $\mathbf{R}^5$ and $f : A \rightarrow \mathbf{R}^2$ is a function of class C^r. Suppose one wishes to "solve" the equation $f(x, y, z, u, v) = 0$ for the two unknowns y and u in terms of the other three. In this case, the implicit function theorem tells us that if $\mathbf{a}$ is a point of A such that $f(\mathbf{a}) = 0$ and

$$\det \frac{\partial f}{\partial(y,u)}(\mathbf{a}) \neq 0,$$

then one can solve for y and u locally near that point, say $y = \phi(x, z, v)$ and $u = \psi(x, z, v)$. Furthermore, the derivatives of ϕ and ψ satisfy the formula

$$\frac{\partial(\phi,\psi)}{\partial(x,z,v)} = -\left[\frac{\partial f}{\partial(y,u)}\right]^{-1} \cdot \left[\frac{\partial f}{\partial(x,z,v)}\right].$$

EXAMPLE 1. Let $f : \mathbf{R}^2 \rightarrow \mathbf{R}$ be given by the equation

$$f(x,y) = x^2 + y^2 - 5.$$

Then the point $(x, y) = (1, 2)$ satisfies the equation $f(x, y) = 0$. Both $\partial f/\partial x$ and $\partial f/\partial y$ are non-zero at (1,2), so we can solve this equation locally for either variable in terms of the other. In particular, we can solve for y in terms of x, obtaining the function

$$y = g(x) = [5 - x^2]^{1/2}.$$

Note that this solution is not unique in a neighborhood of $x = 1$ unless we specify that g is continuous. For instance, the function

$$h(x) = \begin{cases} [5 - x^2]^{1/2} & \text{for } x \geq 1, \\ -[5 - x^2]^{1/2} & \text{for } x < 1 \end{cases}$$

satisfies the same conditions, but is not continuous. See Figure 9.2.

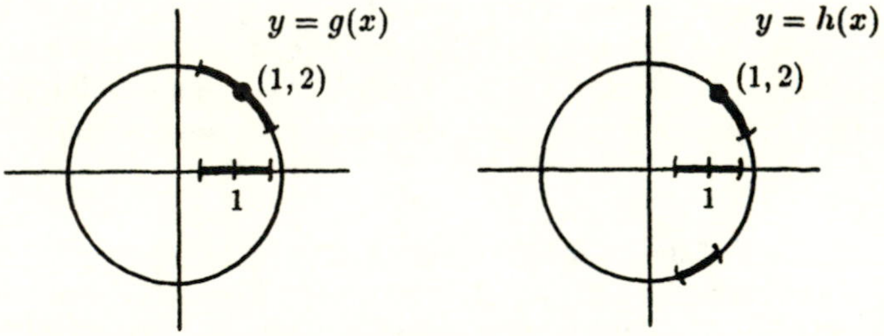

Figure 9.2

EXAMPLE 2. Let f be the function of Example 1. The point $(x, y) = (\sqrt{5}, 0)$ also satisfies the equation $f(x, y) = 0$. The derivative $\partial f/\partial y$ vanishes at $(\sqrt{5}, 0)$, so we do not expect to be able to solve for y in terms of x near this point. And, in fact, there is no neighborhood B of $\sqrt{5}$ on which we can solve for y in terms of x. See Figure 9.3.

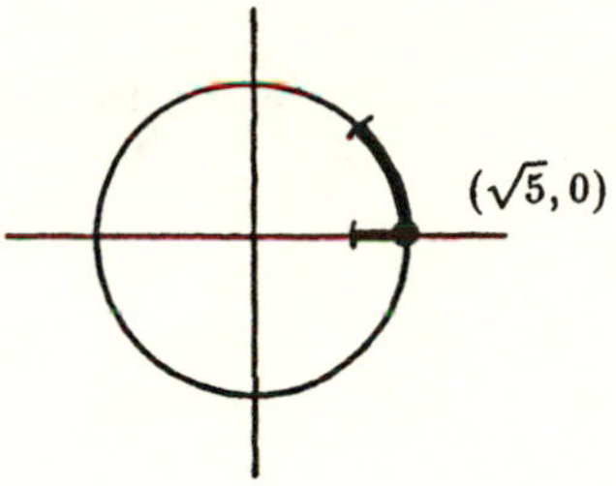

Figure 9.3

EXAMPLE 3. Let $f : \mathbb{R}^2 \to \mathbb{R}$ be given by the equation

$$f(x, y) = x^2 - y^3.$$

Then (0,0) is a solution of the equation $f(x, y) = 0$. Because $\partial f/\partial y$ vanishes at (0,0), we do not expect to be able to solve this equation for y in terms of x near (0,0). But in fact, we can; and furthermore, the solution is unique! However, the function we obtain is not differentiable at $x = 0$. See Figure 9.4.

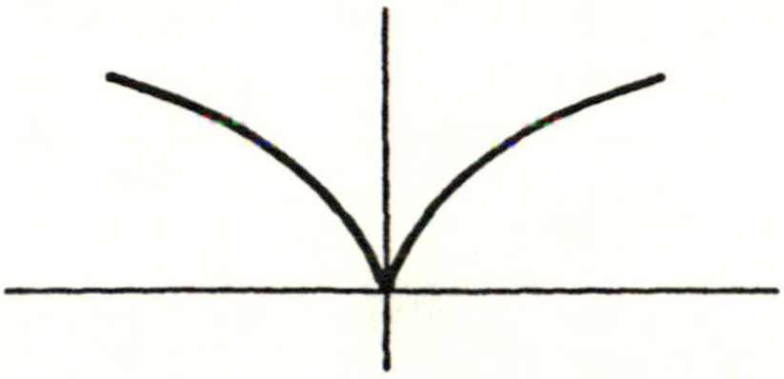

Figure 9.4

EXAMPLE 4. Let $f : \mathbb{R}^2 \to \mathbb{R}$ be given by the equation

$$f(x, y) = y^2 - x^4.$$

Then (0,0) is a solution of the equation $f(x, y) = 0$. Because $\partial f/\partial y$ vanishes at (0,0), we do not expect to be able to solve for y in terms of x near (0,0). In fact, however, we can do so, and we can do so in such a way that the resulting function is differentiable. However, the solution is not unique.

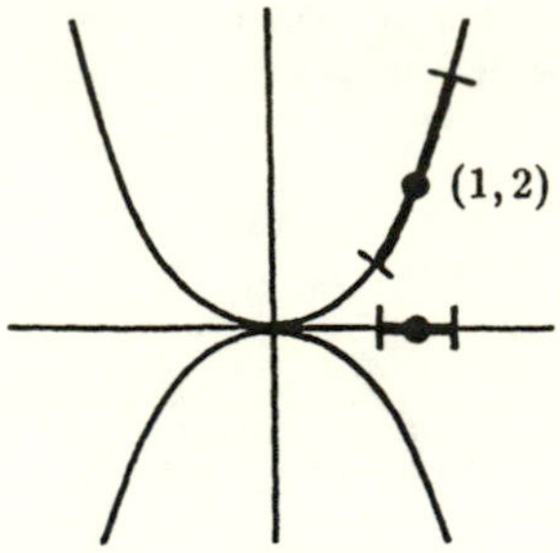

Figure 9.5

Now the point (1,2) also satisfies the equation $f(x,y) = 0$. Because $\partial f/\partial y$ is non-zero at (1,2), one can solve this equation for y as a continuous function of x in a neighborhood of $x = 1$. See Figure 9.5. One can in fact express y as a continuous function of x on a larger neighborhood than the one pictured, but if the neighborhood is large enough that it contains 0, then the solution is not unique on that larger neighborhood.

EXERCISES

1. Let $f : \mathbf{R}^3 \to \mathbf{R}^2$ be of class C^1; write f in the form $f(x, y_1, y_2)$. Assume that $f(3,-1,2) = 0$ and

$$Df(3,-1,2) = \begin{bmatrix} 1 & 2 & 1 \\ 1 & -1 & 1 \end{bmatrix}.$$

   (a) Show there is a function $g : B \to \mathbf{R}^2$ of class C^1 defined on an open set B in $\mathbf{R}$ such that

$$f\big(x, g_1(x), g_2(x)\big) = 0$$

   for $x \in B$, and $g(3) = (-1, 2)$.

   (b) Find $Dg(3)$.

   (c) Discuss the problem of solving the equation $f(x, y_1, y_2) = 0$ for an arbitrary pair of the unknowns in terms of the third, near the point $(3,-1,2)$.

2. Given $f : \mathbf{R}^5 \to \mathbf{R}^2$, of class C^1. Let $\mathbf{a} = (1,2,-1,3,0)$; suppose that $f(\mathbf{a}) = 0$ and

$$Df(\mathbf{a}) = \begin{bmatrix} 1 & 3 & 1 & -1 & 2 \\ 0 & 0 & 1 & 2 & -4 \end{bmatrix}.$$

(a) Show there is a function $g : B \to \mathbf{R}^2$ of class C^1 defined on an open set B of $\mathbf{R}^3$ such that

$$f(x_1, g_1(\mathbf{x}), g_2(\mathbf{x}), x_2, x_3) = 0$$

for $\mathbf{x} = (x_1, x_2, x_3) \in B$, and $g(1,3,0) = (2,-1)$.

(b) Find $Dg(1,3,0)$.

(c) Discuss the problem of solving the equation $f(\mathbf{x}) = \mathbf{0}$ for an arbitrary pair of the unknowns in terms of the others, near the point $\mathbf{a}$.

3. Let $f : \mathbf{R}^2 \to \mathbf{R}$ be of class C^1, with $f(2,-1) = -1$. Set

$$G(x,y,u) = f(x,y) + u^2,$$
$$H(x,y,u) = ux + 3y^3 + u^3.$$

The equations $G(x,y,u) = 0$ and $H(x,y,u) = 0$ have the solution $(x,y,u) = (2,-1,1)$.

(a) What conditions on Df ensure that there are C^1 functions $x = g(y)$ and $u = h(y)$ defined on an open set in $\mathbf{R}$ that satisfy both equations, such that $g(-1) = 2$ and $h(-1) = 1$?

(b) Under the conditions of (a), and assuming that $Df(2,-1) = [1 \;\; -3]$, find $g'(-1)$ and $h'(-1)$.

4. Let $F : \mathbf{R}^2 \to \mathbf{R}$ be of class C^2, with $F(0,0) = 0$ and $DF(0,0) = [2 \;\; 3]$. Let $G : \mathbf{R}^3 \to \mathbf{R}$ be defined by the equation

$$G(x,y,z) = F(x + 2y + 3z - 1, x^3 + y^2 - z^2).$$

(a) Note that $G(-2,3,-1) = F(0,0) = 0$. Show that one can solve the equation $G(x,y,z) = 0$ for z, say $z = g(x,y)$, for (x,y) in a neighborhood B of $(-2,3)$, such that $g(-2,3) = -1$.

(b) Find $Dg(-2,3)$.

*(c) If $D_1D_1F = 3$ and $D_1D_2F = -1$ and $D_2D_2F = 5$ at (0,0), find $D_2D_1g(-2,3)$.

5. Let $f, g : \mathbf{R}^3 \to \mathbf{R}$ be functions of class C^1. "In general," one expects that each of the equations $f(x,y,z) = 0$ and $g(x,y,z) = 0$ represents a smooth surface in $\mathbf{R}^3$, and that their intersection is a smooth curve. Show that if (x_0, y_0, z_0) satisfies both equations, and if $\partial(f,g)/\partial(x,y,z)$ has rank 2 at (x_0, y_0, z_0), then near (x_0, y_0, z_0), one can solve these equations for two of x, y, z in terms of the third, thus representing the solution set locally as a parametrized curve.

6. Let $f : \mathbf{R}^{k+n} \to \mathbf{R}^n$ be of class C^1; suppose that $f(\mathbf{a}) = \mathbf{0}$ and that $Df(\mathbf{a})$ has rank n. Show that if $\mathbf{c}$ is a point of $\mathbf{R}^n$ sufficiently close to $\mathbf{0}$, then the equation $f(\mathbf{x}) = \mathbf{c}$ has a solution.

3

Integration

In this chapter, we define the integral of real-valued function of several real variables, and derive its properties. The integral we study is called the Riemann integral; it is a direct generalization of the integral usually studied in a first course in single-variable analysis.

§10. THE INTEGRAL OVER A RECTANGLE

We begin by defining the volume of a rectangle. Let

$$Q = [a_1, b_1] \times [a_2, b_2] \times \cdots \times [a_n, b_n]$$

be a rectangle in $\mathbf{R}^n$. Each of the intervals $[a_i, b_i]$ is called a **component interval** of Q. The maximum of the numbers $b_1 - a_1, \ldots, b_n - a_n$ is called the **width** of Q. Their product

$$v(Q) = (b_1 - a_1)(b_2 - a_2)\cdots(b_n - a_n)$$

is called the **volume** of Q.

In the case $n = 1$, the volume and the width of the (1-dimensional) rectangle $[a, b]$ are the same, namely, the number $b - a$. This number is also called the **length** of $[a, b]$.

Definition. Given a closed interval $[a, b]$ of $\mathbf{R}$, a **partition** of $[a, b]$ is a finite collection P of points of $[a, b]$ that includes the points a and b. We usually index the elements of P in increasing order, for notational convenience, as

$$a = t_0 < t_1 < \cdots < t_k = b;$$

each of the intervals $[t_{i-1}, t_i]$, for $i = 1, \ldots, k$, is called a **subinterval determined by** P, of the interval $[a, b]$. More generally, given a rectangle

$$Q = [a_1, b_1] \times \cdots \times [a_n, b_n]$$

in $\mathbf{R}^n$, a **partition** P of Q is an n-tuple $(P_1, \ldots, P_n)$ such that P_j is a partition of $[a_j, b_j]$ for each j. If for each j, I_j is one of the subintervals determined by P_j of the interval $[a_j, b_j]$, then the rectangle

$$R = I_1 \times \cdots \times I_n$$

is called a **subrectangle determined by** P, of the rectangle Q. The maximum width of these subrectangles is called the **mesh** of P.

Definition. Let Q be a rectangle in $\mathbf{R}^n$; let $f : Q \to \mathbf{R}$; assume f is bounded. Let P be a partition of Q. For each subrectangle R determined by P, let

$$m_R(f) = \inf\{f(\mathbf{x}) \mid \mathbf{x} \in R\},$$
$$M_R(f) = \sup\{f(\mathbf{x}) \mid \mathbf{x} \in R\}.$$

We define the **lower sum** and the **upper sum**, respectively, of f, determined by P, by the equations

$$L(f, P) = \sum_R m_R(f) \cdot v(R),$$
$$U(f, P) = \sum_R M_R(f) \cdot v(R),$$

where the summations extend over all subrectangles R determined by P.

Let $P = (P_1, \ldots, P_n)$ be a partition of the rectangle Q. If P'' is a partition of Q obtained from P by adjoining additional points to some or all of the partitions $P_1, \ldots, P_n$, then P'' is called a **refinement** of P. Given two partitions P and $P' = (P'_1, \ldots, P'_n)$ of Q, the partition

$$P'' = (P_1 \cup P'_1, \ldots, P_n \cup P'_n)$$

is a refinement of both P and P'; it is called their **common refinement.**

Passing from P to a refinement of P of course affects lower sums and upper sums; in fact, it tends to increase the lower sums and decrease the upper sums. That is the substance of the following lemma:

Lemma 10.1. *Let P be a partition of the rectangle Q; let $f : Q \to \mathbf{R}$ be a bounded function. If P'' is a refinement of P, then*

$$L(f,P) \leq L(f,P'') \quad \text{and} \quad U(f,P'') \leq U(f,P).$$

Proof. Let Q be the rectangle

$$Q = [a_1, b_1] \times \cdots \times [a_n, b_n].$$

It suffices to prove the lemma when P'' is obtained by adjoining a single additional point to the partition of one of the component intervals of Q. Suppose, to be definite, that P is the partition $(P_1, \ldots, P_n)$ and that P'' is obtained by adjoining the point q to the partition P_1. Further, suppose that P_1 consists of the points

$$a_1 = t_0 < t_1 < \cdots < t_k = b_1$$

and that q lies interior to the subinterval $[t_{i-1}, t_i]$.

We first compare the lower sums $L(f,P)$ and $L(f,P'')$. Most of the subrectangles determined by P are also subrectangles determined by P''. An exception occurs for a subrectangle determined by P of the form

$$R_S = [t_{i-1}, t_i] \times S$$

(where S is one of the subrectangles of $[a_2, b_2] \times \cdots \times [a_n, b_n]$ determined by $(P_2, \ldots, P_n)$). The term involving the subrectangle R_S disappears from the lower sum and is replaced by the terms involving the two subrectangles

$$R'_S = [t_{i-1}, q] \times S \quad \text{and} \quad R''_S = [q, t_i] \times S,$$

which are determined by P''. See Figure 10.1.

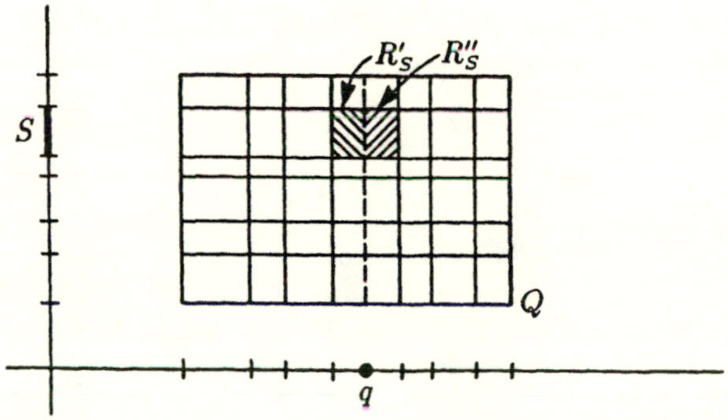

Figure 10.1

Now since $m_{R_S}(f) \leq f(\mathbf{x})$ for each $\mathbf{x} \in R'_S$ and for each $\mathbf{x} \in R''_S$, it follows that

$$m_{R_S}(f) \leq m_{R'_S}(f) \quad \text{and} \quad m_{R_S}(f) \leq m_{R''_S}(f).$$

Because $v(R_S) = v(R'_S) + v(R''_S)$ by direct computation, we have

$$m_{R_S}(f)v(R_S) \leq m_{R'_S}(f)v(R'_S) + m_{R''_S}(f)v(R''_S).$$

Since this inequality holds for each subrectangle of the form R_S, it follows that

$$L(f,P) \leq L(f,P''),$$

as desired.

A similar argument applies to show that $U(f,P) \geq U(f,P'')$. □

Now we explore the relation between upper sums and lower sums. We have the following result:

Lemma 10.2. *Let Q be a rectangle; let $f : Q \to \mathbf{R}$ be a bounded function. If P and P' are any two partitions of Q, then*

$$L(f,P) \leq U(f,P').$$

Proof. In the case where $P = P'$, the result is obvious: For any subrectangle R determined by P, we have $m_R(f) \leq M_R(f)$. Multiplying by $v(R)$ and summing gives the desired inequality.

In general, given partitions P and P' of Q, let P'' be their common refinement. Using the preceding lemma, we conclude that

$$L(f,P) \leq L(f,P'') \leq U(f,P'') \leq U(f,P'). \quad \square$$

Now (finally) we define the integral.

Definition. Let Q be a rectangle; let $f : Q \to \mathbf{R}$ be a bounded function. As P ranges over all partitions of Q, define

$$\underline{\int_Q} f = \sup_P \{L(f,P)\} \qquad \text{and} \qquad \overline{\int_Q} f = \inf_P \{U(f,P)\}.$$

These numbers are called the **lower integral** and **upper integral**, respectively, of f over Q. They exist because the numbers $L(f,P)$ are bounded above by $U(f,P')$ where P' is any fixed partition of Q; and the numbers $U(f,P)$ are bounded below by $L(f,P')$. If the upper and lower integrals of f over Q are equal, we say f is **integrable** over Q, and we define the **integral** of f over Q to equal the common value of the upper and lower integrals. We denote the integral of f over Q by either of the symbols

$$\int_Q f \quad \text{or} \quad \int_{\mathbf{x}\in Q} f(\mathbf{x}).$$

EXAMPLE 1. Let $f : [a,b] \to \mathbf{R}$ be a non-negative bounded function. If P is a partition of $I = [a,b]$, then $L(f,P)$ equals the total area of a bunch of rectangles inscribed in the region between the graph of f and the x-axis, and $U(f,P)$ equals the total area of a bunch of rectangles circumscribed about this region. See Figure 10.2.

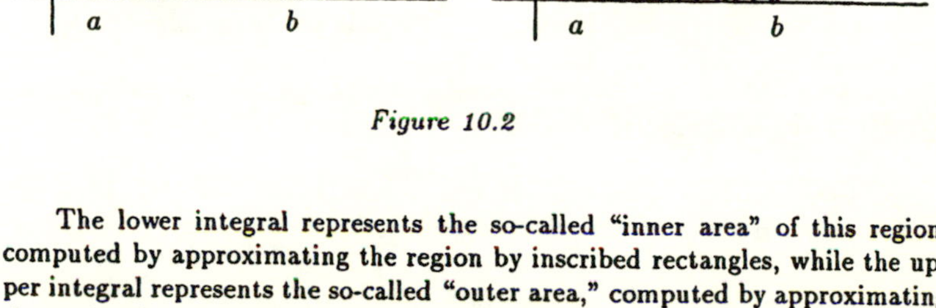

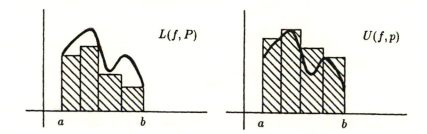

Figure 10.2

The lower integral represents the so-called "inner area" of this region, computed by approximating the region by inscribed rectangles, while the upper integral represents the so-called "outer area," computed by approximating the region by circumscribed rectangles. If the "inner" and "outer" areas are equal, then f is integrable.

Similarly, if Q is a rectangle in $\mathbf{R}^2$ and $f : Q \to \mathbf{R}$ is non-negative and bounded, one can picture $L(f,P)$ as the total volume of a bunch of boxes inscribed in the region between the graph of f and the xy-plane, and $U(f,P)$

as the total volume of a bunch of boxes circumscribed about this region. See Figure 10.3.

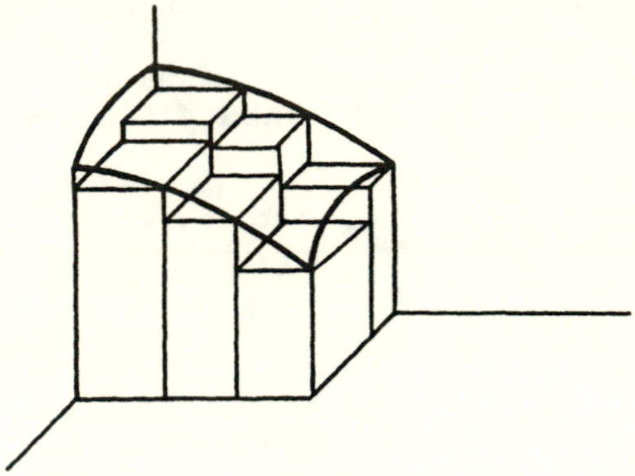

Figure 10.3

EXAMPLE 2. Let $I = [0,1]$. Let $f : I \to \mathbf{R}$ be defined by setting $f(x) = 0$ if x is rational, and $f(x) = 1$ if x is irrational. We show that f is not integrable over I.

Let P be a partition of I. If R is any subinterval determined by P, then $m_R(f) = 0$ and $M_R(f) = 1$, since R contains both rational and irrational numbers. Then

$$L(f,P) = \sum_R 0 \cdot v(R) = 0,$$

and

$$U(f,P) = \sum_R 1 \cdot v(R) = 1.$$

Since P is arbitrary, it follows that the lower integral of f over I equals 0, and the upper integral equals 1. Thus f is not integrable over I.

A condition that is often useful for showing that a given function is integrable is the following:

Theorem 10.3 (The Riemann condition). *Let Q be a rectangle; let $f : Q \to \mathbf{R}$ be a bounded function. Then*

$$\underline{\int_Q} f \le \overline{\int_Q} f;$$

equality holds if and only if given $\epsilon > 0$, there exists a corresponding partition P of Q for which

$$U(f,P) - L(f,P) < \epsilon.$$

Proof. Let P' be a fixed partition of Q. It follows from the fact that $L(f,P) \leq U(f,P')$ for every partition P of Q, that

$$\underline{\int_Q} f \leq U(f,P').$$

Now we use the fact that P' is arbitrary to conclude that

$$\underline{\int_Q} f \leq \overline{\int_Q} f.$$

Suppose now that the upper and lower integrals are equal. Choose a partition P so that $L(f,P)$ is within $\epsilon/2$ of the integral $\int_Q f$, and a partition P' so that $U(f,P')$ is within $\epsilon/2$ of the integral $\int_Q f$. Let P'' be their common refinement. Since

$$L(f,P) \leq L(f,P'') \leq \int_Q f \leq U(f,P'') \leq U(f,P'),$$

the lower and upper sums for f determined by P'' are within ϵ of each other.

Conversely, suppose the upper and lower integrals are not equal. Let

$$\epsilon = \overline{\int_Q} f - \underline{\int_Q} f > 0.$$

Let P be any partition of Q. Then

$$L(f,P) \leq \underline{\int_Q} f < \overline{\int_Q} f \leq U(f,P);$$

hence the upper and lower sums for f determined by P are at least ϵ apart. Thus the Riemann condition does not hold. □

Here is an easy application of this theorem.

Theorem 10.4. *Every constant function $f(x) = c$ is integrable. Indeed, if Q is a rectangle and if P is a partition of Q, then*

$$\int_Q c = c \cdot v(Q) = c \sum_R v(R),$$

where the summation extends over all subrectangles determined by P.

Proof. If R is a subrectangle determined by P, then $m_R(f) = c = M_R(f)$. It follows that

$$L(f,P) = c\sum_R v(R) = U(f,P),$$

so the Riemann condition holds trivially. Thus $\int_Q c$ exists; since it lies between $L(f,P)$ and $U(f,P)$, it must equal $c\sum_R v(R)$.

This result holds for any partition P. In particular, if P is the trivial partition whose only subrectangle is Q itself,

$$\int_Q c = c \cdot v(Q). \quad \square$$

A corollary of this result, which we shall use in the next section, is the following:

Corollary 10.5. *Let Q be a rectangle in $\mathbf{R}^n$; let $\{Q_1,\ldots,Q_k\}$ be a finite collection of rectangles that covers Q. Then*

$$v(Q) \leq \sum_{i=1}^{k} v(Q_i).$$

Proof. Choose a rectangle Q' containing all the rectangles $Q_1,\ldots,Q_k$. Use the end points of the component intervals of the rectangles $Q,Q_1,\ldots,Q_k$ to define a partition P of Q'. Then each of the rectangles $Q,Q_1,\ldots,Q_k$ is a union of subrectangles determined by P. See Figure 10.4.

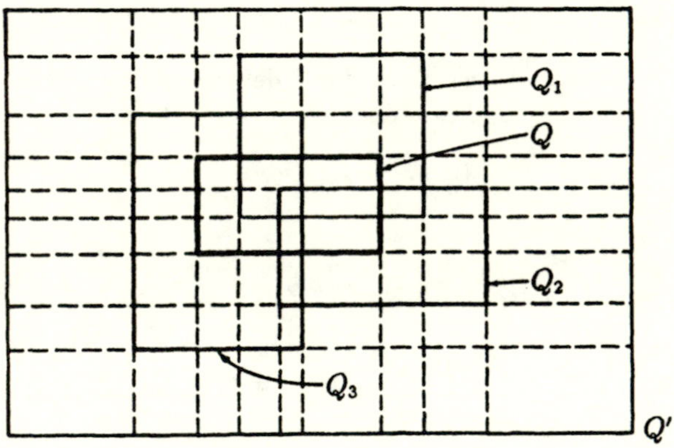

Figure 10.4

From the preceding theorem, we conclude that

$$v(Q) = \sum_{R\subset Q} v(R),$$

where the summation extends over all subrectangles contained in Q. Because each such subrectangle R is contained in at least one of the rectangles $Q_1, \ldots, Q_k$, we have

$$\sum_{R\subset Q} v(R) \leq \sum_{i=1}^{k} \sum_{R\subset Q_i} v(R).$$

Again using Theorem 10.4, we have

$$\sum_{R\subset Q_i} v(R) = v(Q_i);$$

the corollary follows. □

A remark about notation. We shall often use a slightly different notation for the integral in the case $n = 1$. In this case, Q is a closed interval $[a, b]$ in $\mathbf{R}$, and we often denote the integral of f over $[a, b]$ by one of the symbols

$$\int_a^b f \quad \text{or} \quad \int_{x=a}^{x=b} f(x)$$

instead of the symbol $\int_{[a,b]} f$.

Yet another notation is used in calculus for the one-dimensional integral. There it is common to denote this integral by the expression

$$\int_a^b f(x)\,dx,$$

where the symbol "dx" has no independent meaning. We shall avoid this notation for the time being. In a later chapter, we shall give "dx" a meaning and shall introduce this notation.

The definition of the integral we have given is in fact due to Darboux. An equivalent formulation, due to Riemann, is given in Exercise 7. In practice, it has become standard to call this integral the **Riemann integral,** independent of which definition is used.

EXERCISES

1. Let $f, g : Q \to \mathbf{R}$ be bounded functions such that $f(\mathbf{x}) \leq g(\mathbf{x})$ for $\mathbf{x} \in Q$. Show that $\underline{\int}_Q f \leq \underline{\int}_Q g$ and $\overline{\int}_Q f \leq \overline{\int}_Q g$.
2. Suppose $f : Q \to \mathbf{R}$ is continuous. Show f is integrable over Q. [*Hint:* Use uniform continuity of f.]
3. Let $[0,1]^2 = [0,1] \times [0,1]$. Let $f : [0,1]^2 \to \mathbf{R}$ be defined by setting $f(x,y) = 0$ if $y \neq x$, and $f(x,y) = 1$ if $y = x$. Show that f is integrable over $[0,1]^2$.
4. We say $f : [0,1] \to \mathbf{R}$ is increasing if $f(x_1) \leq f(x_2)$ whenever $x_1 < x_2$. If $f, g : [0,1] \to \mathbf{R}$ are increasing and non-negative, show that the function $h(x,y) = f(x)g(y)$ is integrable over $[0,1]^2$.
5. Let $f : \mathbf{R} \to \mathbf{R}$ be defined by setting $f(x) = 1/q$ if $x = p/q$, where p and q are positive integers with no common factor, and $f(x) = 0$ otherwise. Show f is integrable over $[0,1]$.

*6. Prove the following:

Theorem. *Let $f : Q \to \mathbf{R}$ be bounded. Then f is integrable over Q if and only if given $\epsilon > 0$, there is a $\delta > 0$ such that $U(f,P) - L(f,P) < \epsilon$ for every partition P of mesh less than δ.*

Proof. (a) Verify the "if" part of the theorem.

(b) Suppose $|f(\mathbf{x})| \leq M$ for $\mathbf{x} \in Q$. Let P be a partition of Q. Show that if P'' is obtained by adjoining a single point to the partition of one of the component intervals of Q, then

$$0 \leq L(f,P'') - L(f,P) \leq 2M(\text{mesh } P)(\text{width } Q)^{n-1}.$$

Derive a similar result for upper sums.

(c) Prove the "only if" part of the theorem: Suppose f is integrable over Q. Given $\epsilon > 0$, choose a partition P' such that $U(f,P') - L(f,P') < \epsilon/2$. Let N be the number of partition points in P'; then let

$$\delta = \epsilon/8MN\ (\text{width } Q)^{n-1}.$$

Show that if P has mesh less than δ, then $U(f,P) - L(f,P) < \epsilon$. [*Hint:* The common refinement of P and P' is obtained by adjoining at most N points to P.]

7. Use Exercise 6 to prove the following:

Theorem. *Let $f : Q \to \mathbf{R}$ be bounded. Then the statement that f is integrable over Q, with $\int_Q f = A$, is equivalent to the statement that given $\epsilon > 0$, there is a $\delta > 0$ such that if P is any partition of mesh less than δ, and if, for each subrectangle R determined by P, $\mathbf{x}_R$ is a point of R, then*

$$\left|\sum_R f(\mathbf{x}_R)v(R) - A\right| < \epsilon.$$

§11. EXISTENCE OF THE INTEGRAL

In this section, we derive a necessary and sufficient condition for the existence of the integral $\int_Q f$. It involves the notion of a "set of measure zero."

Definition. Let A be a subset of $\mathbf{R}^n$. We say A has **measure zero** in $\mathbf{R}^n$ if for every $\epsilon > 0$, there is a covering $Q_1, Q_2, \ldots$ of A by countably many rectangles such that

$$\sum_{i=1}^{\infty} v(Q_i) < \epsilon.$$

If this inequality holds, we often say that the **total volume** of the rectangles $Q_1, Q_2, \ldots$ is less than ϵ.

We derive some properties of sets of measure zero.

Theorem 11.1. (a) *If $B \subset A$ and A has measure zero in $\mathbf{R}^n$, then so does B.*

(b) *Let A be the union of the countable collection of sets $A_1, A_2, \ldots$. If each A_i has measure zero in $\mathbf{R}^n$, so does A.*

(c) *A set A has measure zero in $\mathbf{R}^n$ if and only if for every $\epsilon > 0$, there is a countable covering of A by open rectangles* Int Q_1, Int $Q_2, \ldots$ *such that*

$$\sum_{i=1}^{\infty} v(Q_i) < \epsilon.$$

(d) *If Q is a rectangle in $\mathbf{R}^n$, then* Bd Q *has measure zero in $\mathbf{R}^n$ but Q does not.*

Proof. (a) is immediate. To prove (b), cover the set A_j by countably many rectangles

$$Q_{1j},\ Q_{2j},\ Q_{3j},\ \ldots$$

of total volume less than $\epsilon/2^j$. Do this for each j. Then the collection of rectangles $\{Q_{ij}\}$ is countable, it covers A, and it has total volume less than

$$\sum_{j=1}^{\infty} \epsilon/2^j = \epsilon.$$

(c) If the open rectangles Int Q_1, Int $Q_2, \ldots$ cover A, then so do the rectangles $Q_1, Q_2, \ldots$. Thus the given condition implies that A has measure zero. Conversely, suppose A has measure zero. Cover A by rectangles

$Q_1', Q_2', \ldots$ of total volume less than $\epsilon/2$. For each i, choose a rectangle Q_i such that

$$Q_i' \subset \text{Int } Q_i \quad \text{and} \quad v(Q_i) \leq 2v(Q_i').$$

(This we can do because $v(Q)$ is a continuous function of the end points of the component intervals of Q.) Then the open rectangles Int Q_1, Int $Q_2, \ldots$ cover A, and $\sum v(Q_i) < \epsilon$.

(d) Let

$$Q = [a_1, b_1] \times \cdots \times [a_n, b_n].$$

The subset of Q consisting of those points $\mathbf{x}$ of Q for which $x_i = a_i$ is called one of the ith faces of Q. The other i^{th} face consists of those $\mathbf{x}$ for which $x_i = b_i$. Each face of Q has measure zero in $\mathbf{R}^n$; for instance, the face for which $x_i = a_i$ can be covered by the single rectangle

$$[a_1, b_1] \times \cdots \times [a_i, a_i + \delta] \times \cdots \times [a_n, b_n],$$

whose volume may be made as small as desired by taking δ small. Now Bd Q is the union of the faces of Q, which are finite in number. Therefore Bd Q has measure zero in $\mathbf{R}^n$.

Now we suppose Q has measure zero in $\mathbf{R}^n$, and derive a contradiction. Set $\epsilon = v(Q)$. We can by (c) cover Q by open rectangles Int Q_1, Int $Q_2, \ldots$ with $\sum v(Q_i) < \epsilon$. Because Q is compact, we can cover Q by finitely many of these open sets, say Int $Q_1, \ldots,$ Int Q_k. But

$$\sum_{i=1}^{k} v(Q_i) < \epsilon,$$

a result that contradicts Corollary 10.5. □

EXAMPLE 1. Allowing for a countably infinite collection of rectangles is an essential part of the definition of a set of measure zero. One would obtain a different notion if one allowed only finite collections. For instance, the set A of rational numbers in $I = [0, 1]$ is a countable union of one-point sets, so that A has measure zero in $\mathbf{R}$ by (b) of the preceding theorem. But A cannot be covered by *finitely* many intervals of total length less than ϵ if $\epsilon < 1$. For suppose $I_1, \ldots I_k$ is a finite collection of intervals covering A. Then the set B which is their union is a finite union of closed sets and therefore closed. Since B contains all rationals in I, it contains all limit points of these rationals; that is, it contains all of I. But this implies that the intervals $I_1, \ldots I_k$ cover I, whence by Corollary 10.5,

$$\sum_{i=1}^{k} v(I_i) \geq v(I) = 1.$$

Now we prove our main theorem.

Theorem 11.2. *Let Q be a rectangle in $\mathbf{R}^n$; let $f : Q \to \mathbf{R}$ be a bounded function. Let D be the set of points of Q at which f fails to be continuous. Then $\int_Q f$ exists if and only if D has measure zero in $\mathbf{R}^n$.*

Proof. Choose M so that $|f(\mathbf{x})| \leq M$ for $\mathbf{x} \in Q$.

Step 1. We prove the "if" part of the theorem. Assume D has measure zero in $\mathbf{R}^n$. We show that f is integrable over Q by showing that given $\epsilon > 0$, there is a partition P of Q for which $U(f,P) - L(f,P) < \epsilon$.

Given ϵ, let ϵ' be the strange number

$$\epsilon' = \epsilon/(2M + 2v(Q)).$$

First, we cover D by countably many open rectangles Int Q_1, Int $Q_2, \dots$ of total volume less than ϵ', using (c) of the preceding theorem. Second, for each point $\mathbf{a}$ of Q not in D, we choose an open rectangle Int $Q_\mathbf{a}$ containing $\mathbf{a}$ such that

$$|f(\mathbf{x}) - f(\mathbf{a})| < \epsilon' \quad \text{for} \quad \mathbf{x} \in Q_\mathbf{a} \cap Q.$$

(This we can do because f is continuous at $\mathbf{a}$.) Then the open sets Int Q_i and Int $Q_\mathbf{a}$, for $i = 1, 2, \dots$ and for $\mathbf{a} \in Q - D$, cover all of Q. Since Q is compact, we can choose a finite subcollection

$$\text{Int } Q_1, \dots, \text{Int } Q_k, \text{Int } Q_{\mathbf{a}_1}, \dots, \text{Int } Q_{\mathbf{a}_\ell}$$

that covers Q. (The open rectangles Int $Q_1, \dots,$ Int Q_k may not cover D, but that does not matter.)

Denote $Q_{\mathbf{a}_j}$ by Q'_j for convenience. Then the rectangles

$$Q_1, \dots, Q_k, Q'_1, \dots, Q'_\ell$$

cover Q, where the rectangles Q_i satisfy the condition

$$\sum_{i=1}^{k} v(Q_i) < \epsilon', \tag{1}$$

and the rectangles Q'_j satisfy the condition

$$|f(\mathbf{x}) - f(\mathbf{y})| \leq 2\epsilon' \quad \text{for} \quad \mathbf{x}, \mathbf{y} \in Q'_j \cap Q. \tag{2}$$

Without change of notation, let us replace each rectangle Q_i by its intersection with Q, and each rectangle Q'_j by its intersection with Q. The new rectangles $\{Q_i\}$ and $\{Q'_j\}$ still cover Q and satisfy conditions (1) and (2).

Now let us use the end points of the component intervals of the rectangles $Q_1, \dots, Q_k, Q'_1, \dots, Q'_\ell$ to define a partition P of Q. Then each of the rectangles Q_i and Q'_j is a union of subrectangles determined by P. We compute the upper and lower sums of f relative to P.

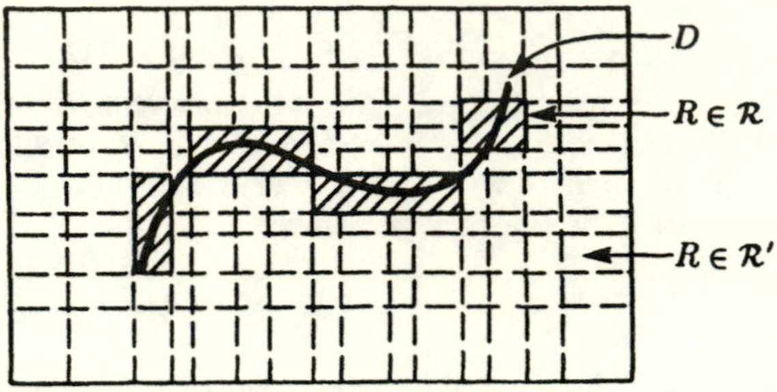

Figure 11.1

Divide the collection of all subrectangles R determined by P into two disjoint subcollections $\mathcal{R}$ and $\mathcal{R}'$, so that each rectangle $R \in \mathcal{R}$ lies in one of the rectangles Q_i, and each rectangle $R \in \mathcal{R}'$ lies in one of the rectangles Q'_j. See Figure 11.1. We have

$$\sum_{R\in\mathcal{R}} (M_R(f) - m_R(f))v(R) \leq 2M \sum_{R\in\mathcal{R}} v(R), \quad \text{and}$$

$$\sum_{R\in\mathcal{R}'} (M_R(f) - m_R(f))v(R) \leq 2\epsilon' \sum_{R\in\mathcal{R}'} v(R);$$

these inequalities follow from the fact that

$$|f(\mathbf{x}) - f(\mathbf{y})| \leq 2M$$

for any two points $\mathbf{x},\mathbf{y}$ belonging to a rectangle $R \in \mathcal{R}$, and

$$|f(\mathbf{x}) - f(\mathbf{y})| \leq 2\epsilon'$$

for any two points $\mathbf{x},\mathbf{y}$ belonging to a rectangle $R \in \mathcal{R}'$. Now

$$\sum_{R\in\mathcal{R}} v(R) \leq \sum_{i=1}^{k} \sum_{R\subset Q_i} v(R) = \sum_{i=1}^{k} v(Q_i) < \epsilon', \quad \text{and}$$

$$\sum_{R\in\mathcal{R}'} v(R) \leq \sum_{R\subset Q} v(R) = v(Q).$$

Thus

$$U(f,P) - L(f,P) < 2M\epsilon' + 2\epsilon' v(Q) = \epsilon.$$

Step 2. We now define what we mean by the "oscillation" of a function f at a point $\mathbf{a}$ of its domain, and relate it to continuity of f at $\mathbf{a}$.

Given $\mathbf{a} \in Q$ and given $\delta > 0$, let A_δ denote the set of values of $f(\mathbf{x})$ at points $\mathbf{x}$ within δ of $\mathbf{a}$. That is,

$$A_\delta = \{f(\mathbf{x}) \mid \mathbf{x} \in Q \quad \text{and} \quad |\mathbf{x} - \mathbf{a}| < \delta\}.$$

Let $M_\delta(f) = \sup A_\delta$, and let $m_\delta(f) = \inf A_\delta$. We define the **oscillation** of f at $\mathbf{a}$ by the equation

$$\nu(f;\mathbf{a}) = \inf_{\delta>0} [M_\delta(f) - m_\delta(f)].$$

Then $\nu(f;\mathbf{a})$ is non-negative; we show that f is continuous at $\mathbf{a}$ if and only if $\nu(f;\mathbf{a}) = 0$.

If f is continuous at $\mathbf{a}$, then, given $\epsilon > 0$, we can choose $\delta > 0$ so that $|f(\mathbf{x}) - f(\mathbf{a})| < \epsilon$ for all $\mathbf{x} \in Q$ with $|\mathbf{x} - \mathbf{a}| < \delta$. It follows that

$$M_\delta(f) \leq f(\mathbf{a}) + \epsilon \quad \text{and} \quad m_\delta(f) \geq f(\mathbf{a}) - \epsilon.$$

Hence $\nu(f;\mathbf{a}) \leq 2\epsilon$. Since ϵ is arbitrary, $\nu(f;\mathbf{a}) = 0$.

Conversely, suppose $\nu(f;\mathbf{a}) = 0$. Given $\epsilon > 0$, there is a $\delta > 0$ such that

$$M_\delta(f) - m_\delta(f) < \epsilon.$$

Now if $\mathbf{x} \in Q$ and $|\mathbf{x} - \mathbf{a}| < \delta$,

$$m_\delta(f) \leq f(\mathbf{x}) \leq M_\delta(f).$$

Since $f(\mathbf{a})$ also lies between $m_\delta(f)$ and $M_\delta(f)$, it follows that $|f(\mathbf{x}) - f(\mathbf{a})| < \epsilon$. Thus f is continuous at $\mathbf{a}$.

Step 3. We prove the "only if" part of the theorem. Assume f is integrable over Q. We show that the set D of discontinuities of f has measure zero in $\mathbf{R}^n$.

For each positive integer m, let

$$D_m = \{\mathbf{a} \mid \nu(f;\mathbf{a}) \geq 1/m\}.$$

Then by Step 2, D equals the union of the sets D_m. We show that each set D_m has measure zero; this will suffice.

Let m be fixed. Given $\epsilon > 0$, we shall cover D_m by countably many rectangles of total volume less than ϵ.

First choose a partition P of Q for which $U(f,P) - L(f,P) < \epsilon/2m$. Then let D'_m consist of those points of D_m that belong to Bd R for some subrectangle R determined by P; and let D''_m consist of the remainder of D_m. We cover each of D'_m and D''_m by rectangles having total volume less than $\epsilon/2$.

For D'_m, this is easy. Given R, the set Bd R has measure zero in $\mathbf{R}^n$; then so does the union $\bigcup_R \text{Bd}\, R$. Since D'_m is contained in this union, it may be covered by countably many rectangles of total volume less than $\epsilon/2$.

Now we consider D''_m. Let $R_1, \ldots, R_k$ be those subrectangles determined by P that contain points of D''_m. We show that these subrectangles have total volume less than $\epsilon/2$. Given i, the rectangle R_i contains a point $\mathbf{a}$ of D''_m. Since $\mathbf{a} \notin \text{Bd}\, R_i$, there is a $\delta > 0$ such that R_i contains the cubical neighborhood of radius δ centered at $\mathbf{a}$. Then

$$1/m \leq \nu(f;\mathbf{a}) \leq M_\delta(f) - m_\delta(f) \leq M_{R_i}(f) - m_{R_i}(f).$$

Multiplying by $v(R_i)$ and summing, we have

$$\sum_{i=1}^{k}(1/m)v(R_i) \leq U(f,P) - L(f,P) < \epsilon/2m.$$

Then the rectangles $R_1, \ldots, R_k$ have total volume less than $\epsilon/2$. □

We give an application of this theorem:

Theorem 11.3. *Let Q be a rectangle in $\mathbf{R}^n$; let $f : Q \to \mathbf{R}$; assume f is integrable over Q.*

(a) *If f vanishes except on a set of measure zero, then $\int_Q f = 0$.*

(b) *If f is non-negative and if $\int_Q f = 0$, then f vanishes except on a set of measure zero.*

Proof. (a) Suppose f vanishes except on a set E of measure zero. Let P be a partition of Q. If R is a subrectangle determined by P, then R is not contained in E, so that f vanishes at some point of R. Then $m_R(f) \leq 0$ and $M_R(f) \geq 0$. It follows that $L(f,P) \leq 0$ and $U(f,P) \geq 0$. Since these inequalities hold for all P,

$$\underline{\int_Q} f \leq 0 \quad \text{and} \quad \overline{\int_Q} f \geq 0.$$

Since $\int_Q f$ exists, it must equal zero.

(b) Suppose $f(\mathbf{x}) \geq 0$ and $\int_Q f = 0$. We show that if f is continuous at $\mathbf{a}$, then $f(\mathbf{a}) = 0$. It follows that f must vanish except possibly at points where f fails to be continuous; the set of such points has measure zero by the preceding theorem.

We suppose that f is continuous at $\mathbf{a}$ and that $f(\mathbf{a}) > 0$ and derive a contradiction. Set $\epsilon = f(\mathbf{a})$. Since f is continuous at $\mathbf{a}$, there is a $\delta > 0$ such that

$$f(\mathbf{x}) > \epsilon/2 \quad \text{for} \quad |\mathbf{x} - \mathbf{a}| < \delta \quad \text{and} \quad \mathbf{x} \in Q.$$

Choose a partition P of Q of mesh less than δ. If R_0 is a subrectangle determined by P that contains $\mathbf{a}$, then $m_{R_0}(f) \geq \epsilon/2$. On the other hand, $m_R(f) \geq 0$ for all R. It follows that

$$L(f,P) = \sum_R m_R(f)\, v(R) \geq (\epsilon/2) v(R_0) > 0.$$

But

$$L(f,P) \leq \int_Q f = 0. \quad \square$$

EXAMPLE 2. The assumption that $\int_Q f$ exists is necessary for the truth of this theorem. For example, let $I = [0,1]$ and let $f(x) = 1$ for x rational and $f(x) = 0$ for x irrational. Then f vanishes except on a set of measure zero. But it is not true that $\int_I f = 0$, for the integral of f over I does not even exist.

EXERCISES

1. Show that if A has measure zero in $\mathbf{R}^n$, the sets $\bar{A}$ and Bd A need not have measure zero.
2. Show that no non-trivial open set in $\mathbf{R}^n$ has measure zero in $\mathbf{R}^n$.
3. Show that the set $\mathbf{R}^{n-1} \times 0$ has measure zero in $\mathbf{R}^n$.
4. Show that the set of irrationals in $[0,1]$ does not have measure zero in $\mathbf{R}$.
5. Show that if A is a *compact* subset of $\mathbf{R}^n$ and A has measure zero in $\mathbf{R}^n$, then given $\epsilon > 0$, there is a finite collection of rectangles of total volume less than ϵ covering A.
6. Let $f : [a,b] \to \mathbf{R}$. The graph of f is the subset

$$G_f = \{(x,y) \mid y = f(x)\}$$

of $\mathbf{R}^2$. Show that if f is continuous, G_f has measure zero in $\mathbf{R}^2$. [*Hint:* Use uniform continuity of f.]

7. Consider the function f defined in Example 2. At what points of $[0,1]$ does f fail to be continuous? Answer the same question for the function defined in Exercise 5 of §10.

8. Let Q be a rectangle in $\mathbf{R}^n$; let $f : Q \to \mathbf{R}$ be a bounded function. Show that if f vanishes except on a *closed* set B of measure zero, then $\int_Q f$ exists and equals zero.
9. Let Q be a rectangle in $\mathbf{R}^n$; let $f : Q \to \mathbf{R}$; assume f is integrable over Q.
   (a) Show that if $f(\mathbf{x}) \geq 0$ for $\mathbf{x} \in Q$, then $\int_Q f \geq 0$.
   (b) Show that if $f(\mathbf{x}) > 0$ for $\mathbf{x} \in Q$, then $\int_Q f > 0$.
10. Show that if $Q_1, Q_2, \ldots$ is a countable collection of rectangles covering Q, then $v(Q) \leq \sum v(Q_i)$.

§12. EVALUATION OF THE INTEGRAL

Given that a function $f : Q \to \mathbf{R}$ is integrable, how does one evaluate its integral?

Even in the case of a function $f : [a,b] \to \mathbf{R}$ of a single variable, the problem is not easy. One tool is provided by the fundamental theorem of calculus, which is applicable when f is continuous. This theorem is familiar to you from single-variable analysis. For reference, we state it here:

Theorem 12.1 (Fundamental theorem of calculus). (a) *If f is continuous on $[a,b]$, and if*

$$F(x) = \int_a^x f$$

for $x \in [a,b]$, then $F'(x)$ exists and equals $f(x)$.

(b) *If f is continuous on $[a,b]$, and if g is a function such that $g'(x) = f(x)$ for $x \in [a,b]$, then*

$$\int_a^b f = g(b) - g(a). \qquad \square$$

(When one refers to the derivatives F' and g' at the end points of the interval $[a,b]$, one means of course the appropriate "one-sided" derivatives.)

The conclusions of this theorem are summarized in the two equations

$$D\int_a^x f = f(x) \quad \text{and} \quad \int_a^x Dg = g(x) - g(a).$$

In each case, the integrand is required to be continuous on the interval in question.

Part (b) of this theorem tells us we can calculate the integral of a continuous function f if we can find an **antiderivative** of f, that is, a function g such that $g' = f$. Part (a) of the theorem tells us that such an antiderivative always exists (in theory), since F is such an antiderivative. The problem, of course, is to find such an antiderivative in practice. That is what the so-called "Technique of Integration," as studied in calculus, is about.

The same difficulties of evaluating the integral occur with n-dimensional integrals. One way of approaching the problem is to attempt to reduce the computation of an n-dimensional integral to the presumably simpler problem of computing a sequence of lower-dimensional integrals. One might even be able to reduce the problem to computing a sequence of one-dimensional integrals, to which, if the integrand is continuous, one could apply the fundamental theorem of calculus.

This is the approach used in calculus to compute a double integral. To integrate the continuous function $f(x,y)$ over the rectangle $Q = [a,b] \times [c,d]$, for example, one integrates f first with respect to y, holding x fixed, and then integrates the resulting function with respect to x. (Or the other way around.) In doing so, one is using the formula

$$\int_Q f = \int_{x=a}^{x=b} \int_{y=c}^{y=d} f(x,y)$$

or its reverse. (In calculus, one usually inserts the meaningless symbols "dx" and "dy," but we are avoiding this notation here.) These formulas are not usually proved in calculus. In fact, it is seldom mentioned that a proof is needed; they are taken as "obvious." We shall prove them, and their appropriate n-dimensional versions, in this section.

These formulas hold when f is continuous. But when f is integrable but not continuous, difficulties can arise concerning the existence of the various integrals involved. For instance, the integral

$$\int_{y=c}^{y=d} f(x,y)$$

may not exist for all x even though $\int_Q f$ exists, for the function f can behave badly along a single vertical line without that behavior affecting the existence of the double integral.

One could avoid the problem by simply assuming that all the integrals involved exist. What we shall do instead is to replace the inner integral in the statement of the formula by the corresponding lower integral (or upper integral), which we know exists. When we do this, a correct general theorem results; it includes as a special case the case where all the integrals exist.

Theorem 12.2 (Fubini's theorem). *Let $Q = A \times B$, where A is a rectangle in $\mathbf{R}^k$ and B is a rectangle in $\mathbf{R}^n$. Let $f : Q \to \mathbf{R}$ be a bounded function; write f in the form $f(\mathbf{x},\mathbf{y})$ for $\mathbf{x} \in A$ and $\mathbf{y} \in B$. For each $\mathbf{x} \in A$, consider the lower and upper integrals*

$$\underline{\int}_{\mathbf{y}\in B} f(\mathbf{x},\mathbf{y}) \quad \textit{and} \quad \overline{\int}_{\mathbf{y}\in B} f(\mathbf{x},\mathbf{y}).$$

If f is integrable over Q, then these two functions of $\mathbf{x}$ are integrable over A, and

$$\int_Q f = \int_{\mathbf{x}\in A} \underline{\int}_{\mathbf{y}\in B} f(\mathbf{x},\mathbf{y}) = \int_{\mathbf{x}\in A} \overline{\int}_{\mathbf{y}\in B} f(\mathbf{x},\mathbf{y}).$$

Proof. For purposes of this proof, define

$$\underline{I}(\mathbf{x}) = \underline{\int}_{\mathbf{y}\in B} f(\mathbf{x},\mathbf{y}) \quad \text{and} \quad \overline{I}(\mathbf{x}) = \overline{\int}_{\mathbf{y}\in B} f(\mathbf{x},\mathbf{y})$$

for $\mathbf{x} \in A$. Assuming $\int_Q f$ exists, we show that $\underline{I}$ and $\overline{I}$ are integrable over A, and that their integrals equal $\int_Q f$.

Let P be a partition of Q. Then P consists of a partition P_A of A, and a partition P_B of B. We write $P = (P_A, P_B)$. If R_A is the general subrectangle of A determined by P_A, and if R_B is the general subrectangle of B determined by P_B, then $R_A \times R_B$ is the general subrectangle of Q determined by P.

We begin by comparing the lower and upper sums for f with the lower and upper sums for $\underline{I}$ and $\overline{I}$.

Step 1. We first show that

$$L(f,P) \le L(\underline{I}, P_A);$$

that is, the lower sum for f is no larger than the lower sum for the lower integral, $\underline{I}$.

Consider the general subrectangle $R_A \times R_B$ determined by P. Let $\mathbf{x}_0$ be a point of R_A. Now

$$m_{R_A\times R_B}(f) \le f(\mathbf{x}_0,\mathbf{y})$$

for all $\mathbf{y} \in R_B$; hence

$$m_{R_A\times R_B}(f) \le m_{R_B}(f(\mathbf{x}_0,\mathbf{y})).$$

See Figure 12.1. Holding $\mathbf{x}_0$ and R_A fixed, multiply by $v(R_B)$ and sum over all subrectangles R_B. One obtains the inequalities

$$\sum_{R_B} m_{R_A\times R_B}(f)v(R_B) \le L(f(\mathbf{x}_0,\mathbf{y}), P_B) \le \underline{\int}_{\mathbf{y}\in B} f(\mathbf{x}_0,\mathbf{y}) = \underline{I}(\mathbf{x}_0).$$

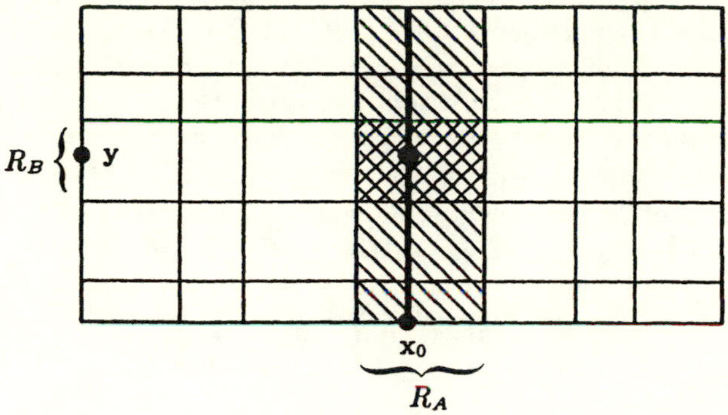

Figure 12.1

This result holds for each $x_0 \in R_A$. We conclude that

$$\sum_{R_B} m_{R_A \times R_B}(f) v(R_B) \le m_{R_A}(\underline{I}).$$

Now multiply through by $v(R_A)$ and sum. Since $v(R_A)v(R_B) = v(R_A \times R_B)$, one obtains the desired inequality

$$L(f,P) \le L(\underline{I}, P_A).$$

Step 2. An entirely similar proof shows that

$$U(f,P) \ge U(\overline{I}, P_A);$$

that is, the upper sum for f is no smaller than the upper sum for the upper integral, $\overline{I}$. The proof is left as an exercise.

Step 3. We summarize the relations that hold among the upper and lower sums of $f, \underline{I}$, and $\overline{I}$ in the following diagram:

$$L(f,P) \le L(\underline{I}, P_A) \begin{matrix} \le U(\underline{I}, P_A) \le \\ \\ \le L(\overline{I}, P_A) \le \end{matrix} U(\overline{I}, P_A) \le U(f,P).$$

The first and last inequalities in this diagram come from Steps 1 and 2. Of the remaining inequalities, the two on the upper left and lower right follow from the fact that $L(h,P) \le U(h,P)$ for any h and P. The ones on the lower left and upper right follow from the fact that $\underline{I}(\mathbf{x}) \le \overline{I}(\mathbf{x})$ for all $\mathbf{x}$. This diagram contains all the information we shall need.

Step 4. We prove the theorem. Because f is integrable over Q, we can, given $\epsilon > 0$, choose a partition $P = (P_A, P_B)$ of Q so that the numbers at the extreme ends of the diagram in Step 3 are within ϵ of each other. Then the upper and lower sums for $\underline{I}$ are within ϵ of each other, and so are the upper and lower sums for $\overline{I}$. It follows that both $\underline{I}$ and $\overline{I}$ are integrable over A.

Now we note that by definition the integral $\int_A \underline{I}$ lies between the upper and lower sums of $\underline{I}$. Similarly, the integral $\int_A \overline{I}$ lies between the upper and lower sums for $\overline{I}$. Hence all three numbers

$$\int_A \underline{I} \quad \text{and} \quad \int_A \overline{I} \quad \text{and} \quad \int_Q f$$

lie between the numbers at the extreme ends of the diagram. Because ϵ is arbitrary, we must have

$$\int_A \underline{I} = \int_A \overline{I} = \int_Q f. \quad \square$$

This theorem expresses $\int_Q f$ as an iterated integral. To compute $\int_Q f$, one first computes the lower integral (or upper integral) of f with respect to $\mathbf{y}$, and then one integrates the resulting function with respect to $\mathbf{x}$. There is nothing special about the order of integration; a similar proof shows that one can compute $\int_Q f$ by first taking the lower integral (or upper integral) of f with respect to $\mathbf{x}$, and then integrating this function with respect to $\mathbf{y}$.

Corollary 12.3. *Let $Q = A \times B$, where A is a rectangle in $\mathbf{R}^k$ and B is a rectangle in $\mathbf{R}^n$. Let $f : Q \to \mathbf{R}$ be a bounded function. If $\int_Q f$ exists, and if $\int_{\mathbf{y}\in B} f(\mathbf{x},\mathbf{y})$ exists for each $\mathbf{x} \in A$, then*

$$\int_Q f = \int_{\mathbf{x}\in A} \int_{\mathbf{y}\in B} f(\mathbf{x},\mathbf{y}). \quad \square$$

Corollary 12.4. *Let $Q = I_1 \times \cdots \times I_n$, where I_j is a closed interval in $\mathbf{R}$ for each j. If $f : Q \to \mathbf{R}$ is continuous, then*

$$\int_Q f = \int_{x_1 \in I_1} \cdots \int_{x_n \in I_n} f(x_1, \ldots, x_n). \quad \square$$

EXERCISES

1. Carry out Step 2 of the proof of Theorem 12.2.
2. Let $I = [0,1]$; let $Q = I \times I$. Define $f : Q \to \mathbf{R}$ by letting $f(x,y) = 1/q$ if y is rational and $x = p/q$, where p and q are positive integers with no common factor; let $f(x,y) = 0$ otherwise.
   (a) Show that $\int_Q f$ exists.
   (b) Compute
   $$\underline{\int}_{y\in I} f(x,y) \quad \text{and} \quad \overline{\int}_{y\in I} f(x,y).$$
   (c) Verify Fubini's theorem.
3. Let $Q = A \times B$, where A is a rectangle in $\mathbf{R}^k$ and B is a rectangle in $\mathbf{R}^n$. Let $f : Q \to \mathbf{R}$ be a bounded function.
   (a) Let g be a function such that
   $$\underline{\int}_{\mathbf{y}\in B} f(\mathbf{x},\mathbf{y}) \le g(\mathbf{x}) \le \overline{\int}_{\mathbf{y}\in B} f(\mathbf{x},\mathbf{y})$$
   for all $\mathbf{x} \in A$. Show that if f is integrable over Q, then g is integrable over A, and $\int_Q f = \int_A g$. [*Hint:* Use Exercise 1 of §10.]
   (b) Give an example where $\int_Q f$ exists and one of the iterated integrals
   $$\int_{\mathbf{x}\in A} \int_{\mathbf{y}\in B} f(\mathbf{x},\mathbf{y}) \quad \text{and} \quad \int_{\mathbf{y}\in B} \int_{\mathbf{x}\in A} f(\mathbf{x},\mathbf{y})$$
   exists, but the other does not.
   *(c) Find an example where both the iterated integrals of (b) exist, but the integral $\int_Q f$ does not. [*Hint:* One approach is to find a subset S of Q whose closure equals Q, such that S contains at most one point on each vertical line and at most one point on each horizontal line.]
4. Let A be open in $\mathbf{R}^2$; let $f : A \to \mathbf{R}$ be of class C^2. Let Q be a rectangle contained in A.
   (a) Use Fubini's theorem and the fundamental theorem of calculus to show that
   $$\int_Q D_2 D_1 f = \int_Q D_1 D_2 f.$$
   (b) Give a proof, independent of the one given in §6, that $D_2 D_1 f(\mathbf{x}) = D_1 D_2 f(\mathbf{x})$ for each $\mathbf{x} \in A$.

§13. THE INTEGRAL OVER A BOUNDED SET

In the applications of integration theory, one usually wishes to integrate functions over sets that are *not* rectangles. The problem of finding the mass of a circular plate of variable density, for instance, involves integrating a function over a circular region. So does the problem of finding the center of gravity of a spherical cap. Therefore we seek to generalize our definition of the integral. That is not in fact difficult.

Definition. Let S be a bounded set in $\mathbf{R}^n$; let $f : S \to \mathbf{R}$ be a bounded function. Define $f_S : \mathbf{R}^n \to \mathbf{R}$ by the equation

$$f_S(\mathbf{x}) = \begin{cases} f(\mathbf{x}) & \text{for } \mathbf{x} \in S, \\ 0 & \text{otherwise.} \end{cases}$$

Choose a rectangle Q containing S. We define the **integral** of f over S by the equation

$$\int_S f = \int_Q f_S,$$

provided the latter integral exists.

We must show this definition is independent of the choice of Q. That is the substance of the following lemma:

Lemma 13.1. *Let Q and Q' be two rectangles in $\mathbf{R}^n$. If $f : \mathbf{R}^n \to \mathbf{R}$ is a bounded function that vanishes outside $Q \cap Q'$, then*

$$\int_Q f = \int_{Q'} f;$$

one integral exists if and only if the other does.

Proof. We consider first the case where $Q \subset Q'$. Let E be the set of points of Int Q at which f fails to be continuous. Then both the maps $f : Q \to \mathbf{R}$ and $f : Q' \to \mathbf{R}$ are continuous except at points of E and possibly at points of Bd Q. Existence of each integral is thus equivalent to the requirement that E have measure zero.

Now suppose both integrals exist. Let P be a partition of Q', and let P'' be the refinement of P obtained from P by adjoining the end points of the component intervals of Q. Then Q is a union of subrectangles R determined by P''. See Figure 13.1. If R is a subrectangle determined by P'' that is not contained in Q, then f vanishes at some point of R, whence $m_R(f) \leq 0$. It follows that

$$L(f, P'') \leq \sum_{R \subset Q} m_R(f)\, v(R) \leq \int_Q f.$$

We conclude that $L(f,P) \leq \int_Q f$.

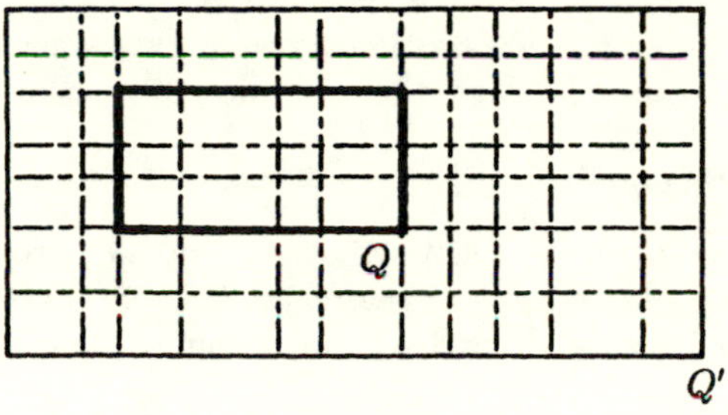

Figure 13.1

An entirely similar argument shows that $U(f,P) \geq \int_Q f$. Since P is an arbitrary partition of Q', it follows that $\int_Q f = \int_{Q'} f$.

The proof for an arbitrary pair of rectangles Q, Q' involves choosing a rectangle Q'' containing them both, and noting that $\int_Q f = \int_{Q''} f = \int_{Q'} f$. □

In the remainder of this section, we study the basic *properties* of this integral, and we obtain conditions for its *existence*. In the next section, we derive (as far as we are able) a method for its *evaluation*.

Lemma 13.2. *Let S be a subset of $\mathbf{R}^n$; let $f, g : S \to \mathbf{R}^n$. Let $F, G : S \to \mathbf{R}^n$ be defined by the equations*

$$F(\mathbf{x}) = \max\{f(\mathbf{x}), g(\mathbf{x})\} \quad \textit{and} \quad G(\mathbf{x}) = \min\{f(\mathbf{x}), g(\mathbf{x})\}.$$

(a) *If f and g are continuous at $\mathbf{x}_0$, so are F and G.*

(b) *If f and g are integrable over S, so are F and G.*

Proof. (a) Suppose f and g are continuous at $\mathbf{x}_0$. Consider first the case in which $f(\mathbf{x}_0) = g(\mathbf{x}_0) = r$. Then $F(\mathbf{x}_0) = G(\mathbf{x}_0) = r$. By continuity, given $\epsilon > 0$, we can choose $\delta > 0$ so that

$$|f(\mathbf{x}) - r| < \epsilon \quad \text{and} \quad |g(\mathbf{x}) - r| < \epsilon$$

for $|\mathbf{x} - \mathbf{x}_0| < \delta$ and $\mathbf{x} \in S$; for such values of $\mathbf{x}$, it follows automatically that

$$|F(\mathbf{x}) - F(\mathbf{x}_0)| < \epsilon \quad \text{and} \quad |G(\mathbf{x}) - G(\mathbf{x}_0)| < \epsilon.$$

On the other hand, suppose $f(\mathbf{x}_0) > g(\mathbf{x}_0)$. By continuity, we can find a neighborhood U of $\mathbf{x}_0$ such that $f(\mathbf{x}) - g(\mathbf{x}) > 0$ for $\mathbf{x} \in U$ and $\mathbf{x} \in S$.

Then $F(\mathbf{x}) = f(\mathbf{x})$ and $G(\mathbf{x}) = g(\mathbf{x})$ on $U \cap S$; it follows that F and G are continuous at $\mathbf{x}_0$. A similar argument holds if $f(\mathbf{x}_0) < g(\mathbf{x}_0)$.

(b) Suppose f and g are integrable over S. Let Q be a rectangle containing S. Then f_S and g_S are continuous on Q except on subsets D and E, respectively, of Q, each of measure zero. Now

$$F_S(\mathbf{x}) = \max\{f_S(\mathbf{x}), g_S(\mathbf{x})\} \quad \text{and} \quad G_S(\mathbf{x}) = \min\{f_S(\mathbf{x}), g_S(\mathbf{x})\},$$

as you can easily check. It follows that F_S and G_S are continuous on Q except on the set $D \cup E$, which has measure zero. Furthermore, F_S and G_S are bounded because f_S and g_S are. Then F_S and G_S are integrable over Q. □

Theorem 13.3 (Properties of the integral). *Let S be a bounded set in $\mathbf{R}^n$; let $f, g : S \to \mathbf{R}$ be bounded functions.*

(a) *(Linearity). If f and g are integrable over S, so is $af + bg$, and*

$$\int_S (af + bg) = a\int_S f + b\int_S g.$$

(b) *(Comparison). Suppose f and g are integrable over S. If $f(\mathbf{x}) \leq g(\mathbf{x})$ for $\mathbf{x} \in S$, then*

$$\int_S f \leq \int_S g.$$

Furthermore, $|f|$ is integrable over S and

$$\left|\int_S f\right| \leq \int_S |f|.$$

(c) *(Monotonicity). Let $T \subset S$. If f is non-negative on S and integrable over T and S, then*

$$\int_T f \leq \int_S f.$$

(d) *(Additivity). If $S = S_1 \cup S_2$ and f is integrable over S_1 and S_2, then f is integrable over S and $S_1 \cap S_2$; furthermore*

$$\int_S f = \int_{S_1} f + \int_{S_2} f - \int_{S_1 \cap S_2} f.$$

Proof. (a) It suffices to prove this result for the integral over a rectangle, since

$$(af + bg)_S = af_S + bg_S.$$

So suppose f and g are integrable over Q. Then f and g are continuous except on sets D, E, respectively, of measure zero. It follows that the function $af+bg$ is continuous except on the set $D \cup E$, so it is integrable over Q.

We consider first the case where $a, b \geq 0$. Let P'' be an arbitrary partition of Q. If R is a subrectangle determined by P'', then

$$a\, m_R(f) + b\, m_R(g) \leq a\, f(\mathbf{x}) + b\, g(\mathbf{x})$$

for all $\mathbf{x} \in R$. It follows that

$$a\, m_R(f) + b\, m_R(g) \leq m_R(af + bg),$$

so that

$$a\, L(f, P'') + b\, L(g, P'') \leq L(af + bg, P'') \leq \int_Q (af + bg).$$

A similar argument shows that

$$a\, U(f, P'') + b\, U(g, P'') \geq \int_Q (af + bg).$$

Now let P and P' be any two partitions of Q, and let P'' be their common refinement. It follows from what have just proved that

$$a\, L(f, P) + b\, L(g, P') \leq \int_Q (af + bg) \leq a\, U(f, P) + b\, U(g, P').$$

Now by definition the number $a \int_Q f + b \int_Q g$ also lies between the numbers at the ends of this sequence of inequalities. Since P and P' are arbitrary, we conclude that

$$\int_Q (af + bg) \quad = \quad a \int_Q f \quad + \quad b \int_Q g.$$

Now we complete the proof by showing that

$$\int_Q (-f) = -\int_Q f.$$

Let P be a partition of Q; let R be a subrectangle determined by P. For $\mathbf{x} \in R$, we have

$$-M_R(f) \leq -f(\mathbf{x}) \leq -m_R(f),$$

so that

$$-M_R(f) \leq m_R(-f) \quad \text{and} \quad M_R(-f) \leq -m_R(f).$$

Multiplying by $v(R)$ and summing, we obtain the inequalities

$$-U(f,P) \leq L(-f,P) \leq \int_Q (-f) \leq U(-f,P) \leq -L(f,P).$$

By definition, the number $-\int_Q f$ also lies between the numbers at the extreme ends of this sequence of inequalities. Since P is arbitrary, our result follows.

(b) It suffices to prove the comparison property for the integral over a rectangle. So suppose $f(\mathbf{x}) \leq g(\mathbf{x})$ for $\mathbf{x} \in Q$. If R is any rectangle contained in Q, then

$$m_R(f) \leq f(\mathbf{x}) \leq g(\mathbf{x})$$

for each $\mathbf{x} \in R$. Then $m_R(f) \leq m_R(g)$. It follows that if P is any partition of Q,

$$L(f,P) \leq L(g,P) \leq \int_Q g.$$

Since P is arbitrary, we conclude that

$$\int_Q f \leq \int_Q g.$$

The fact that $|f|$ is integrable over S follows from the equation

$$|f(\mathbf{x})| = \max\{f(\mathbf{x}), -f(\mathbf{x})\}.$$

The desired inequality follows by applying the comparison property to the inequalities

$$-|f(\mathbf{x})| \leq f(\mathbf{x}) \leq |f(\mathbf{x})|.$$

(c) If f is non-negative and if $T \subset S$, then $f_T(\mathbf{x}) \leq f_S(\mathbf{x})$ for all $\mathbf{x}$. One then applies the comparison property.

(d) Let $T = S_1 \cap S_2$. We prove f is integrable over S and T. Consider first the special case where f is non-negative on S. Let Q be a rectangle containing S. Then both f_{S_1} and f_{S_2} are integrable over Q by hypothesis. It follows from the equations

$$f_S(\mathbf{x}) = \max\{f_{S_1}(\mathbf{x}), f_{S_2}(\mathbf{x})\} \quad \text{and} \quad f_T(\mathbf{x}) = \min\{f_{S_1}(\mathbf{x}), f_{S_2}(\mathbf{x})\}$$

that f_S and f_T are integrable over Q.

In the general case, set

$$f_+(\mathbf{x}) = \max\{f(\mathbf{x}), 0\} \quad \text{and} \quad f_-(\mathbf{x}) = \max\{-f(\mathbf{x}), 0\}.$$

Since f is integrable over S_1 and S_2, so are f_+ and f_-. By the special case already considered, f_+ and f_- are integrable over S and T. Because

$$f(\mathbf{x}) = f_+(\mathbf{x}) - f_-(\mathbf{x}),$$

it follows from linearity that f is integrable over S and T.

The desired additivity formula follows by applying linearity to the equation

$$f_S(\mathbf{x}) = f_{S_1}(\mathbf{x}) + f_{S_2}(\mathbf{x}) - f_T(\mathbf{x}). \quad \square$$

Corollary 13.4. *Let $S_1, \ldots, S_k$ be bounded sets in $\mathbf{R}^n$; assume $S_i \cap S_j$ has measure zero whenever $i \neq j$. Let $S = S_1 \cup \cdots \cup S_k$. If $f : S \to \mathbf{R}$ is integrable over each set S_i, then f is integrable over S and*

$$\int_S f = \int_{S_1} f + \cdots + \int_{S_k} f.$$

Proof. The case $k = 2$ follows from additivity, since the integral of f over $S_1 \cap S_2$ vanishes by Theorem 11.3. The general case follows by induction. $\square$

Up to this point, we have made no *a priori* restrictions on the functions f we deal with in integration theory, other than that they be bounded. In particular, we have not required f to be continuous. The reason is obvious; in order to define the integral $\int_S f$, even in the case where f is continuous on S, we needed to deal with the function f_S, which need not be continuous at points of Bd S.

However, our primary interest in this book is in integrals of the form $\int_S f$, where f is continuous on S. Therefore we make the following:

Convention. *Henceforth, we restrict ourselves in studying integration theory to the integration of continuous functions $f : S \to \mathbf{R}$.*

Now we consider conditions under which the integral $\int_S f$ exists. Even if we assume f is bounded and continuous on S, we need some sort of condition involving the set S to ensure that $\int_S f$ exists. That condition is the following:

Theorem 13.5. *Let S be a bounded set in $\mathbf{R}^n$; let $f : S \to \mathbf{R}$ be a bounded continuous function. Let E be the set of points $\mathbf{x}_0$ of* Bd S *for which the condition*

$$\lim_{\mathbf{x} \to \mathbf{x}_0} f(\mathbf{x}) = 0$$

fails to hold. If E has measure zero, then f is integrable over S.

The converse of this theorem also holds; since we shall not need it, we leave the proof to the exercises.

Proof. Let $\mathbf{x}_0$ be a point of $\mathbf{R}^n$ not in E. We show that the function f_S is continuous at $\mathbf{x}_0$; the theorem follows.

If $\mathbf{x}_0 \in \text{Int } S$, then the functions f and f_S agree in a neighborhood of $\mathbf{x}_0$; since f is continuous at $\mathbf{x}_0$, so is f_S. If $\mathbf{x}_0 \in \text{Ext } S$, then f_S vanishes in a neighborhood of $\mathbf{x}_0$. Suppose $\mathbf{x}_0 \in \text{Bd } S$; then $\mathbf{x}_0$ may or may not belong to S. See Figure 13.2. Since $\mathbf{x}_0 \notin E$, we know that $f(\mathbf{x}) \to 0$ as $\mathbf{x}$ approaches $\mathbf{x}_0$ through points of S. Since f is continuous, it follows that $f(\mathbf{x}_0) = 0$ if $\mathbf{x}_0$ belongs to S. It also follows, since $f_S(\mathbf{x})$ equals either $f(\mathbf{x})$ or 0, that $f_S(\mathbf{x}) \to 0$, as $\mathbf{x}$ approaches $\mathbf{x}_0$ through points of $\mathbf{R}^n$. To show that f_S is continuous at $\mathbf{x}_0$, we must show that $f_S(\mathbf{x}_0) = 0$. If $\mathbf{x}_0 \notin S$, this follows by definition. If $\mathbf{x}_0 \in S$, then $f_S(\mathbf{x}_0) = f(\mathbf{x}_0)$, which vanishes, as noted earlier. □

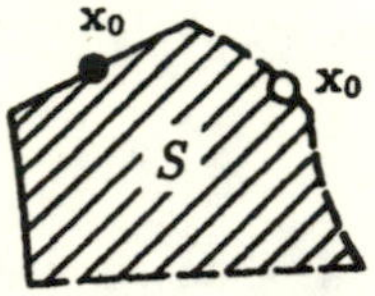

Figure 13.2

The same techniques may be used to prove the following theorem, which is sometimes useful:

Theorem 13.6. *Let S be a bounded set in $\mathbf{R}^n$; let $f : S \to \mathbf{R}$ be a bounded continuous function; let $A = \text{Int } S$. If f is integrable over S, then f is integrable over A, and $\int_S f = \int_A f$.*

Proof. Step 1. We show that if f_S is continuous at $\mathbf{x}_0$, then f_A is continuous, and agrees with f_S, at $\mathbf{x}_0$. The proof is easy. If $\mathbf{x}_0 \in \text{Int } S$ or $\mathbf{x}_0 \in \text{Ext } S$, then f_S and f_A agree in a neighborhood of $\mathbf{x}_0$, and the result is trivial. Let $\mathbf{x}_0 \in \text{Bd } S$. Continuity of f_S at $\mathbf{x}_0$ implies that $f_S(\mathbf{x}) \to f_S(\mathbf{x}_0)$ as $\mathbf{x} \to \mathbf{x}_0$. Arbitrarily near $\mathbf{x}_0$ are points $\mathbf{x}$ not in S, for which $f_S(\mathbf{x}) = 0$; hence this limit must be 0. Thus $f_S(\mathbf{x}_0) = 0$. Since $f_A(\mathbf{x})$ equals either $f_S(\mathbf{x})$ or 0, we have $f_A(\mathbf{x}) \to 0$ also as $\mathbf{x} \to \mathbf{x}_0$. Furthermore, $f_A(\mathbf{x}_0) = 0$ because $\mathbf{x}_0 \notin A$. Thus f_A is continuous at $\mathbf{x}_0$ and agrees with f_S at $\mathbf{x}_0$.

Step 2. We prove the theorem. If f is integrable over S, then f_S is continuous except on a set D of measure zero. Then f_A is continuous at points not in D, so f is integrable over A. Since $f_S - f_A$ vanishes at points not in D, we have $\int_Q (f_S - f_A) = 0$, where Q is a rectangle containing S. Then $\int_S f = \int_A f$. □

EXERCISES

1. Let $f, g : S \to \mathbf{R}$; assume f and g are integrable over S.
   (a) Show that if f and g agree except on a set of measure zero, then $\int_S f = \int_S g$.
   (b) Show that if $f(\mathbf{x}) \leq g(\mathbf{x})$ for $\mathbf{x} \in S$ and $\int_S f = \int_S g$, then f and g agree except on a set of measure zero.
2. Let A be a rectangle in $\mathbf{R}^k$; let B be a rectangle in $\mathbf{R}^n$; let $Q = A \times B$. Let $f : Q \to \mathbf{R}$ be a bounded function. Show that if $\int_Q f$ exists, then
$$\int_{\mathbf{y} \in B} f(\mathbf{x}, \mathbf{y})$$
exists for $\mathbf{x} \in A - D$, where D is a set of measure zero in $\mathbf{R}^k$.
3. Complete the proof of Corollary 13.4.
4. Let S_1 and S_2 be bounded sets in $\mathbf{R}^n$; let $f : S_1 \cup S_2 \to \mathbf{R}$ be a bounded function. Show that if f is integrable over S_1 and S_2, then f is integrable over $S_1 - S_2$, and
$$\int_{S_1 - S_2} f = \int_{S_1} f - \int_{S_1 \cap S_2} f.$$
5. Let S be a bounded set in $\mathbf{R}^n$; let $f : S \to \mathbf{R}$ be a bounded continuous function; let $A = \text{Int } S$. Give an example where $\int_A f$ exists and $\int_S f$ does not.
6. Show that Theorem 13.6 holds without the hypothesis that f is continuous on S.

*7. Prove the following:

Theorem. *Let S be a bounded set in $\mathbf{R}^n$; let $f : S \to \mathbf{R}$ be a bounded function. Let D be the set of points of S at which f fails to be continuous. Let E be the set of points of Bd S at which the condition*
$$\lim_{\mathbf{x} \to \mathbf{x}_0} f(\mathbf{x}) = 0$$
fails to hold. Then $\int_S f$ exists if and only if D and E have measure zero.

Proof. (a) Show that f_S is continuous at each point $\mathbf{x}_0 \notin D \cup E$.
(b) Let B be the set of isolated points of S; then $B \subset E$ because the limit cannot be defined if $\mathbf{x}_0$ is not a limit point of S. Show that if f_S is continuous at $\mathbf{x}_0$, then $\mathbf{x}_0 \notin D \cup (E - B)$.
(c) Show that B is countable.
(d) Complete the proof.

§14. RECTIFIABLE SETS

We now extend the volume function, defined for rectangles, to more general subsets of $\mathbf{R}^n$. Then we relate this notion to integration theory, and extend the Fubini theorem to certain integrals of the form $\int_S f$.

Definition. Let S be a bounded set in $\mathbf{R}^n$. If the constant function 1 is integrable over S, we say that S is **rectifiable**, and we define the (n-dimensional) **volume** of S by the equation

$$v(S) = \int_S 1.$$

Note that this definition agrees with our previous definition of volume when S is a rectangle.

Theorem 14.1. *A subset S of $\mathbf{R}^n$ is rectifiable if and only if S is bounded and* $\operatorname{Bd} S$ *has measure zero.*

Proof. The function 1_S that equals 1 on S and 0 outside S is continuous on the open sets Ext S and Int S. It fails to be continuous at each point of $\operatorname{Bd} S$. By Theorem 11.2, the function 1_S is integrable over a rectangle Q containing S if and only if $\operatorname{Bd} S$ has measure zero. □

We list some properties of rectifiable sets.

Theorem 14.2. (a) *(Positivity). If S is rectifiable, $v(S) \geq 0$.*

(b) *(Monotonicity). If S_1 and S_2 are rectifiable and if $S_1 \subset S_2$, then $v(S_1) \leq v(S_2)$.*

(c) *(Additivity). If S_1 and S_2 are rectifiable, so are $S_1 \cup S_2$ and $S_1 \cap S_2$, and*

$$v(S_1 \cup S_2) = v(S_1) + v(S_2) - v(S_1 \cap S_2).$$

(d) *Suppose S is rectifiable. Then $v(S) = 0$ if and only if S has measure zero.*

(e) *If S is rectifiable, so is the set $A =$ Int S, and $v(S) = v(A)$.*

(f) *If S is rectifiable, and if $f : S \to \mathbf{R}$ is a bounded continuous function, then f is integrable over S.*

Proof. Parts (a), (b), and (c) follow from Theorem 13.3. Part (d) follows by applying Theorem 11.3 to the non-negative function 1_S. Part (e) follows from Theorem 13.6, and (f) from Theorem 13.5. □

Let us make a remark on terminology. The concept of volume, as we have defined it, was called classically the theory of *content* (or Jordan content). This terminology distinguishes this concept from a more general one, called *measure* (or Lebesgue measure). This concept is important in the development of an integral called the Lebesgue integral, which is a generalization of the Riemann integral.

Measure is defined for a larger class of sets than content is, but the two concepts agree when both are defined. A "set of measure zero" as we have defined it is in fact just a set whose Lebesgue measure exists and equals zero. Such a set need not of course be rectifiable.

A set whose Lebesgue measure is defined is usually called *measurable.* But there is no universally accepted corresponding term for a set whose Jordan content is defined. Some call such sets "Jordan-measurable"; others refer to such sets as "domains of integration," because bounded continuous functions are integrable over such sets. One student suggested to me that a set whose Jordan content is defined should be called "contented"! I have taken the term *rectifiable,* which is commonly used to refer to a curve whose length is defined, and have adopted it to refer to any set having volume (content).

The class of rectifiable sets in $\mathbb{R}^n$ is not easy to describe other than by the condition stated in Theorem 14.1. It is tempting to think, for instance, that any bounded open set in $\mathbb{R}^n$, or any bounded closed set in $\mathbb{R}^n$, should be rectifiable. That is not the case, as the following example shows:

EXAMPLE 1. We construct a bounded open set A in $\mathbb{R}$ such that Bd A does *not* have measure zero.

The rational numbers in the open interval (0,1) are countable; let us arrange them in a sequence $q_1, q_2, \ldots$. Let $0 < a < 1$ be fixed. For each i, choose an open interval (a_i, b_i) of length less than $a/2^i$ that contains q_i and is contained in (0,1). These intervals will overlap, of course, but that doesn't matter. Let A be the following open set of $\mathbb{R}$:

$$A = (a_1, b_1) \cup (a_2, b_2) \cup \cdots.$$

We assume Bd A has measure zero and derive a contradiction. Set $\epsilon = 1 - a$. Since Bd A has measure zero, we may cover Bd A by countably many open intervals of total length less than ϵ. Because A is a subset of [0,1] that contains each rational in (0,1), we have $\bar{A} = [0,1]$. Since $\bar{A} = A \cup \text{Bd}\, A$, the open intervals covering Bd A, along with the open intervals (a_i, b_i) whose union is A, give an open covering of the interval [0,1]. The total length of the intervals covering Bd A is less than ϵ, and the total length of the intervals covering A is less than $\sum a/2^i = a$. Because [0,1] is compact, it can be covered by finitely many of these intervals; the total length of these intervals is less than $\epsilon + a = 1$. This contradicts Corollary 10.5.

We conclude this section by discussing certain rectifiable sets that are especially useful; they are called the "simple regions." For these sets, a version of the Fubini theorem holds, as we shall see. We shall use these results only in the examples and the exercises.

Definition. Let C be a compact rectifiable set in $\mathbf{R}^{n-1}$; let $\phi, \psi : C \to \mathbf{R}$ be continuous functions such that $\phi(\mathbf{x}) \leq \psi(\mathbf{x})$ for $\mathbf{x} \in C$. The subset S of $\mathbf{R}^n$ defined by the equation

$$S = \{(\mathbf{x},t) \mid \mathbf{x} \in C \quad \text{and} \quad \phi(\mathbf{x}) \leq t \leq \psi(\mathbf{x})\}$$

is called a **simple region** in $\mathbf{R}^n$.

There is nothing special about the last coordinate here. If $k + \ell = n - 1$, and if $\mathbf{y}$ and $\mathbf{z}$ denote the general points of $\mathbf{R}^k$ and $\mathbf{R}^\ell$, respectively, then the set

$$S' = \{(\mathbf{y},t,\mathbf{z}) \mid (\mathbf{y},\mathbf{z}) \in C \quad \text{and} \quad \phi(\mathbf{y},\mathbf{z}) \leq t \leq \psi(\mathbf{y},\mathbf{z})\}$$

is also called a simple region in $\mathbf{R}^n$.

***Lemma 14.3.** *If S is a simple region in $\mathbf{R}^n$, then S is compact and rectifiable.*

Proof. Let S be a simple region, as in the definition. We show that S is compact and that $\operatorname{Bd} S$ has measure zero.

Step 1. The graph of ϕ is the subset of $\mathbf{R}^n$ defined by the equation

$$G_\phi = \{(\mathbf{x},t) \mid \mathbf{x} \in C \quad \text{and} \quad t = \phi(\mathbf{x})\}.$$

We show that $\operatorname{Bd} S$ lies in the union of the three sets G_ϕ and G_ψ and

$$D = \{(\mathbf{x},t) \mid \mathbf{x} \in \operatorname{Bd} C \quad \text{and} \quad \phi(\mathbf{x}) \leq t \leq \psi(\mathbf{x})\}.$$

Since each of these sets is contained in S, it follows that $\operatorname{Bd} S \subset S$, so that S is closed. Being bounded, S is thus compact. See Figure 14.1.

Suppose that $(\mathbf{x}_0, t_0)$ belongs to none of the sets G_ϕ, G_ψ, or D. We show that $(\mathbf{x}_0, t_0)$ lies either in $\operatorname{Int} S$ or $\operatorname{Ext} S$. As you can check, there are three possibilities:

(1) $\mathbf{x}_0 \notin C$,

(2) $\mathbf{x}_0 \in C$ and either $t_0 < \phi(\mathbf{x}_0)$ or $t_0 > \psi(\mathbf{x}_0)$,

(3) $\mathbf{x}_0 \in \operatorname{Int} C$ and $\phi(\mathbf{x}_0) < t_0 < \psi(\mathbf{x}_0)$.

In case (1), there is a neighborhood U of $\mathbf{x}_0$ disjoint from C. Then $U \times \mathbf{R}$ is disjoint from S, so that $(\mathbf{x}_0, t_0) \in \operatorname{Ext} S$.

Consider case (2). Suppose that $t_0 < \phi(\mathbf{x}_0)$. By continuity of ϕ, we can choose a neighborhood W of $(\mathbf{x}_0, t_0)$ such that the function $\phi(\mathbf{x}) - t$ is positive

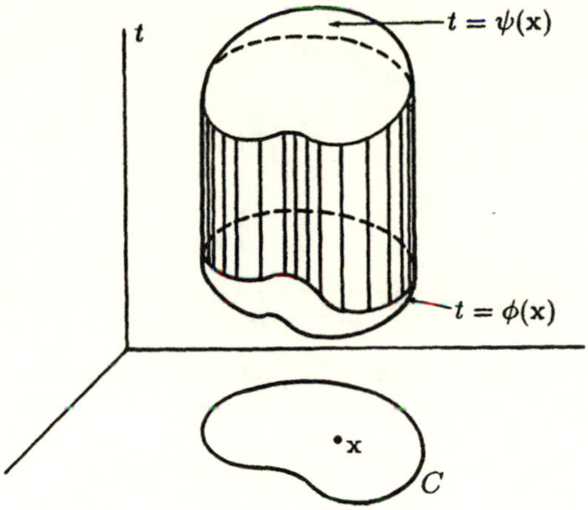

Figure 14.1

for $\mathbf{x} \in C$ and $(\mathbf{x},t) \in W$. Then W is disjoint from S, so that $(\mathbf{x}_0, t_0) \in \text{Ext}\ S$. A similar argument applies if $t_0 > \psi(\mathbf{x}_0)$.

Consider case (3). By continuity, there is a neighborhood $U \times V$ of $(\mathbf{x}_0, t_0)$ in $\mathbf{R}^n$ such that $U \subset C$ and both functions $t - \phi(\mathbf{x})$ and $\psi(\mathbf{x}) - t$ are positive on $U \times V$. Then $U \times V$ is contained in S, so that $(\mathbf{x}_0, t_0) \in \text{Int}\ S$.

Step 2. We show that G_ϕ and G_ψ have measure zero.

It suffices to consider the case of G_ϕ. Choose a rectangle Q in $\mathbf{R}^{n-1}$ containing the set C. Given $\epsilon > 0$, let ϵ' be the number $\epsilon' = \epsilon/2v(Q)$. Because ϕ is continuous and C is compact, there is, by the theorem on uniform continuity, a $\delta > 0$ such that $|\phi(\mathbf{x}) - \phi(\mathbf{y})| < \epsilon'$ whenever $\mathbf{x}, \mathbf{y} \in C$ and $|\mathbf{x} - \mathbf{y}| < \delta$. Choose a partition P of Q of mesh less than δ. If R is a subrectangle determined by P, and if R intersects C, then $|\phi(\mathbf{x}) - \phi(\mathbf{y})| < \epsilon'$ for $\mathbf{x}, \mathbf{y} \in R \cap C$. For each such R, choose a point $\mathbf{x}_R$ of $R \cap C$ and define I_R to be the interval

$$I_R = [\phi(\mathbf{x}_R) - \epsilon', \phi(\mathbf{x}_R) + \epsilon'].$$

Then the n-dimensional rectangle $R \times I_R$ contains every point of the form $(\mathbf{x}, \phi(\mathbf{x}))$ for which $\mathbf{x} \in C \cap R$. See Figure 14.2.

The rectangles $R \times I_R$, as R ranges over all subrectangles that intersect C, thus cover G_ϕ. Their total volume is

$$\sum_R v(R \times I_R) = \sum_R v(R)\,(2\epsilon') \leq 2\epsilon' v(Q) = \epsilon.$$

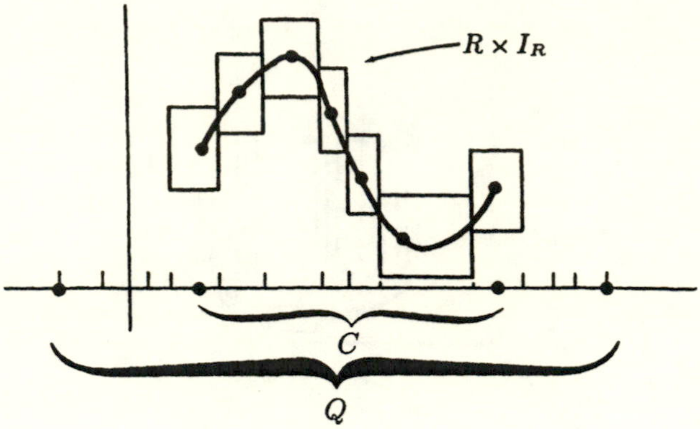

Figure 14.2

Step 3. We show the set D has measure zero; then the proof is complete. Because ϕ and ψ are continuous and C is compact, there is a number M such that

$$-M \leq \phi(\mathbf{x}) \leq \psi(\mathbf{x}) \leq M$$

for $\mathbf{x} \in C$. Given $\epsilon > 0$, cover Bd C by rectangles $Q_1, Q_2, \ldots$ in $\mathbb{R}^{n-1}$ of total volume less than $\epsilon/2M$. Then the rectangles $Q_i \times [-M, M]$ in $\mathbb{R}^n$ cover D and have total volume less than ϵ. □

***Theorem 14.4 (Fubini's theorem for simple regions).** *Let*

$$S = \{(\mathbf{x},t) |\ \mathbf{x} \in C \quad and \quad \phi(\mathbf{x}) \leq t \leq \psi(\mathbf{x})\}$$

be a simple region in $\mathbb{R}^n$*. Let* $f : S \to \mathbb{R}$ *be a continuous function. Then* f *is integrable over* S*, and*

$$\int_S f \quad = \quad \int_{\mathbf{x} \in C} \int_{t=\phi(\mathbf{x})}^{t=\psi(\mathbf{x})} f(\mathbf{x},t).$$

Proof. Let $Q \times [-M, M]$ be a rectangle in $\mathbb{R}^n$ containing S. Because f is continuous and bounded on S and S is rectifiable, f is integrable over S. Furthermore, for fixed $\mathbf{x}_0 \in Q$, the function $f_S(\mathbf{x}_0, t)$ is either identically zero (if $\mathbf{x}_0 \notin C$), or it is continuous at all but two points of **R**. We conclude from Fubini's theorem that

$$\int_S f = \int_{\mathbf{x}\in Q} \int_{t=-M}^{t=M} f_S(\mathbf{x},t).$$

Since the inner integral vanishes if $\mathbf{x} \notin C$, we can write this equation as

$$\int_S f = \int_{\mathbf{x}\in C} \int_{t=-M}^{t=M} f_S(\mathbf{x},t).$$

Furthermore, the number $f_S(\mathbf{x},t)$ vanishes unless $\phi(\mathbf{x}) \leq t \leq \psi(\mathbf{x})$, in which case it equals $f(\mathbf{x},t)$. Therefore we can write

$$\int_S f = \int_{\mathbf{x}\in C} \int_{t=\phi(\mathbf{x})}^{t=\psi(\mathbf{x})} f(\mathbf{x},t). \quad \square$$

The preceding theorem gives us a reasonable method for reducing the n-dimensional integral $\int_S f$ to lower-dimensional integrals, at least if the integrand is continuous and the set S is a simple region.

If the set S is *not* a simple region, one can often in practice express S as a union of simple regions that overlap in sets of measure zero. Additivity of the integral tells us that we can evaluate the integral $\int_S f$ by integrating over each of these regions separately and adding the results together. Just as in calculus, the procedure can be reasonably laborious. But at least it is straightforward.

Of course, there are rectifiable sets that cannot be broken up in this way into simple regions. Computing integrals over such sets is more difficult. One way of proceeding is to approximate S by a union of simple regions and follow a limiting procedure.

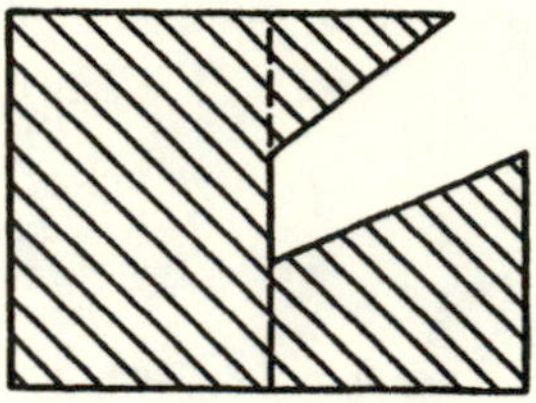

Figure 14.3

EXAMPLE 2. Suppose one wishes to integrate a continuous function f over the set S in $\mathbf{R}^2$ pictured in Figure 14.3. While S is not a simple region, it is easy to break S up into simple regions that overlap in sets of measure zero, as indicated by the dotted lines.

EXAMPLE 3. Consider the set S in $\mathbf{R}^2$ given by

$$S = \{(x, y) \mid 1 \leq x^2 + y^2 \leq 4\};$$

it is pictured in Figure 14.4. While S is not a simple region, one can evaluate an integral over S by breaking S up into two simple regions that overlap in a set of measure zero, as indicated, and integrating over each of these regions separately. The limits of integration will be rather unpleasant, of course.

Now if one were actually assigned a problem like this in a calculus course, one would do no such thing! What one would do instead would be to express the integral in terms of polar coordinates, thereby obtaining an integral with much simpler limits of integration.

Expressing a two-dimensional integral in terms of polar coordinates is a special case of a quite general method for evaluating integrals, which is called "substitution" or "change of variables." We shall deal with it in the next chapter.

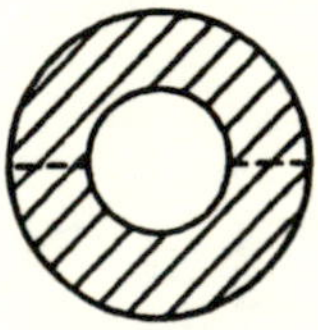

Figure 14.4

Let us make one final remark. There is one thing lacking in our discussion of the notion of volume. How do we know that the volume of a set is independent of its position in space? Said differently, if S is a rectifiable set, and if $h : \mathbf{R}^n \to \mathbf{R}^n$ is a rigid motion (whatever that means), how do we know that the sets S and $h(S)$ have the same volume?

For example, each of the sets S and T pictured in Figure 14.5 represents a square with edge length 5; in fact T is obtained by rotating S through the angle $\theta = \arctan 3/4$. It is immediate from the definition that S has volume 25. It is clear that T is rectifiable, for it is a simple region. *But how do we know T has volume* 25?

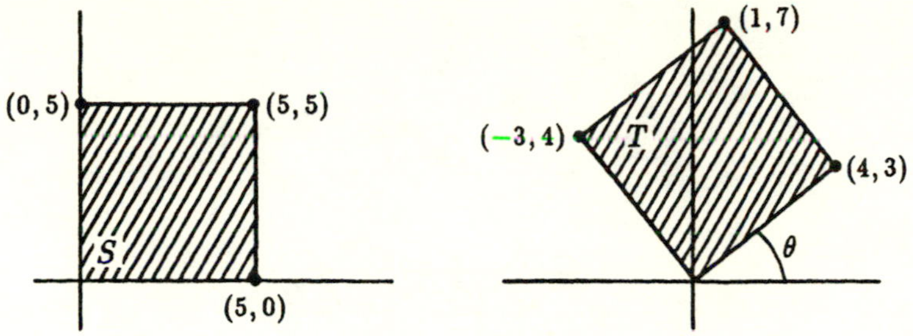

Figure 14.5

One can of course simply calculate $v(T)$. One way to proceed is to write equations for the functions $\psi(x)$ and $\phi(x)$ whose graphs bound T above and below respectively, and to integrate the function $\psi(x) - \phi(x)$ over the interval $[-3, 4]$. See Figure 14.6.

Another way to proceed is to enclose T in a rectangle Q, take a partition P of Q, and calculate the upper and lower sums of the function 1_T with respect to P. The lower sum equals the total area of all subrectangles contained in T,

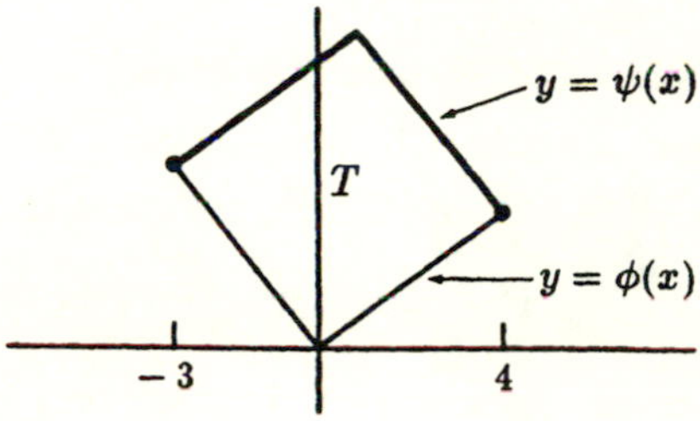

Figure 14.6

while the upper sum equals the total area of all subrectangles that intersect T. One needs to show that

$$L(1_T, P) \leq 25 \leq U(1_T, P)$$

for all P. See Figure 14.7.

Neither of these procedures is especially appealing! What one needs is a general theorem. In the next chapter, we shall prove the following result:

Suppose $h : \mathbf{R}^n \to \mathbf{R}^n$ is a function satisfying the condition

$$\|h(\mathbf{x}) - h(\mathbf{y})\| = \|\mathbf{x} - \mathbf{y}\|$$

for all $\mathbf{x}, \mathbf{y} \in \mathbf{R}^n$; such a function is called an **isometry.** If S is a rectifiable set in $\mathbf{R}^n$, then the set $T = h(S)$ is also rectifiable, and $v(T) = v(S)$.

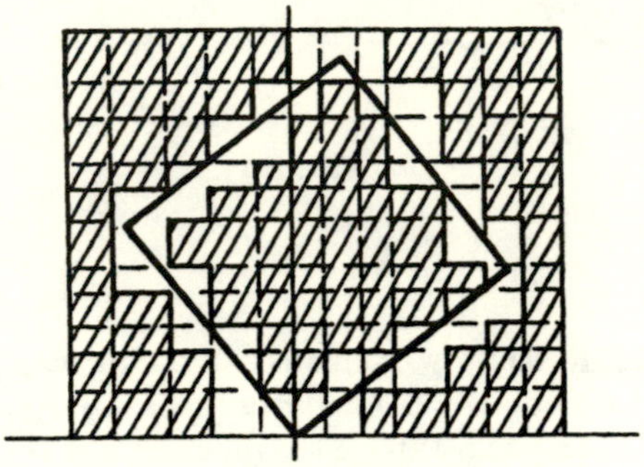

Figure 14.7

EXERCISES

1. Let S be a bounded set in $\mathbf{R}^n$ that is the union of the countable collection of rectifiable sets $S_1, S_2, \ldots$.
   (a) Show that $S_1 \cup \cdots \cup S_n$ is rectifiable.
   (b) Give an example showing that S need not be rectifiable.
2. Show that if S_1 and S_2 are rectifiable, so is $S_1 - S_2$, and

$$v(S_1 - S_2) = v(S_1) - v(S_1 \cap S_2).$$

3. Show that if A is a nonempty, rectifiable open set in $\mathbf{R}^n$, then $v(A) > 0$.
4. Give an example of a bounded set of measure zero that is rectifiable, and an example of a bounded set of measure zero that is not rectifiable.
5. Find a bounded closed set in $\mathbf{R}$ that is not rectifiable.

6. Let A be a bounded open set in $\mathbf{R}^n$; let $f : \mathbf{R}^n \to \mathbf{R}$ be a bounded continuous function. Give an example where $\int_{\bar{A}} f$ exists but $\int_A f$ does not.
7. Let S be a bounded set in $\mathbf{R}^n$.
   (a) Show that if S is rectifiable, then so is the set $\bar{S}$, and $v(S) = v(\bar{S})$.
   (b) Give an example where $\bar{S}$ and Int S are rectifiable, but S is not.
8. Let A and B be rectangles in $\mathbf{R}^k$ and $\mathbf{R}^n$, respectively. Let S be a set contained in $A \times B$. For each $\mathbf{y} \in B$, let

$$S_{\mathbf{y}} = \{\mathbf{x} \mid \mathbf{x} \in A \quad \text{and} \quad (\mathbf{x}, \mathbf{y}) \in S\}.$$

   We call $S_{\mathbf{y}}$ a cross-section of S. Show that if S is rectifiable, and if $S_{\mathbf{y}}$ is rectifiable for each $\mathbf{y} \in B$, then

$$v(S) = \int_{\mathbf{y} \in B} v(S_{\mathbf{y}}).$$

§15. IMPROPER INTEGRALS

We now extend our notion of the integral. We define the integral $\int_S f$ in the case where S is not necessarily bounded and f is not necessarily bounded. Such an integral is sometimes called an *improper integral.*

We shall define our extended notion of the integral only in the case where S is open in $\mathbf{R}^n$.

Definition. Let A be an open set in $\mathbf{R}^n$; let $f : A \to \mathbf{R}$ be a continuous function. If f is non-negative on A, we define the (extended) **integral** of f over A, denoted $\int_A f$, to be the supremum of the numbers $\int_D f$, as D ranges over all compact rectifiable subsets of A, provided this supremum exists. In this case, we say that f is **integrable** over A (in the extended sense). More generally, if f is an arbitrary continuous function on A, set

$$f_+(\mathbf{x}) = \max\{f(\mathbf{x}), 0\} \quad \text{and} \quad f_-(\mathbf{x}) = \max\{-f(\mathbf{x}), 0\}.$$

We say that f is **integrable** over A (in the extended sense) if both f_+ and f_- are; and in this case we set

$$\int_A f = \int_A f_+ - \int_A f_-,$$

where $\int_A$ denotes the extended integral throughout.

If A is open in $\mathbf{R}^n$ and both f and A are bounded, we now have two different meanings for the symbol $\int_A f$. It could mean the extended integral, or it could mean the ordinary integral. It turns out that if the ordinary integral exists, then so does the extended integral and the two integrals are equal. Nevertheless, some ambiguity persists, because the extended integral may exist when the ordinary integral does not. To avoid ambiguity, we make the following convention:

Convention. *If A is an open set in $\mathbf{R}^n$, then $\int_A f$ will denote the extended integral unless specifically stated otherwise.*

Of course, if A is not open, there is no ambiguity; $\int_A f$ must denote the ordinary integral in this case.

We now give a reformulation of the definition of the extended integral that is convenient for many purposes. It is related to the way improper integrals are defined in calculus. We begin with a preliminary lemma:

Lemma 15.1. *Let A be an open set in $\mathbf{R}^n$. Then there exists a sequence C_1, C_2, ... of compact rectifiable subsets of A whose union is A, such that $C_N \subset \text{Int } C_{N+1}$ for each N.*

Proof. Let d denote the sup metric $d(\mathbf{x}, \mathbf{y}) = |\mathbf{x} - \mathbf{y}|$ on $\mathbf{R}^n$. If $B \subset \mathbf{R}^n$, let $d(\mathbf{x}, B)$ denote the distance from $\mathbf{x}$ to B, as usual. (See §4.)

Now set $B = \mathbf{R}^n - A$. Then given a positive integer N, let D_N denote the set

$$D_N = \{\mathbf{x} \mid d(\mathbf{x},B) \geq 1/N \quad \text{and} \quad d(\mathbf{x},\mathbf{0}) \leq N\}.$$

Since $d(\mathbf{x},B)$ and $d(\mathbf{x},\mathbf{0})$ are continuous functions of $\mathbf{x}$ (see the proof of Theorem 4.6), D_N is a closed subset of $\mathbf{R}^n$. Because D_N is contained in the cube of radius N centered at $\mathbf{0}$, it is bounded and thus compact. Also, D_N is contained in A, since the inequality $d(\mathbf{x},B) \geq 1/N$ implies that $\mathbf{x}$ cannot be in B. To show the sets D_N cover A, let $\mathbf{x}$ be a point of A. Since A is open, $d(\mathbf{x},B) > 0$; then there is an N such that $d(\mathbf{x},B) \geq 1/N$ and $d(\mathbf{x},\mathbf{0}) \leq N$, so that $\mathbf{x} \in D_N$. Finally, we note that the set

$$A_{N+1} = \{\mathbf{x} \mid d(\mathbf{x},B) > 1/(N+1) \quad \text{and} \quad d(\mathbf{x},\mathbf{0}) < N+1\}$$

is open (because $d(\mathbf{x},B)$ and $d(\mathbf{x},\mathbf{0})$ are continuous). Since A_{N+1} is contained in D_{N+1} and contains D_N by definition, it follows that $D_N \subset \text{Int } D_{N+1}$. See Figure 15.1.

The sets D_N are not quite the sets we want, since they may not be rectifiable. We construct the sets C_N as follows: For each $\mathbf{x} \in D_N$, choose a closed cube that is centered at $\mathbf{x}$ and is contained in Int D_{N+1}. The interiors of these cubes cover D_N; choose finitely many of them whose interiors cover

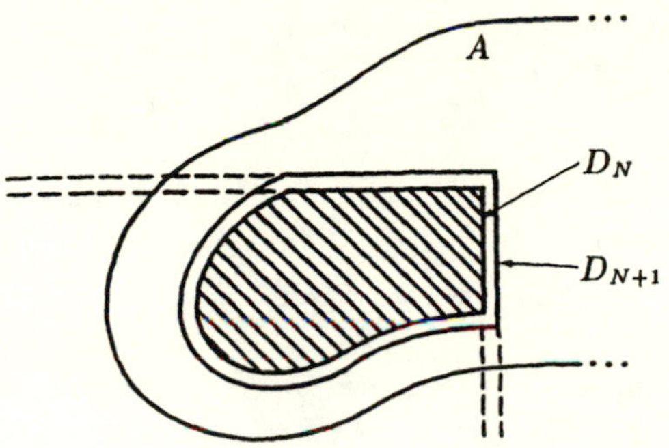

Figure 15.1

D_N and let their union be C_N. Since C_N is a finite union of rectangles, it is compact and rectifiable. Then

$$D_N \subset \text{Int } C_N \subset C_N \subset \text{Int } D_{N+1}.$$

It follows that the union of the sets C_N equals A and that $C_N \subset \text{Int } C_{N+1}$ for each N. □

Now we obtain our alternate formulation of the definition:

Theorem 15.2. *Let A be open in $\mathbf{R}^n$; let $f : A \to \mathbf{R}$ be continuous. Choose a sequence C_N of compact rectifiable subsets of A whose union is A such that $C_N \subset \text{Int } C_{N+1}$ for each N. Then f is integrable over A if and only if the sequence $\int_{C_N} |f|$ is bounded. In this case,*

$$\int_A f = \lim_{N \to \infty} \int_{C_N} f.$$

It follows from this theorem that f is integrable over A if and only if $|f|$ is integrable over A.

Proof. Step 1. We prove the theorem first in the case where f is non-negative. Here $f = |f|$. Since the sequence $\int_{C_N} f$ is increasing (by monotonicity), it converges if and only if it is bounded.

Suppose first that f is integrable over A. If we let D range over all compact rectifiable subsets of A, then

$$\int_{C_N} f \leq \sup_D \left\{ \int_D f \right\} = \int_A f,$$

since C_N is itself a compact rectifiable subset of A. It follows that the sequence $\int_{C_N} f$ is bounded, and

$$\lim_{N\to\infty} \int_{C_N} f \leq \int_A f.$$

Conversely, suppose the sequence $\int_{C_N} f$ is bounded. Let D be an arbitrary compact rectifiable subset of A. Then D is covered by the open sets

$$\text{Int } C_1 \subset \text{Int } C_2 \subset \cdots,$$

hence by finitely many of them, and hence by one of them, say Int C_M. Then

$$\int_D f \leq \int_{C_M} f \leq \lim_{N\to\infty} \int_{C_N} f.$$

Since D is arbitrary, it follows that f is integrable over A, and

$$\int_A f \leq \lim_{N\to\infty} \int_{C_N} f.$$

Step 2. Now let $f : A \to \mathbf{R}$ be an arbitrary continuous function. By definition, f is integrable over A if and only if f_+ and f_- are integrable over A; this occurs if and only if the sequences $\int_{C_N} f_+$ and $\int_{C_N} f_-$ are bounded, by Step 1. Note that

$$0 \leq f_+(\mathbf{x}) \leq |f(\mathbf{x})| \quad \text{and} \quad 0 \leq f_-(\mathbf{x}) \leq |f(\mathbf{x})|,$$

while

$$|f(\mathbf{x})| = f_+(\mathbf{x}) + f_-(\mathbf{x}).$$

It follows that the sequences $\int_{C_N} f_+$ and $\int_{C_N} f_-$ are bounded if and only if the sequence $\int_{C_N} |f|$ is bounded. In this case, the former two sequences converge to $\int_A f_+$ and $\int_A f_-$, respectively. Since convergent sequences can be added term-by-term, the sequence

$$\int_{C_N} f = \int_{C_N} f_+ - \int_{C_N} f_-$$

converges to $\int_A f_+ - \int_A f_-$; and the latter equals $\int_A f$ by definition. □

We now verify the properties of the extended integral; many are analogous to those of the ordinary integral. Then we relate the extended integral to the ordinary integral in the case where both are defined.

Theorem 15.3. *Let A be an open set in $\mathbf{R}^n$. Let $f, g : A \to \mathbf{R}$ be continuous functions.*

(a) *(Linearity). If f and g are integrable over A, so is $af + bg$; and*

$$\int_A (af + bg) \quad = \quad a\int_A f \quad + \quad b\int_A g.$$

(b) *(Comparison). Let f and g be integrable over A. If $f(\mathbf{x}) \leq g(\mathbf{x})$ for $\mathbf{x} \in A$, then*

$$\int_A f \leq \int_A g.$$

In particular,

$$|\int_A f| \leq \int_A |f|.$$

(c) *(Monotonicity). Assume B is open and $B \subset A$. If f is non-negative on A and integrable over A, then f is integrable over B and*

$$\int_B f \leq \int_A f.$$

(d) *(Additivity). Suppose A and B are open in $\mathbf{R}^n$ and f is continuous on $A \cup B$. If f is integrable on A and B, then f is integrable on $A \cup B$ and $A \cap B$, and*

$$\int_{A \cup B} f = \int_A f + \int_B f - \int_{A \cap B} f.$$

Note that by our convention, the integral symbol denotes the extended integral throughout the statement of this theorem.

Proof. Let C_N be a sequence of compact rectifiable sets whose union is A, such that $C_N \subset \text{Int}\, C_{N+1}$ for all N.

(a) We have

$$\int_{C_N} |af + bg| \quad \leq \quad |a|\int_{C_N} |f| \quad + \quad |b|\int_{C_N} |g|,$$

by the comparison and linearity properties of the ordinary integral. Since both sequences $\int_{C_N} |f|$ and $\int_{C_N} |g|$ are bounded, so is $\int_{C_N} |af + bg|$. Linearity now follows by taking limits in the equation

$$\int_{C_N} (af + bg) \quad = \quad a\int_{C_N} f \quad + \quad b\int_{C_N} g.$$

(b) If $f(\mathbf{x}) \leq g(\mathbf{x})$, one takes limits in the inequality

$$\int_{C_N} f \leq \int_{C_N} g.$$

(c) If D is a compact rectifiable subset of B, then D is also a compact rectifiable subset of A, so that

$$\int_D f \leq \int_A f,$$

by definition. Since D is arbitrary, f is integrable over B and $\int_B f \leq \int_A f$.

(d) Let D_N be a sequence of compact rectifiable sets whose union is B such that $D_N \subset \text{Int } D_{N+1}$ for each N. Let

$$E_N = C_N \cup D_N \quad \text{and} \quad F_N = C_N \cap D_N.$$

Then E_N and F_N are sequences of compact rectifiable sets whose unions equal $A \cup B$ and $A \cap B$, respectively. See Figure 15.2.

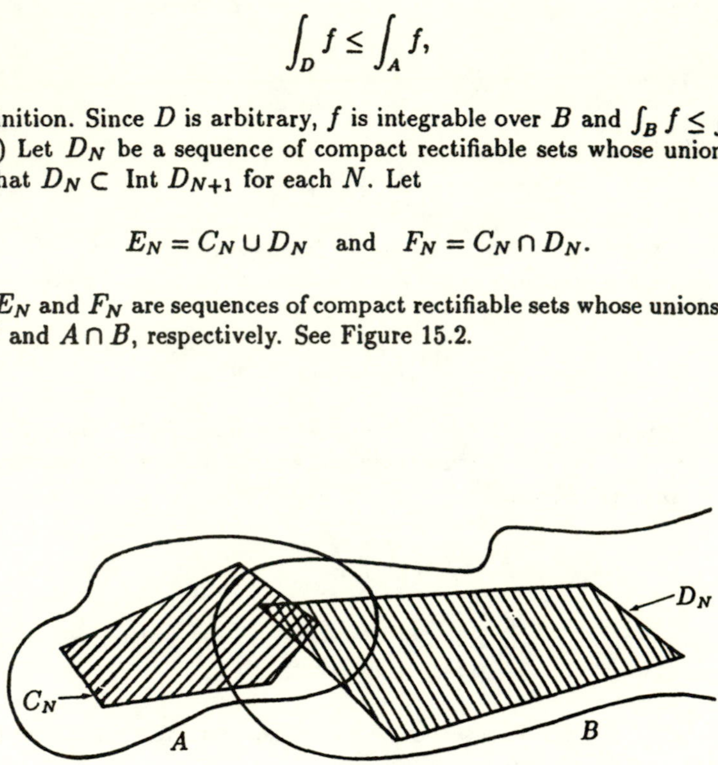

Figure 15.2

We show $E_N \subset \text{Int } E_{N+1}$ and $F_N \subset \text{Int } F_{N+1}$. If $\mathbf{x} \in E_N$, then $\mathbf{x}$ is in either C_N or D_N. If the former, then some neighborhood of $\mathbf{x}$ is contained in C_{N+1}. If the latter, some neighborhood of $\mathbf{x}$ is contained in D_{N+1}. In either case, this neighborhood of $\mathbf{x}$ is contained in E_{N+1}, so that $\mathbf{x} \in \text{Int } E_{N+1}$.

Similarly, if $\mathbf{x} \in F_N$, then some neighborhood U of $\mathbf{x}$ is contained in C_{N+1}, and some neighborhood of V of $\mathbf{x}$ is contained in D_{N+1}. The neighborhood $U \cap V$ of $\mathbf{x}$ is thus contained in F_{N+1}, so that $\mathbf{x} \in \text{Int } F_{N+1}$.

Additivity of the ordinary integral tells us that

$$(*) \qquad \int_{E_N} f = \int_{C_N} f + \int_{D_N} f - \int_{F_N} f.$$

Applying this equation to the function $|f|$, we see that $\int_{E_N} |f|$ and $\int_{F_N} |f|$ are bounded above by

$$\int_{C_N} |f| + \int_{D_N} |f|.$$

Thus f is integrable over $A \cup B$ and $A \cap B$. The desired equation now follows by taking limits in $(*)$. $\square$

Now we relate the extended integral to the ordinary integral.

Theorem 15.4. *Let A be a bounded open set in $\mathbf{R}^n$; let $f : A \to \mathbf{R}$ be a bounded continuous function. Then the extended integral $\int_A f$ exists. If the ordinary integral $\int_A f$ also exists, then these two integrals are equal.*

Proof. Let Q be a rectangle containing A.

Step 1. We show the extended integral of f exists. Choose M so that $|f(x)| \leq M$ for $\mathbf{x} \in A$. Then for any compact rectifiable subset D of A,

$$\int_D |f| \leq \int_D M \leq M \cdot v(Q).$$

Thus f is integrable over A in the extended sense.

Step 2. We consider the case where f is non-negative. Suppose the ordinary integral of f over A exists. It equals, by definition, the integral over Q of the function f_A. If D is a compact rectifiable subset of A, then

$$\begin{aligned} \int_D f &= \int_D f_A \quad \text{because} \quad f = f_A \quad \text{on} \quad D, \\ &\leq \int_Q f_A \quad \text{by monotonicity}, \\ &= (\text{ordinary}) \int_A f. \end{aligned}$$

Since D is arbitrary, it follows that

$$(\text{extended}) \int_A f \leq (\text{ordinary}) \int_A f.$$

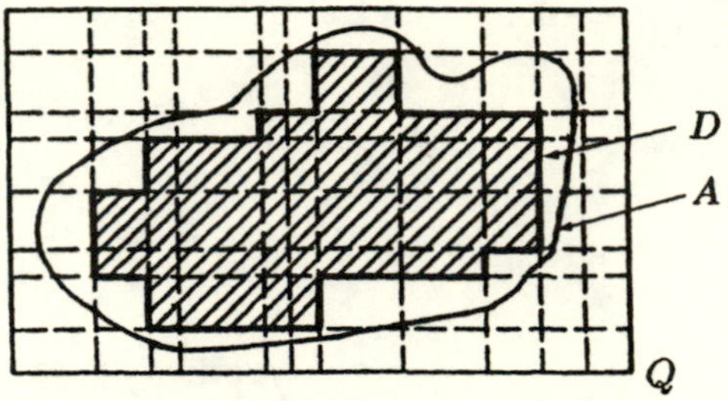

Figure 15.3

On the other hand, let P be a partition of Q, and let R denote the general subrectangle determined by P. Denote by $R_1, \ldots, R_k$ those subrectangles that lie in A, and let $D = R_1 \cup \cdots \cup R_k$. See Figure 15.3. Now

$$L(f_A, P) = \sum_{i=1}^{k} m_{R_i}(f) \cdot v(R_i),$$

because $m_R(f_A) = m_R(f)$ if R is contained in A and $m_R(f_A) = 0$ if R is not contained in A. On the other hand,

$$\begin{aligned} \sum_{i=1}^{k} m_{R_i}(f) \cdot v(R_i) &\leq \sum_{i=1}^{k} \int_{R_i} f \qquad \text{by the comparison property,} \\ &= \int_D f \qquad \text{by additivity,} \\ &\leq (\text{extended}) \int_A f \qquad \text{by definition.} \end{aligned}$$

Since P is arbitrary, we conclude that

$$(\text{ordinary}) \int_A f \leq (\text{extended}) \int_A f.$$

Step 3. Now we consider the general case. Write $f = f_+ - f_-$, as usual. Since f is integrable over A in the ordinary sense, so are f_+ and f_-, by

Lemma 13.2. Then

$$
\begin{aligned}
\text{(ordinary)} \int_A f &= \text{(ordinary)} \int_A f_+ - \text{(ordinary)} \int_A f_- && \text{by linearity,} \\
&= \text{(extended)} \int_A f_+ - \text{(extended)} \int_A f_- && \text{by Step 2,} \\
&= \text{(extended)} \int_A f && \text{by definition. } \square
\end{aligned}
$$

EXAMPLE 1. If A is a bounded open set in $\mathbf{R}^n$ and $f : A \to \mathbf{R}$ is a bounded continuous function, then the extended integral $\int_A f$ exists, but the ordinary integral $\int_A f$ may not. For example, let A be the open subset of $\mathbf{R}$ constructed in Example 1 of §14. The set A is bounded, but $\text{Bd } A$ does not have measure zero. Then the ordinary integral $\int_A 1$ does not exist, although the extended integral $\int_A 1$ does.

A consequence of the preceding theorem is the following:

Corollary 15.5. *Let S be a bounded set in $\mathbf{R}^n$; let $f : S \to \mathbf{R}$ be a bounded continuous function. If f is integrable over S in the ordinary sense, then*

$$\textit{(ordinary)} \int_S f \quad = \quad \textit{(extended)} \int_{\text{Int } S} f.$$

Proof. One applies Theorems 13.6 and 15.4. □

This corollary tells us that any theorem we prove about extended integrals has implications for ordinary integrals. The change of variables theorem, which we prove in the next chapter, is an important example.

We have already given two formulations of the definition of the extended integral, and we will give another in the next chapter. All these versions of the definition are useful for different theoretical purposes. Actually applying them to computational problems can be a bit awkward, however. Here is a formulation that is useful in many practical situations. We shall use it in some of the examples and exercises:

***Theorem 15.6.** *Let A be open in $\mathbf{R}^n$; let $f : A \to \mathbf{R}$ be continuous. Let $U_1 \subset U_2 \subset \cdots$ be a sequence of open sets whose union is A. Then $\int_A f$ exists if and only if the sequence $\int_{U_N} |f|$ exists and is bounded; in this case,*

$$\int_A f = \lim_{N\to\infty} \int_{U_N} f.$$

Proof. It suffices, as usual, to consider the case where f is non-negative.

Suppose the integral $\int_A f$ exists. Monotonicity of the extended integral implies that f is integrable over U_N and that for each N,

$$\int_{U_N} f \leq \int_A f.$$

It follows that the increasing sequence $\int_{U_N} f$ converges, and that

$$\lim_{N\to\infty} \int_{U_N} f \leq \int_A f.$$

Conversely, suppose the sequence $\int_{U_N} f$ exists and is bounded. Let D be a compact rectifiable subset of A. Since D is covered by the open sets $U_1 \subset U_2 \subset \cdots$, it is covered by finitely many of them, and hence by one of them, say U_M. Then, by definition,

$$\int_D f \leq \int_{U_M} f \leq \lim_{N\to\infty} \int_{U_N} f.$$

Since D is arbitrary,

$$\int_A f \leq \lim_{N\to\infty} \int_{U_N} f. \quad \square$$

In applying this theorem, we usually choose U_N so that it is rectifiable and f is bounded on U_N; then the integral $\int_{U_N} f$ exists as an ordinary integral (and hence as an extended integral) and can be computed by familiar techniques. See the examples following.

EXAMPLE 2. Let A be the open set in $\mathbf{R}^2$ defined by the equation

$$A = \{(x,y) \mid x > 1 \quad \text{and} \quad y > 1\}.$$

Let $f(x,y) = 1/x^2y^2$. Then f is bounded on A, but A is unbounded. We could use Theorem 15.2 to calculate $\int_A f$, by setting $C_N = [(N+1)/N, N]^2$ and integrating f over C_N. It is a bit easier to use Theorem 15.6, setting $U_N = (1,N)^2$ and integrating f over U_N. See Figure 15.4. The set U_N is rectifiable; f is bounded on U_N because $\overline{U}_N$ is compact and f is continuous on $\overline{U}_N$. Thus $\int_{U_N} f$ exists as an ordinary integral, so we can apply the Fubini theorem. We compute

$$\int_{U_N} f = \int_{x=1}^{x=N} \int_{y=1}^{y=N} 1/x^2y^2 = \left((N-1)/N\right)^2.$$

We conclude that $\int_A f = 1$.

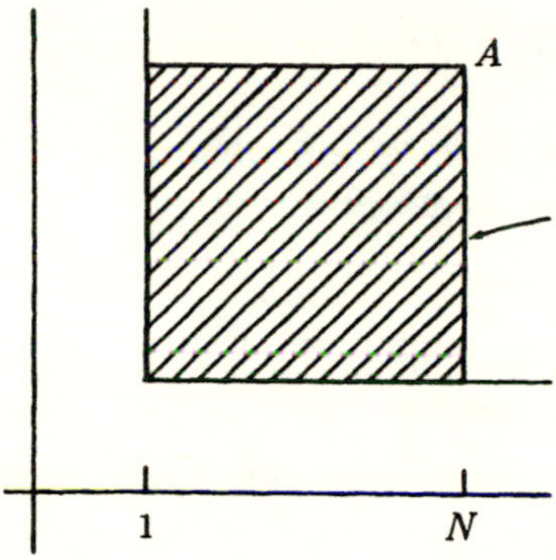

Figure 15.4

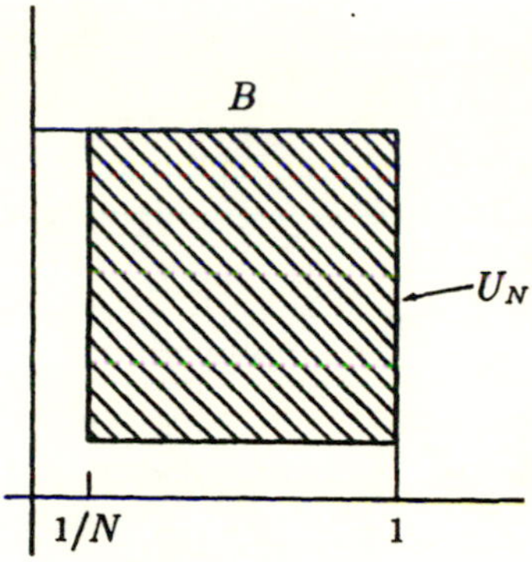

Figure 15.5

EXAMPLE 3. Let $B = (0,1)^2$; let $f(x,y) = 1/x^2y^2$, as before. Here B is bounded but f is not bounded on B; indeed, f is unbounded near each point of the x and y axes. However, if we set $U_N = (1/N,1)^2$, then f is bounded on U_N. See Figure 15.5. We compute

$$\int_{U_N} f = (-1+N)^2.$$

We conclude that $\int_B f$ does not exist.

EXERCISES

1. Let $f : \mathbf{R} \rightarrow \mathbf{R}$ be the function $f(x) = x$. Show that, given $\lambda \in \mathbf{R}$, there exists a sequence C_N of compact rectifiable subsets of $\mathbf{R}$ whose union is $\mathbf{R}$, such that $C_N \subset \text{Int } C_{N+1}$ for each N and

$$\lim_{N\to\infty} \int_{C_N} f = \lambda.$$

Does the extended integral $\int_{\mathbf{R}} f$ exist?

2. Let A be open in $\mathbf{R}^n$; let $f, g : A \rightarrow \mathbf{R}$ be continuous; suppose that $|f(\mathbf{x})| \leq g(\mathbf{x})$ for $\mathbf{x} \in A$. Show that if $\int_A g$ exists, so does $\int_A f$. (This result is analogous to the so-called "comparison test" for the convergence of an infinite series.)

3. (a) Let A and B be the sets of Examples 2 and 3; let $f(x, y) = 1/(xy)^{1/2}$. Determine whether $\int_A f$ and $\int_B f$ exist; if either does, calculate it.

   (b) Let $C = \{(x, y) \mid x > 0 \text{ and } y > 0\}$. Let

$$f(x, y) = 1/(x^2 + \sqrt{x})(y^2 + \sqrt{y}).$$

   Show that $\int_C f$ exists; do not attempt to calculate it.

4. Let $f(x, y) = 1/(y + 1)^2$. Let A and B be the open sets

$$A = \{(x, y) \mid x > 0 \quad \text{and} \quad x < y < 2x\},$$
$$B = \{(x, y) \mid x > 0 \quad \text{and} \quad x^2 < y < 2x^2\},$$

of $\mathbf{R}^2$. Show that $\int_A f$ does not exist; show that $\int_B f$ does exist and calculate it. See Figure 15.6.

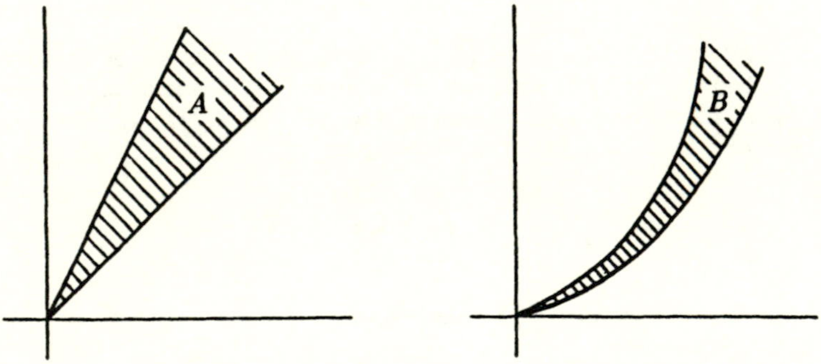

Figure 15.6

5. Let $f(x,y) = 1/x(xy)^{1/2}$ for $x > 0$ and $y > 0$. Let

$$A_0 = \{(x,y) \mid 0 < x < 1 \quad \text{and} \quad x < y < 2x\},$$
$$B_0 = \{(x,y) \mid 0 < x < 1 \quad \text{and} \quad x^2 < y < 2x^2\}.$$

Determine whether $\int_{A_0} f$ and $\int_{B_0} f$ exist; if so, calculate.

6. Let A be the set in $\mathbf{R}^2$ defined by the equation

$$A = \{(x,y) \mid x > 1 \quad \text{and} \quad 0 < y < 1/x\}.$$

Calculate $\int_A 1/xy^{1/2}$ if it exists.

*7. Let A be a bounded open set in $\mathbf{R}^n$; let $f : A \to \mathbf{R}$ be a bounded continuous function. Let Q be a rectangle containing A. Show that

$$\int_A f = \underline{\int_Q} (f_+)_A - \underline{\int_Q} (f_-)_A.$$

*8. Let A be open in $\mathbf{R}^n$. We say $f : A \to \mathbf{R}$ is locally bounded on A if each $\mathbf{x}$ in A has a neighborhood on which f is bounded. Let $\mathcal{F}(A)$ be the set of all functions $f : A \to \mathbf{R}$ that are locally bounded on A and continuous on A except on a set of measure zero.

(a) Show that if f is continuous on A, then $f \in \mathcal{F}(A)$.

(b) Show that if f is in $\mathcal{F}(A)$, then f is bounded on each compact subset of A and the definition of the extended integral $\int_A f$ goes through without change.

(c) Show that Theorem 15.3 holds for functions f in $\mathcal{F}(A)$.

(d) Show that Theorem 15.4 holds if the word "continuous" in the hypothesis is replaced by "continuous except on a set of measure zero."

4

Change of Variables

In evaluating the integral of a function of a single variable, one of the most useful tools is the so-called "substitution rule." It is used in calculus, for example, to evaluate such an integral as

$$\int_0^1 (2x^2 + 1)^{10}(4x)\, dx;$$

one makes the substitution $y = 2x^2 + 1$, reducing this integral to the integral

$$\int_1^3 y^{10}\, dy,$$

which is easy to evaluate. (Here we use the "dx" and "dy" notation of calculus.)

Our intention in this chapter is to generalize the substitution rule in two ways:

(1) We shall deal with n-dimensional integrals rather than one-dimensional integrals.

(2) We shall prove it for the extended integral, rather than merely for integrals of bounded functions over bounded sets. This will require us to limit ourselves to integrals over *open* sets in $\mathbf{R}^n$, but, as Corollary 15.5 shows, this is not a serious restriction.

We call the generalized version of the substitution rule the *change of variables theorem.*

§16. PARTITIONS OF UNITY

In order to prove the change of variables theorem, we need to reformulate the definition of the extended integral $\int_A f$. This integral is obtained by breaking the *set* A up into compact rectifiable sets C_N, and taking the limit of the corresponding integrals $\int_{C_N} f$. In our new approach, we instead break the *function* f up into functions f_N, each of which vanishes outside a compact set, and we take the limit of the corresponding integrals $\int_A f_N$. This approach has many advantages, especially for theoretical purposes; it will recur throughout the rest of the book.

This approach involves a notion of comparatively recent origin in mathematics, called a "partition of unity," which we define in this section.

We begin with several lemmas.

Lemma 16.1. *Let Q be a rectangle in $\mathbf{R}^n$. There is a C^∞ function $\phi : \mathbf{R}^n \to \mathbf{R}$ such that $\phi(\mathbf{x}) > 0$ for $\mathbf{x} \in \text{Int } Q$ and $\phi(\mathbf{x}) = 0$ otherwise.*

Proof. Let $f : \mathbf{R} \to \mathbf{R}$ be defined by the equation

$$f(x) = \begin{cases} e^{-1/x} & \text{if } x > 0, \\ 0 & \text{otherwise.} \end{cases}$$

Then $f(x) > 0$ for $x > 0$. It is a standard result of single-variable analysis that f is of class C^∞. (A proof is outlined in the exercises.) Define

$$g(x) = f(x) \cdot f(1 - x).$$

Then g is of class C^∞; furthermore, g is positive for $0 < x < 1$ and vanishes otherwise. See Figure 16.1. Finally, if

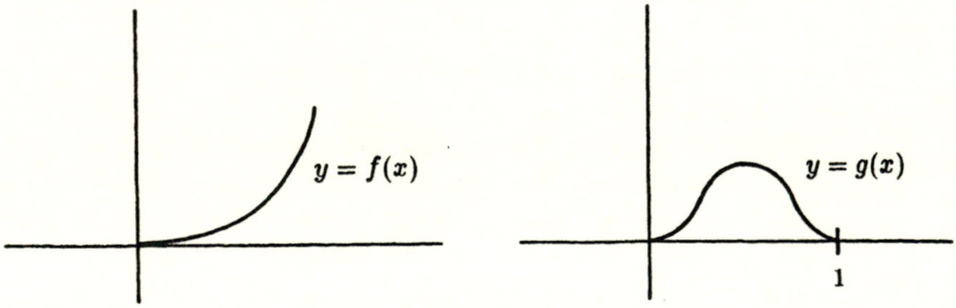

Figure 16.1

$$Q = [a_1, b_1] \times \cdots \times [a_n, b_n],$$

define

$$\phi(\mathbf{x}) = g\left(\frac{x_1 - a_1}{b_1 - a_1}\right) \cdot g\left(\frac{x_2 - a_2}{b_2 - a_2}\right) \cdots g\left(\frac{x_n - a_n}{b_n - a_n}\right). \quad \square$$

Lemma 16.2. *Let $\mathcal{A}$ be a collection of open sets in $\mathbf{R}^n$; let A be their union. Then there exists a countable collection $Q_1, Q_2, \ldots$ of rectangles contained in A such that:*

(1) *The sets* Int Q_i *cover A.*

(2) *Each Q_i is contained in an element of $\mathcal{A}$.*

(3) *Each point of A has a neighborhood that intersects only finitely many of the sets Q_i.*

Proof. It is not difficult to find rectangles Q_i satisfying (1) and (2). Choosing them so they also satisfy (3), the so-called "local finiteness condition," is more difficult.

Step 1. Let $D_1, D_2, \ldots$ be a sequence of compact subsets of A whose union is A, such that $D_i \subset \text{Int } D_{i+1}$ for each i. For convenience in notation, let D_i denote the empty set for $i \leq 0$. Then for each i, define

$$B_i = D_i - \text{Int } D_{i-1}.$$

The set B_i is bounded, being a subset of D_i; and it is closed, being the intersection of the closed sets D_i and $\mathbf{R}^n - \text{Int } D_{i-1}$. Thus B_i is compact. Also, B_i is disjoint from the closed set D_{i-2}, since $D_{i-2} \subset \text{Int } D_{i-1}$. For each $\mathbf{x} \in B_i$, we choose a closed cube $C_\mathbf{x}$ centered at $\mathbf{x}$ that is contained in A and is disjoint from D_{i-2}; also choose $C_\mathbf{x}$ small enough that it is contained in an element of the collection of open sets $\mathcal{A}$. See Figure 16.2.

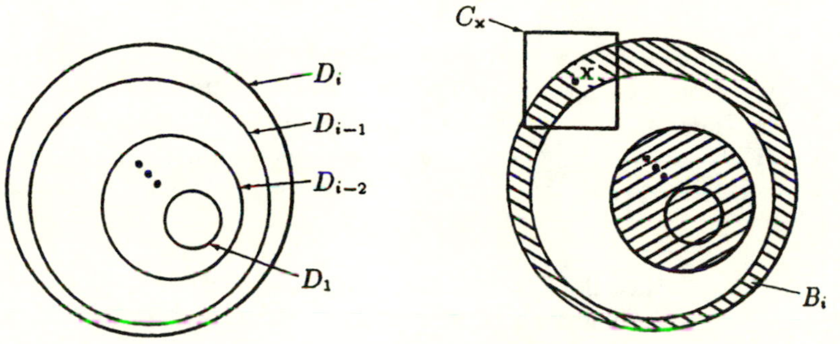

Figure 16.2

The interiors of the cubes $C_{\mathbf{x}}$ cover B_i; choose finitely many of these cubes whose interiors cover B_i; let $\mathcal{C}_i$ denote this finite collection of cubes. See Figure 16.3.

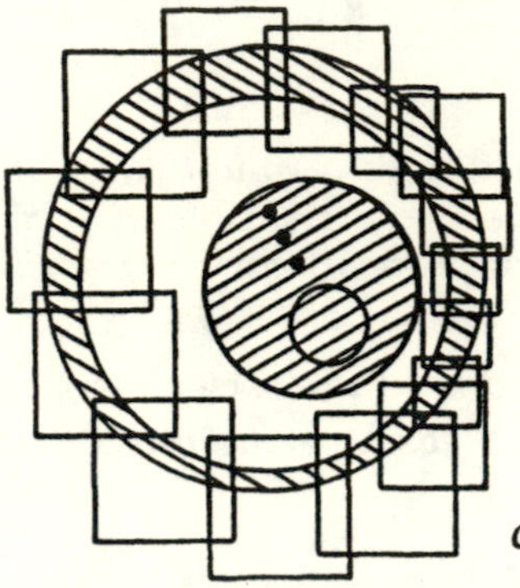

Figure 16.3

Step 2. Let $\mathcal{C}$ be the collection

$$\mathcal{C} = \mathcal{C}_1 \cup \mathcal{C}_2 \cup \cdots ;$$

then $\mathcal{C}$ is a countable collection of rectangles (in fact, of cubes). We show this collection satisfies the requirements of the lemma.

By construction, each element of $\mathcal{C}$ is a rectangle contained in an element of the collection $\mathcal{A}$. We show that the interiors of these rectangles cover A. Given $\mathbf{x} \in A$, let i be the smallest integer such that $\mathbf{x} \in \text{Int } D_i$. Then $\mathbf{x}$ is an element of the set $B_i = D_i - \text{Int } D_{i-1}$. Since the interiors of the cubes belonging to the collection $\mathcal{C}_i$ cover B_i, the point $\mathbf{x}$ lies interior to one of these cubes.

Finally, we check the local finiteness condition. Given $\mathbf{x}$, we have $\mathbf{x} \in \text{Int } D_i$ for some i. Each cube belonging to one of the collections $\mathcal{C}_{i+2}, \mathcal{C}_{i+3}, \ldots$ is disjoint from D_i, by construction. Therefore the open set Int D_i can intersect only the cubes belonging to one of the collections $\mathcal{C}_1, \ldots, \mathcal{C}_{i+1}$. Thus $\mathbf{x}$ has a neighborhood that intersects only finitely many cubes from the collection $\mathcal{C}$. □

We remark that the local finiteness condition holds for each point $\mathbf{x}$ of A, but it does *not* hold for a point $\mathbf{x}$ of Bd A. Each neighborhood of such a point necessarily intersects infinitely many of the cubes from the collection $\mathcal{C}$, as you can check.

Definition. If $\phi : \mathbf{R}^n \to \mathbf{R}$, then the support of ϕ is defined to be the closure of the set $\{\mathbf{x} \mid \phi(\mathbf{x}) \neq 0\}$. Said differently, the support of ϕ is characterized by the property that if $\mathbf{x} \notin \text{Support } \phi$, then there is some neighborhood of $\mathbf{x}$ on which the function ϕ vanishes identically.

Theorem 16.3 (Existence of a partition of unity). *Let $\mathcal{A}$ be a collection of open sets in $\mathbf{R}^n$; let A be their union. There exists a sequence $\phi_1, \phi_2, \ldots$ of continuous functions $\phi_i : \mathbf{R}^n \to \mathbf{R}$ such that:*

(1) $\phi_i(\mathbf{x}) \geq 0$ *for all* $\mathbf{x}$.

(2) *The set* $S_i = \text{Support } \phi_i$ *is contained in* A.

(3) *Each point of A has a neighborhood that intersects only finitely many of the sets S_i.*

(4) $\sum_{i=1}^{\infty} \phi_i(\mathbf{x}) = 1$ *for each* $\mathbf{x} \in A$.

(5) *The functions ϕ_i are of class C^∞.*

(6) *The sets S_i are compact.*

(7) *For each i, the set S_i is contained in an element of $\mathcal{A}$.*

A collection of functions $\{\phi_i\}$ satisfying conditions (1)–(4) is called a **partition of unity** on A. If it satisfies (5), it is said to be **of class** C^∞; if it satisfies (6), it is said to have **compact supports**; if it satisfies (7), it said to be **dominated by the collection** $\mathcal{A}$.

Proof. Given $\mathcal{A}$ and A, let $Q_1, Q_2, \ldots$ be a sequence of rectangles in A satisfying the conditions stated in Lemma 16.2. For each i, let $\psi_i : \mathbf{R}^n \to \mathbf{R}$ be a C^∞ function that is positive on Int Q_i and zero elsewhere. Then $\psi_i(\mathbf{x}) \geq 0$ for all $\mathbf{x}$. Furthermore, Support $\psi_i = Q_i$; the latter is a compact subset of A that is contained in an element of $\mathcal{A}$. Finally, each point of A has a neighborhood that intersects only finitely many of the sets Q_i. The collection $\{\psi_i\}$ thus satisfies all the conditions of our theorem except for (4).

Condition (3) tells us that for $\mathbf{x} \in A$, only finitely many of the numbers $\psi_1(\mathbf{x}), \psi_2(\mathbf{x}), \ldots$ are non-zero. Thus the series

$$\lambda(\mathbf{x}) = \sum_{i=1}^{\infty} \psi_i(\mathbf{x})$$

converges trivially. Because each $\mathbf{x} \in A$ has a neighborhood on which $\lambda(\mathbf{x})$ equals a finite sum of C^∞ functions, $\lambda(\mathbf{x})$ is of class C^∞. Finally, $\lambda(\mathbf{x}) > 0$ for each $\mathbf{x} \in A$; given $\mathbf{x}$, there is a rectangle Q_i whose interior contains $\mathbf{x}$, whence $\psi_i(\mathbf{x}) > 0$. We now define

$$\phi_i(\mathbf{x}) = \psi_i(\mathbf{x})/\lambda(\mathbf{x});$$

the functions ϕ_i satisfy all of the conditions of our theorem. □

Conditions (1) and (4) imply that, for each $\mathbf{x} \in A$, the numbers $\phi_i(\mathbf{x})$ actually "partition unity," that is, they express the unity element 1 as a sum of non-negative numbers. The local finiteness condition (3) has the consequence that for any compact set C contained in A, there is an open set about C on which ϕ_i vanishes identically except for finitely many i. To find such an open set, one covers C by finitely many neighborhoods, on each of which ϕ_i vanishes except for finitely many i; then one takes the union of this finite collection of neighborhoods.

EXAMPLE 1. Let $f : \mathbf{R} \to \mathbf{R}$ be defined by the equation

$$f(x) = \begin{cases} (1+\cos x)/2 & \text{for } -\pi \leq x \leq \pi, \\ 0 & \text{otherwise.} \end{cases}$$

Then f is of class C^1. For each integer $m \geq 0$, set $\phi_{2m+1}(x) = f(x - m\pi)$. For each integer $m \geq 1$, set $\phi_{2m}(x) = f(x + m\pi)$. Then the collection $\{\phi_i\}$ forms a partition of unity on $\mathbf{R}$. The support S_i of ϕ_i is a closed interval of the form $[k\pi, (k+2)\pi]$, which is compact, and each point of $\mathbf{R}$ has a neighborhood that intersects at most three of the sets S_i. We leave it to you to check that $\sum \phi_i(x) = 1$. Thus $\{\phi_i\}$ is a partition of unity on $\mathbf{R}$. See Figure 16.4.

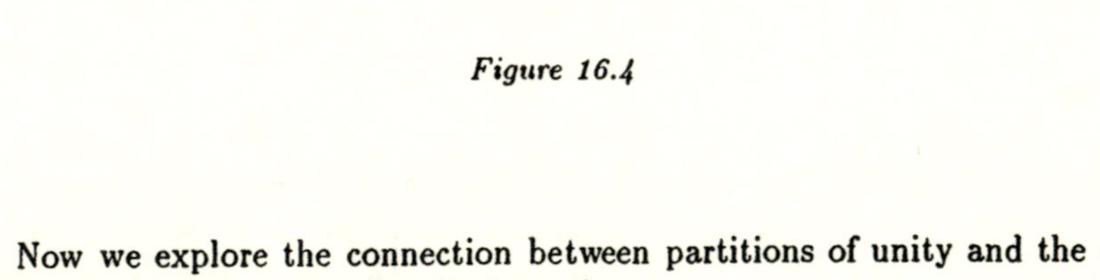

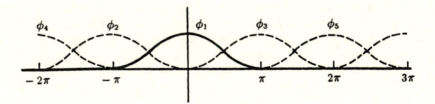

Figure 16.4

Now we explore the connection between partitions of unity and the extended integral. We need a preliminary lemma:

Lemma 16.4. *Let A be open in $\mathbf{R}^n$; let $f : A \to \mathbf{R}$ be continuous. If f vanishes outside the compact subset C of A, then the integrals $\int_A f$ and $\int_C f$ exist and are equal.*

Proof. The integral $\int_C f$ exists because C is bounded, and the function f_C, which equals f on A and vanishes outside C, is continuous and bounded on all of $\mathbf{R}^n$.

Let C_i be a sequence of compact rectifiable sets whose union is A, such that $C_i \subset \text{Int } C_{i+1}$ for each i. Then C is covered by finitely many sets Int C_i, and hence by one of them, say Int C_M. Since f vanishes outside C,

$$\int_C f = \int_{C_N} f$$

for all $N \geq M$. Applying this fact to the function $|f|$ shows that $\lim \int_{C_N} |f|$ exists, so that f is integrable over A; applying it to f shows that $\int_C f = \lim \int_{C_N} f = \int_A f$. □

Theorem 16.5. *Let A be open in $\mathbf{R}^n$; let $f : A \to \mathbf{R}$ be continuous. Let $\{\phi_i\}$ be a partition of unity on A having compact supports. The integral $\int_A f$ exists if and only if the series*

$$\sum_{i=1}^{\infty} [\int_A \phi_i |f|]$$

converges; in this case,

$$\int_A f = \sum_{i=1}^{\infty} [\int_A \phi_i f].$$

Note that the integral $\int_A \phi_i f$ exists and equals the ordinary integral $\int_{S_i} \phi_i f$ (where $S_i = \text{Support } \phi_i$) by the preceding lemma.

Proof. We consider first the case where f is non-negative on A.

Step 1. Suppose f is non-negative on A, and suppose the series $\sum[\int_A \phi_i f]$ converges. We show that $\int_A f$ exists and

$$\int_A f \leq \sum_{i=1}^{\infty} [\int_A \phi_i f].$$

Let D be a compact rectifiable subset of A. There exists an M such that for all $i > M$, the function ϕ_i vanishes identically on D. Then

$$f(\mathbf{x}) = \sum_{i=1}^{M} \phi_i(\mathbf{x}) f(\mathbf{x})$$

for $\mathbf{x} \in D$. We conclude that

$$\begin{aligned}
\int_D f &= \sum_{i=1}^{M} [\int_D \phi_i f] && \text{by linearity,} \\
&\leq \sum_{i=1}^{M} [\int_{D \cup S_i} \phi_i f] && \text{by monotonicity,} \\
&= \sum_{i=1}^{M} [\int_A \phi_i f] && \text{by the preceding lemma,} \\
&\leq \sum_{i=1}^{\infty} [\int_A \phi_i f].
\end{aligned}$$

It follows that f is integrable over A, and

$$\int_A f \leq \sum_{i=1}^{\infty} [\int_A \phi_i f].$$

Step 2. Suppose f is non-negative on A, and suppose f is integrable over A. We show the series $\sum [\int_A \phi_i f]$ converges, and

Given N,

$$\begin{aligned}
\sum_{i=1}^{N} [\int_A \phi_i f] &= \int_A [\sum_{i=1}^{N} \phi_i f] && \text{by linearity,} \\
&\leq \int_A f && \text{by the comparison property.}
\end{aligned}$$

Thus the series $\sum[\int_A \phi_i f]$ converges because its partial sums are bounded, and its sum is less than or equal to $\int_A f$.

The theorem is now proved for non-negative functions f.

Step 3. Consider the case of an arbitrary continuous function $f : A \to \mathbf{R}$. By Theorem 15.2, the integral $\int_A f$ exists if and only if the integral $\int_A |f|$ exists, and this occurs if and only if the series

$$\sum_{i=1}^{\infty}[\int_A \phi_i |f|]$$

converges, by Steps 1 and 2.

On the other hand, if $\int_A f$ exists, then

$$\begin{aligned}\int_A f &= \int_A f_+ - \int_A f_- \quad \text{by definition,}\\ &= \sum_{i=1}^{\infty}[\int_A \phi_i f_+] - \sum_{i=1}^{\infty}[\int_A \phi_i f_-] \quad \text{by Steps 1 and 2,}\\ &= \sum_{i=1}^{\infty}[\int_A \phi_i f] \quad \text{by linearity,}\end{aligned}$$

since convergent series can be added term-by-term. □

EXERCISES

1. Prove that the function f of Lemma 16.1 is of class C^∞ as follows: Given any integer $n \geq 0$, define $f_n : \mathbf{R} \to \mathbf{R}$ by the equation

$$f_n(x) = \begin{cases} (e^{-1/x})/x^n & \text{for } x > 0,\\ 0 & \text{for } x \leq 0.\end{cases}$$

   (a) Show that f_n is continuous at 0. [*Hint:* Show that $a < e^a$ for all a. Then set $a = t/2n$ to conclude that

$$\frac{t^n}{e^t} < \frac{(2n)^n}{e^{t/2}}.$$

   Set $t = 1/x$ and let x approach 0 through positive values.]

   (b) Show that f_n is differentiable at 0.

   (c) Show that $f_n'(x) = f_{n+2}(x) - nf_{n+1}(x)$ for all x.

   (d) Show that f_n is of class C^∞.

2. Show that the functions defined in Example 1 form a partition of unity on $\mathbf{R}$. [*Hint:* Let $f_m(x) = f(x - m\pi)$, for *all* integers m. Show that $\sum f_{2m}(x) = (1 + \cos x)/2$. Then find $\sum f_{2m+1}(x)$.]

3. (a) Let S be an arbitrary subset of $\mathbf{R}^n$; let $\mathbf{x}_0 \in S$. We say that the function $f : S \to \mathbf{R}$ is differentiable at $\mathbf{x}_0$, of class C^r, provided there is a C^r function $g : U \to \mathbf{R}$ defined in a neighborhood U of $\mathbf{x}_0$ in $\mathbf{R}^n$, such that g agrees with f on the set $U \cap S$. In this case, show that if $\phi : \mathbf{R}^n \to \mathbf{R}$ is a C^r function whose support lies in U, then the function

$$h(\mathbf{x}) = \begin{cases} \phi(\mathbf{x})g(\mathbf{x}) & \text{for } \mathbf{x} \in U, \\ 0 & \text{for } \mathbf{x} \notin \text{Support } \phi, \end{cases}$$

is well-defined and of class C^r on $\mathbf{R}^n$.

(b) Prove the following:

Theorem. *If $f : S \to \mathbf{R}$ and f is differentiable of class C^r at each point $\mathbf{x}_0$ of S, then f may be extended to a C^r function $h : A \to \mathbf{R}$ that is defined on an open set A of $\mathbf{R}^n$ containing S.*

[*Hint:* Cover S by appropriately chosen neighborhoods, let A be their union, and take a C^∞ partition of unity on A dominated by this collection of neighborhoods.]

§17. THE CHANGE OF VARIABLES THEOREM

Now we discuss the general change of variables theorem. We begin by reviewing the version of it used in calculus; although this version is usually proved in a first course in single-variable analysis, we reprove it here.

Recall the common convention that if f is integrable over $[a, b]$, then one defines

$$\int_b^a f = -\int_a^b f.$$

Theorem 17.1 (Substitution rule). *Let $I = [a, b]$. Let $g : I \to \mathbf{R}$ be a function of class C^1, with $g'(x) \neq 0$ for $x \in (a, b)$. Then the set $g(I)$ is a closed interval J with end points $g(a)$ and $g(b)$. If $f : J \to \mathbf{R}$ is continuous, then*

$$\int_{g(a)}^{g(b)} f = \int_a^b (f \circ g)g',$$

or equivalently,

$$\int_J f = \int_I (f \circ g)|g'|.$$

Proof. Continuity of g' and the intermediate-value theorem imply that either $g'(x) > 0$ or $g'(x) < 0$ on all of (a, b). Hence g is either strictly increasing or strictly decreasing on I, by the mean-value theorem, so that g is one-to-one. In the case where $g' > 0$, we have $g(a) < g(b)$; in the case where $g' < 0$, we have $g(a) > g(b)$. In either case, let $J = [c, d]$ denote the interval with end points $g(a)$ and $g(b)$. See Figure 17.1. The intermediate-value theorem implies that g carries I *onto* J. Then the composite function $f(g(x))$ is defined for all x in $[a, b]$, so the theorem at least makes sense.

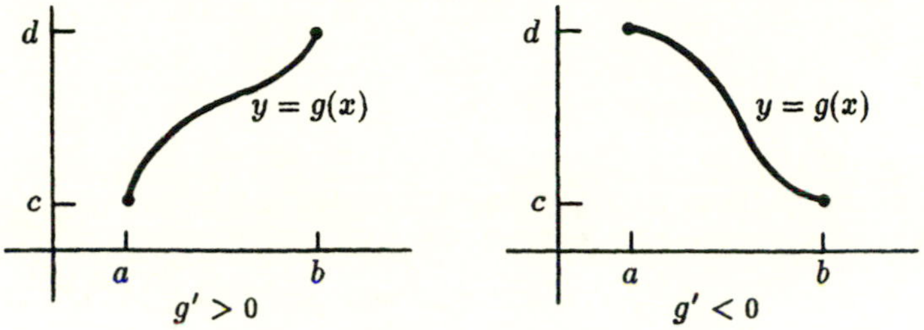

Figure 17.1

Define

$$F(y) = \int_c^y f$$

for y in $[c, d]$. Because f is continuous, the fundamental theorem of calculus implies that $F'(y) = f(y)$. Consider the composite function $h(x) = F(g(x))$; we differentiate it by the chain rule. We have

$$h'(x) = F'(g(x))g'(x) = f(g(x))g'(x).$$

Because the latter function is continuous, we can apply the fundamental theorem of calculus to integrate it. We have

$$\begin{aligned}\int_{x=a}^{x=b} f(g(x))g'(x) &= h(b) - h(a) \\ &= F(g(b)) - F(g(a)) \\ &= \int_c^{g(b)} f \; - \; \int_c^{g(a)} f.\end{aligned}$$

Now c equals either $g(a)$ or $g(b)$. In either case, this equation can be written in the form

$$(*)\qquad \int_a^b (f\circ g)g' = \int_{g(a)}^{g(b)} f.$$

This is the first of our desired formulas.

Now in the case where $g' > 0$, we have $J = [g(a), g(b)]$. Since $|g'| = g'$ in this case, equation $(*)$ can be written in the form

$$(**)\qquad \int_I (f\circ g)|g'| = \int_J f.$$

In the case where $g' < 0$, we have $J = [g(b), g(a)]$. Since $|g'| = -g'$ in this case, equation $(*)$ can again be written in the form $(**)$. □

EXAMPLE 1. Consider the integral

$$\int_{x=0}^{x=1} (2x^2+1)^{10}(4x).$$

Set $f(y) = y^{10}$ and $g(x) = 2x^2+1$. Then $g'(x) = 4x$, which is positive for $0 < x < 1$. See Figure 17.2. The substitution rule implies that

$$\int_{x=0}^{x=1} (2x^2+1)^{10}(4x) = \int_{x=0}^{x=1} f(g(x))g'(x) = \int_{y=1}^{y=3} f(y) = \int_{y=1}^{y=3} y^{10}.$$

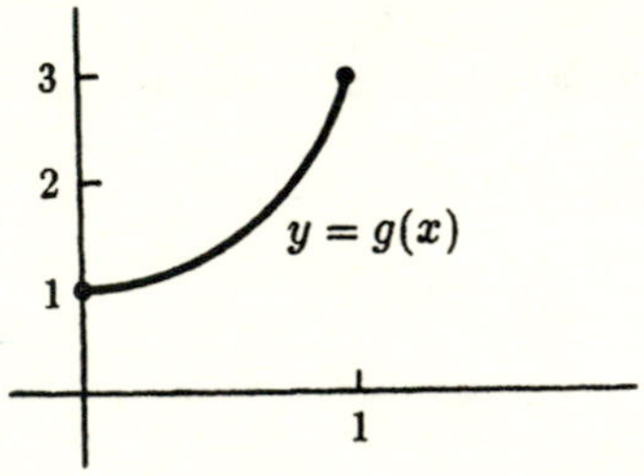

Figure 17.2

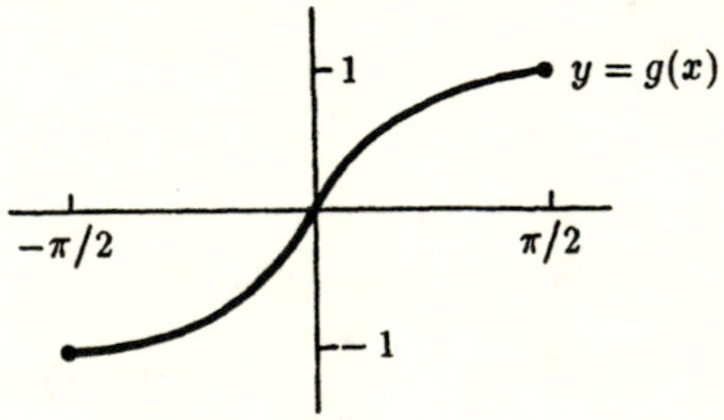

Figure 17.3

EXAMPLE 2. Consider the integral

$$\int_{y=-1}^{y=1} 1/(1-y^2)^{1/2}.$$

In calculus one proceeds as follows: Set $y = g(x) = \sin x$ for $-\pi/2 \leq x \leq \pi/2$. Then $g'(x) = \cos x$, which is positive on $(-\pi/2, \pi/2)$ and satisfies the conditions $g(-\pi/2) = -1$ and $g(\pi/2) = 1$. See Figure 17.3. If $f(y)$ is continuous on the interval $[-1, 1]$, then the substitution rule tells us that

$$\int_{-1}^{1} f = \int_{-\pi/2}^{\pi/2} (f \circ g)g'.$$

Applying this rule to the function $f(y) = 1/(1-y^2)^{1/2}$, we have

$$\int_{-1}^{1} 1/(1-y^2)^{1/2} = \int_{-\pi/2}^{\pi/2} [1/(1-\sin^2 x)^{1/2}] \cos x = \int_{-\pi/2}^{\pi/2} 1 = \pi.$$

Thus the problem seems to be solved.

However, there is a difficulty here. The substitution rule does *not* apply in this case, for the function $f(y)$ is *not* continuous on the interval $-1 \leq y \leq 1$! The integral of f is in fact an improper integral, since f is not even bounded on the interval $(-1, 1)$.

As indicated earlier, we shall generalize the substitution rule to n-dimensional integrals, and we shall prove it for the extended integral rather than merely for the ordinary integral. One reason is that the extended integral is actually easier to work with in this context than the ordinary integral. The other is that even in elementary problems one often needs to use the substitution rule in a situation where Theorem 17.1 does not apply, as Example 2 shows.

If we are to generalize this rule, we need to determine what a "substitution" or a "change of variables" is to be, in an n-dimensional extended integral. It is the following:

Definition. Let A be open in $\mathbf{R}^n$. Let $g : A \to \mathbf{R}^n$ be a one-to-one function of class C^r, such that $\det Dg(\mathbf{x}) \neq 0$ for $\mathbf{x} \in A$. Then g is called a **change of variables in $\mathbf{R}^n$**.

An equivalent notion is the following: If A and B are open sets in $\mathbf{R}^n$ and if $g : A \to B$ is a one-to-one function carrying A onto B such that both g and g^{-1} are of class C^r, then g is called a **diffeomorphism** (of class C^r). Now if g is a diffeomorphism, then the chain rule implies that Dg is non-singular, so that $\det Dg \neq 0$; thus g is also a change of variables. Conversely, if $g : A \to \mathbf{R}^n$ is a change of variables in $\mathbf{R}^n$, then Theorem 8.2 tells us that the set $B = g(A)$ is open in $\mathbf{R}^n$ and the function $g^{-1} : B \to A$ is of class C^r. Thus the terms "diffeomorphism" and "change of variables" are different terms for the same concept.

We now state the general change of variables theorem:

Theorem 17.2 (Change of variables theorem). *Let $g : A \to B$ be a diffeomorphism of open sets in $\mathbf{R}^n$. Let $f : B \to \mathbf{R}$ be a continuous function. Then f is integrable over B if and only if the function $(f \circ g)|\det Dg|$ is integrable over A; in this case,*

$$\int_B f = \int_A (f \circ g)|\det Dg|.$$

Note that in the special case $n = 1$, the derivative Dg is the 1 by 1 matrix whose entry is g'. Thus this theorem includes the classical substitution rule as a special case. It includes more, of course, since the integrals involved are extended integrals. It justifies, for example, the computations made in Example 2.

We shall prove this theorem in a later section. For the present, let us illustrate how it can be used to justify computations commonly made in multivariable calculus.

EXAMPLE 3. Let B be the open set in $\mathbf{R}^2$ defined by the equation

$$B = \{(x,y) \mid x > 0 \quad \text{and} \quad y > 0 \quad \text{and} \quad x^2 + y^2 < a^2\}.$$

One commonly computes an integral over B, such as $\int_B x^2y^2$, by the use of the **polar coordinate transformation.** This is the transformation $g : \mathbf{R}^2 \to \mathbf{R}^2$ defined by the equation

$$g(r,\theta) = (r\cos\theta, r\sin\theta).$$

One checks readily that $\det Dg(r,\theta) = r$, and that the map g carries the open rectangle

$$A = \{(r,\theta) \mid 0 < r < a \quad \text{and} \quad 0 < \theta < \pi/2\}$$

in the (r,θ) plane onto B in a one-to-one fashion. Since $\det Dg = r > 0$ on A, the map $g : A \to B$ is a diffeomorphism. See Figure 17.4.

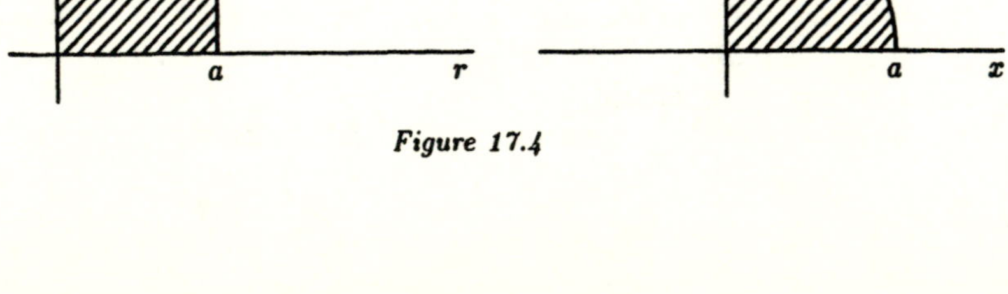

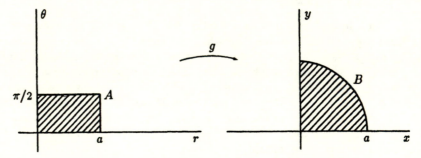

Figure 17.4

The change of variables theorem implies that

$$\int_B x^2y^2 = \int_A (r\cos\theta)^2(r\sin\theta)^2 r;$$

since the latter exists as an ordinary integral as well as an extended integral, it can be evaluated (easily) by use of the Fubini theorem.

EXAMPLE 4. Suppose we wish to integrate the same function x^2y^2 over the open set

$$W = \{(x,y)|x^2 + y^2 < a^2\}.$$

Here the use of polar coordinates is a bit more tricky. The polar coordinate transformation g does *not* in this case define a diffeomorphism of an open set in the (r,θ) plane with W. However, g *does* define a diffeomorphism of the open set $U = (0,a) \times (0,2\pi)$ with the open set

$$V = \{(x,y) \mid x^2 + y^2 < a^2 \quad \text{and} \quad x < 0 \quad \text{if} \quad y = 0\}$$

of $\mathbf{R}^2$. See Figure 17.5; the set V consists of W with the non-negative x-axis deleted. Because the non-negative x-axis has measure zero,

$$\int_W x^2y^2 = \int_V x^2y^2.$$

The latter can be expressed as an integral over U, by use of the polar coordinate transformation.

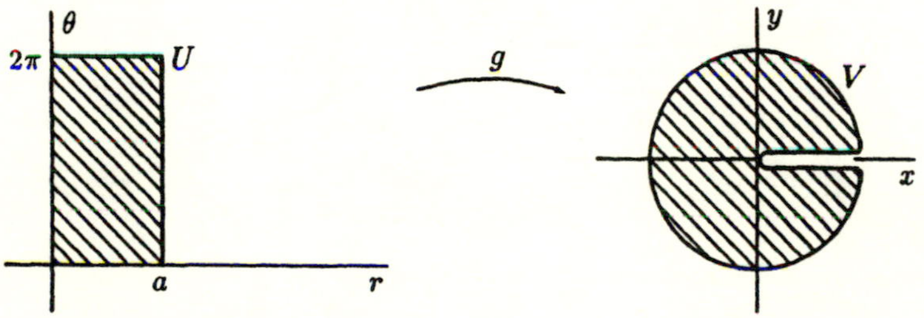

Figure 17.5

EXAMPLE 5. Let B be the open set in $\mathbf{R}^3$ defined by the equation

$$B = \{(x, y, z) \mid x > 0 \quad \text{and} \quad y > 0 \quad \text{and} \quad x^2 + y^2 + z^2 < a^2\}.$$

One commonly evaluates an integral over B, such as $\int_B x^2 z$, by the use of the **spherical coordinate transformation**, which is the transformation $g : \mathbf{R}^3 \to \mathbf{R}^3$ defined by the equation

$$g(\rho, \phi, \theta) = (\rho \sin\phi \cos\theta, \rho \sin\phi \sin\theta, \rho \cos\phi).$$

Now $\det Dg = \rho^2 \sin\phi$, as you can check. Thus $\det Dg$ is positive if $0 < \phi < \pi$ and $\rho \neq 0$. The transformation g carries the open set

$$A = \{(\rho, \phi, \theta) \mid 0 < \rho < a \quad \text{and} \quad 0 < \phi < \pi \quad \text{and} \quad 0 < \theta < \pi/2\}$$

in a one-to-one fashion onto B, as you can check. See Figure 17.6. Since $\det Dg > 0$ on A, the change of variables theorem implies that

$$\int_B x^2 z = \int_A (\rho \sin\phi \cos\theta)^2 (\rho \cos\phi) \rho^2 \sin\phi.$$

The latter can be evaluated by the Fubini theorem.

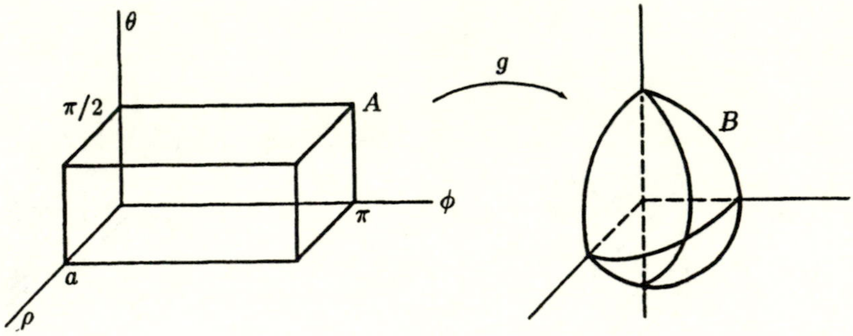

Figure 17.6

EXERCISES

1. Check the computations made in Examples 3 and 5.

2. If
$$V = \{(x,y,z) \mid x^2 + y^2 + z^2 < a^2 \quad \text{and} \quad z > 0\},$$
use the spherical coordinate transformation to express $\int_V z$ as an integral over an appropriate set in (ρ, ϕ, θ) space. Justify your answer.

3. Let U be the open set in $\mathbf{R}^2$ consisting of all $\mathbf{x}$ with $\|\mathbf{x}\| < 1$. Let $f(x,y) = 1/(x^2 + y^2)$ for $(x,y) \neq 0$. Determine whether f is integrable over $U - 0$ and over $\mathbf{R}^2 - \bar{U}$; if so, evaluate.

4. (a) Show that
$$\int_{\mathbf{R}^2} e^{-(x^2+y^2)} = [\int_{\mathbf{R}} e^{-x^2}]^2,$$
provided the first of these integrals exists.

   (b) Show the first of these integrals exists and evaluate it.

5. Let B be the portion of the first quadrant in $\mathbf{R}^2$ lying between the hyperbolas $xy = 1$ and $xy = 2$ and the two straight lines $y = x$ and $y = 4x$. Evaluate $\int_B x^2 y^3$. [*Hint:* Set $x = u/v$ and $y = uv$.]

6. Let S be the tetrahedron in $\mathbf{R}^3$ having vertices (0,0,0), (1,2,3), (0,1,2), and $(-1,1,1)$. Evaluate $\int_S f$, where $f(x,y,z) = x + 2y - z$. [*Hint:* Use a suitable linear transformation g as a change of variables.]

7. Let $0 < a < b$. If one takes the circle in the xz-plane of radius a centered at the point $(b,0,0)$, and if one rotates it about the z-axis, one obtains a surface called the **torus**. If one rotates the corresponding circular disc instead of the circle, one obtains a 3-dimensional solid called the **solid torus**. Find the volume of this solid torus. See Figure 17.7. [*Hint:* One can proceed directly, but it is easier to use the **cylindrical coordinate transformation**
$$g(r,\theta,z) = (r\cos\theta, r\sin\theta, z).$$
The solid torus is the image under g of the set of all (r,θ,z) for which $(r-b)^2 + z^2 \leq a^2$ and $0 \leq \theta \leq 2\pi$.]

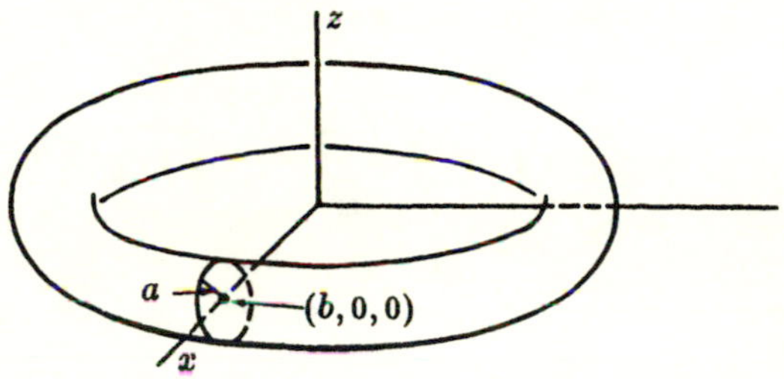

Figure 17.7

§18. DIFFEOMORPHISMS IN R^n

In order to prove the change of variables theorem, we need to obtain some fundamental properties of diffeomorphisms. This we do in the present section. Our first basic result is that the image of a compact rectifiable set under a diffeomorphism is another compact rectifiable set. And the second is that any diffeomorphism can be broken up locally into a composite of diffeomorphisms of a special type, called "primitive diffeomorphisms."

We begin with a preliminary lemma.

Lemma 18.1. *Let A be open in $\mathbf{R}^n$; let $g : A \to \mathbf{R}^n$ be a function of class C^1. If the subset E of A has measure zero in $\mathbf{R}^n$, then the set $g(E)$ also has measure zero in $\mathbf{R}^n$.*

Proof. *Step 1.* Let $\epsilon, \delta > 0$. We first show that if a set S has measure zero in $\mathbf{R}^n$, then S can be covered by countably many closed *cubes,* each of width less than δ, having total volume less than ϵ.

To prove this fact, it suffices to show that if Q is a rectangle

$$Q = [a_1, b_1] \times \cdots \times [a_n, b_n]$$

in $\mathbf{R}^n$, then Q can be covered by finitely many cubes, each of width less than δ, having total volume less than $2v(Q)$. Choose $\lambda > 0$ so that the rectangle

$$Q_\lambda = [a_1 - \lambda, b_1 + \lambda] \times \cdots \times [a_n - \lambda, b_n + \lambda]$$

has volume less than $2v(Q)$.

Then choose N so that $1/N$ is less than the smaller of δ and λ. Consider all rational numbers of the form m/N, where m is an arbitrary integer. Let c_i be the largest such number for which $c_i \leq a_i$, and let d_i be the smallest such number for which $d_i \geq b_i$. Then $[a_i, b_i] \subset [c_i, d_i] \subset [a_i - \lambda, b_i + \lambda]$. See Figure 18.1. Let Q' be the rectangle

$$Q' = [c_1, d_1] \times \cdots \times [c_n, d_n],$$

which contains Q and is contained in Q_λ. Then $v(Q') < 2v(Q)$. Each of the component intervals $[c_i, d_i]$ of Q' is partitioned by points of the form m/N into subintervals of length $1/N$. Then Q' is partitioned into subrectangles that are cubes of width $1/N$ (which is less than δ); these subrectangles cover Q. By Theorem 10.4, the total volume of these cubes equals $v(Q')$.

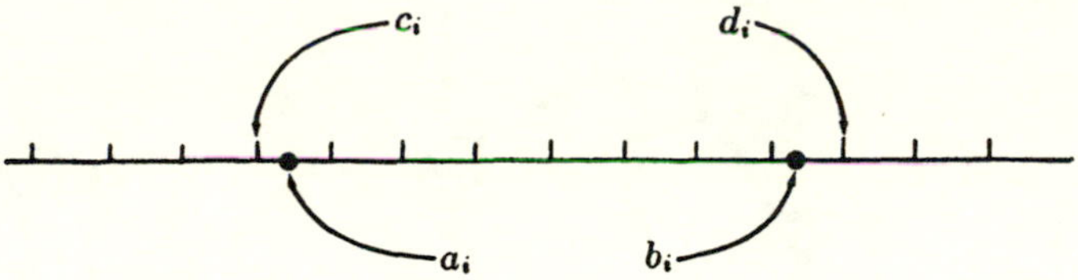

Figure 18.1

Step 2. Let C be a closed cube contained in A. Let

$$|Dg(\mathbf{x})| \leq M \quad \text{for} \quad \mathbf{x} \in C.$$

We show that if C has width w, then $g(C)$ is contained in a closed cube in $\mathbf{R}^n$ of width $(nM)w$.

Let $\mathbf{a}$ be the center of C; then C consists of all points $\mathbf{x}$ of $\mathbf{R}^n$ such that $|\mathbf{x} - \mathbf{a}| \leq w/2$. Now the mean-value theorem implies that given $\mathbf{x} \in C$, there is a point $\mathbf{c}_j$ on the line segment from $\mathbf{a}$ to $\mathbf{x}$ such that

$$g_j(\mathbf{x}) - g_j(\mathbf{a}) = Dg_j(\mathbf{c}_j) \cdot (\mathbf{x} - \mathbf{a}).$$

Then

$$|g_j(\mathbf{x}) - g_j(\mathbf{a})| \leq n|Dg_j(\mathbf{c}_j)| \cdot |\mathbf{x} - \mathbf{a}| \leq nM(w/2).$$

It follows from this inequality that if $\mathbf{x} \in C$, then $g(\mathbf{x})$ lies in the cube consisting of all $\mathbf{y} \in \mathbf{R}^n$ such that

$$|\mathbf{y} - g(\mathbf{a})| \leq nM(w/2).$$

This cube has width $(nM)w$, as desired.

Step 3. Now we prove the theorem. Suppose E is a subset of A and E has measure zero. We show that $g(E)$ has measure zero.

Let C_i be a sequence of compact sets whose union is A, such that $C_i \subset \text{Int}\, C_{i+1}$ for each i. Let $E_k = C_k \cap E$; it suffices to show that $g(E_k)$ has measure zero. Given $\epsilon > 0$, we shall cover $g(E_k)$ by cubes of total volume less than ϵ.

Since C_k is compact, we can choose $\delta > 0$ so that the δ-neighborhood of C_k (in the sup metric) lies in $\text{Int}\, C_{k+1}$, by Theorem 4.6. Choose M so that

$$|Dg(x)| \leq M \quad \text{for} \quad \mathbf{x} \in C_{k+1}.$$

Using Step 1, cover E_k by countably many cubes, each of width less than δ, having total volume less than

$$\epsilon' = \epsilon/(nM)^n.$$

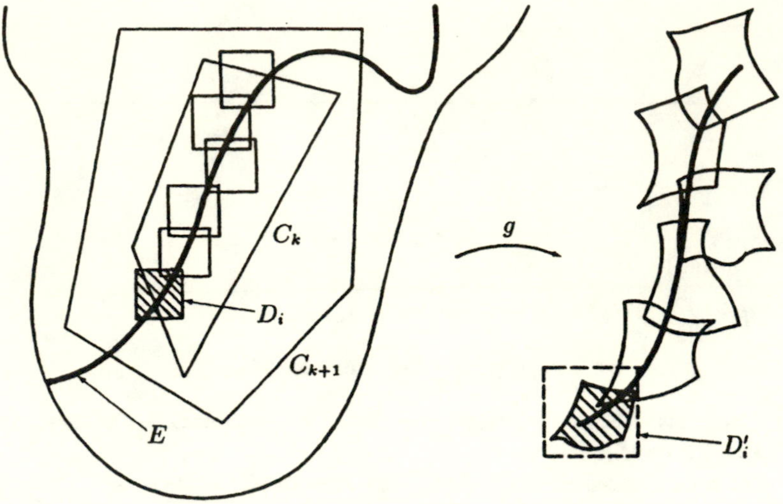

Figure 18.2

Let $D_1, D_2, \ldots$ denote those cubes that actually intersect E_k. Because D_i has width less than δ, it is contained in C_{k+1}. Then $|Dg(\mathbf{x})| \leq M$ for $\mathbf{x} \in D_i$, so that $g(D_i)$ lies in a cube D_i' of width $nM(\text{width } D_i)$, by Step 2. The cube D_i' has volume

$$v(D_i') = (nM)^n(\text{width } D_i)^n = (nM)^n v(D_i).$$

Therefore the cubes D_i', which cover $g(E_k)$, have total volume less than $(nM)^n\epsilon' = \epsilon$, as desired. See Figure 18.2. □

EXAMPLE 1. Differentiability is needed for the truth of the preceding lemma. If g is merely continuous, then the image of a set of measure zero need not have measure zero. This fact follows from the existence of a continuous map $f : [0,1] \to [0,1]^2$ whose image set is the *entire square* $[0,1]^2$! It is called the *Peano space-filling curve;* and it is studied in topology. (See [*M*], for example.)

Theorem 18.2. *Let $g : A \to B$ be a diffeomorphism of class C^r, where A and B are open sets in $\mathbb{R}^n$. Let D be a compact subset of A, and let $E = g(D)$.*

(a) *We have* $g(\text{Int } D) = \text{Int } E$ *and* $g(\text{Bd } D) = \text{Bd } E$.

(b) *If D is rectifiable, so is E.*

These results also hold when D is not compact, provided $\text{Bd } D \subset A$ *and* $\text{Bd } E \subset B$.

Proof. (a) The map g^{-1} is continuous. Therefore, for any open set U contained in A, the set $g(U)$ is an open set contained in B. In particular, $g(\text{Int } D)$ is an open set in $\mathbb{R}^n$ contained in the set $g(D) = E$. Thus

$$(1) \qquad g(\text{Int } D) \subset \text{Int } E.$$

Similarly, g carries the open set $(\text{Ext } D) \cap A$ onto an open set contained in B. Because g is one-to-one, the set $g\big((\text{Ext } D) \cap A\big)$ is disjoint from $g(D) = E$. Thus

$$(2) \qquad g\big((\text{Ext } D) \cap A\big) \subset \text{Ext } E.$$

It follows that

$$(3) \qquad g(\text{Bd } D) \supset \text{Bd } E.$$

For let $\mathbf{y} \in \text{Bd } E$; we show that $\mathbf{y} \in g(\text{Bd } D)$. The set E is compact, since D is compact and g is continuous. Hence E is closed, so it must contain its boundary point $\mathbf{y}$. Then $\mathbf{y} \in B$. Let $\mathbf{x}$ be the point of A such that $g(\mathbf{x}) = \mathbf{y}$. The point $\mathbf{x}$ cannot lie in $\text{Int } D$, by (1), and cannot lie in $\text{Ext } D$, by (2). Therefore $\mathbf{x} \in \text{Bd } D$, so that $\mathbf{y} \in g(\text{Bd } D)$, as desired. See Figure 18.3.

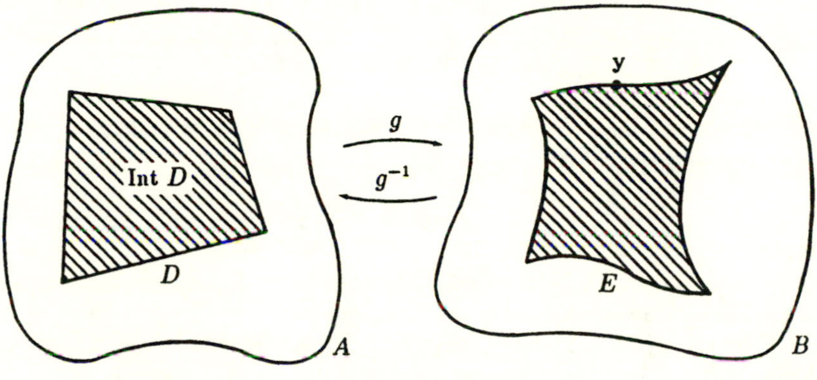

Figure 18.3

Symmetry implies that these same results hold for the map $g^{-1} : B \to A$. In particular,

$$(1') \qquad g^{-1}(\text{Int } E) \subset \text{Int } D,$$

$$(3') \qquad g^{-1}(\text{Bd } E) \supset \text{Bd } D.$$

Combining (1) and (1′) we see that $g(\text{Int } D) = \text{Int } E$; combining (3) and (3′) gives the equation $g(\text{Bd } D) = \text{Bd } E$.

(b) If D is rectifiable, then Bd D has measure zero. By the preceding lemma, $g(\text{Bd } D)$ also has measure zero. But $g(\text{Bd } D) = \text{Bd } E$. Thus E is rectifiable. □

Now we show that an arbitrary diffeomorphism of open sets in $\mathbf{R}^n$ can be "factored" locally into diffeomorphisms of a certain special type. This technical result will be crucial in the proof of the change of variables theorem.

Definition. Let $h : A \to B$ be a diffeomorphism of open sets in $\mathbf{R}^n$ (where $n \geq 2$), given by the equation

$$h(\mathbf{x}) = (h_1(\mathbf{x}), \ldots, h_n(\mathbf{x})).$$

Given i, we say that h **preserves the ith coordinate** if $h_i(\mathbf{x}) = x_i$ for all $\mathbf{x} \in A$. If h preserves the ith coordinate for some i, then h is called a **primitive diffeomorphism.**

Theorem 18.3. *Let $g : A \to B$ be a diffeomorphism of open sets in $\mathbf{R}^n$, where $n \geq 2$. Given $\mathbf{a} \in A$, there is a neighborhood U_0 of $\mathbf{a}$ contained in A, and a sequence of diffeomorphisms of open sets in $\mathbf{R}^n$,*

$$U_0 \xrightarrow{h_1} U_1 \xrightarrow{h_2} U_2 \longrightarrow \cdots \xrightarrow{h_k} U_k,$$

such that the composite $h_k \circ \cdots \circ h_2 \circ h_1$ equals $g|U_0$, and such that each h_i is a primitive diffeomorphism.

Proof. *Step 1.* We first consider the special case of a linear transformation. Let $T : \mathbf{R}^n \to \mathbf{R}^n$ be the linear transformation $T(\mathbf{x}) = C \cdot \mathbf{x}$, where C is a non-singular n by n matrix. We show that T factors into a sequence of primitive non-singular linear transformations.

This is easy. The matrix C equals a product of elementary matrices, by Theorem 2.4. The transformation corresponding to an elementary matrix may either (1) switch two coordinates, or (2) replace the ith coordinate by itself plus a multiple of another coordinate, or (3) multiply the ith coordinate by a non-zero scalar. Transformations of types (2) and (3) are clearly primitive, since they leave all but the ith coordinate fixed. We show that a transformation of type (1) is a composite of transformation of types (2) and (3), and our result follows. Indeed, it is easy to check that the following sequence of elementary operations has the effect of exchanging rows i and j:

	Row i	Row j
Initial state	a	b
Replace (row i) by (row i) − (row j)	a−b	b
Replace (row j) by (row j) + (row i)	a−b	a
Replace (row i) by (row i) − (row j)	−b	a
Multiply (row i) by −1	b	a

Step 2. We next consider the case where g is a translation. Let $t : \mathbf{R}^n \to \mathbf{R}^n$ be the map $t(\mathbf{x}) = \mathbf{x} + \mathbf{c}$. Then t is the composite of the translations

$$t_1(\mathbf{x}) = \mathbf{x} + (0, c_2, \ldots, c_n),$$

$$t_2(\mathbf{x}) = \mathbf{x} + (c_1, 0, \ldots, 0),$$

both of which are primitive.

Step 3. We now consider the special case where $\mathbf{a} = \mathbf{0}$ and $g(\mathbf{0}) = \mathbf{0}$ and $Dg(\mathbf{0}) = I_n$. We show that in this case, g factors locally as a composite of two primitive diffeomorphisms.

Let us write g in components as

$$g(\mathbf{x}) = (g_1(\mathbf{x}), \ldots, g_n(\mathbf{x})) = (g_1(x_1, \ldots, x_n), \ldots, g_n(x_1, \ldots, x_n)).$$

Define $h : A \to \mathbf{R}^n$ by the equation

$$h(\mathbf{x}) = (g_1(\mathbf{x}), \ldots, g_{n-1}(\mathbf{x}), x_n).$$

Now $h(\mathbf{0}) = \mathbf{0}$, because $g_i(\mathbf{0}) = 0$ for all i; and

$$Dh(\mathbf{x}) = \begin{bmatrix} \partial(g_1, & \ldots, & g_{n-1})/\partial\mathbf{x} & \\ & & & \\ 0 & \ldots & 0 & 1 \end{bmatrix}.$$

Since the matrix $\partial(g_1, \ldots, g_{n-1})/\partial\mathbf{x}$ equals the first $n-1$ rows of the matrix Dg, and $Dg(\mathbf{0}) = I_n$, we have $Dh(\mathbf{0}) = I_n$. It follows from the inverse function theorem that h is a diffeomorphism of a neighborhood V_0 of $\mathbf{0}$ with an open set V_1 in $\mathbf{R}^n$. See Figure 18.4. Now we define $k : V_1 \to \mathbf{R}^n$ by the equation

$$k(\mathbf{y}) = (y_1, \ldots, y_{n-1}, g_n(h^{-1}(\mathbf{y}))).$$

Then $k(\mathbf{0}) = \mathbf{0}$ (since $h^{-1}(\mathbf{0}) = \mathbf{0}$ and $g_n(\mathbf{0}) = 0$). Furthermore,

$$Dk(\mathbf{y}) = \begin{bmatrix} I_{n-1} & 0 \\ & \\ D(g_n \circ h^{-1})(\mathbf{y}) & \end{bmatrix}.$$

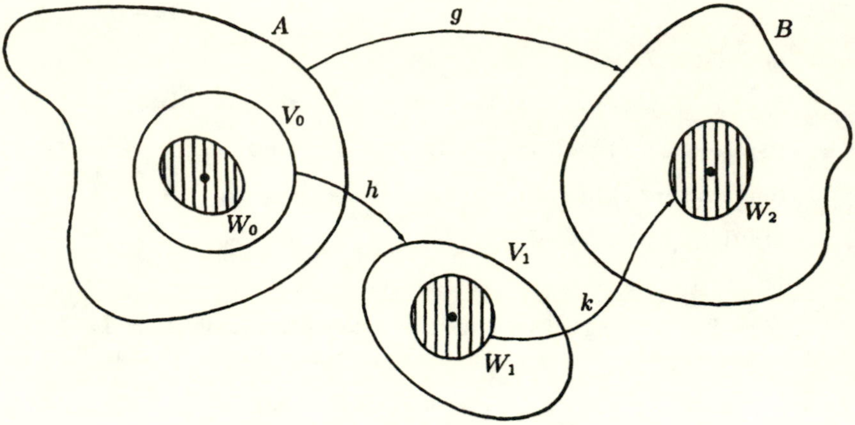

Figure 18.4

Applying the chain rule, we compute

$$\begin{aligned} D(g_n \circ h^{-1})(0) &= Dg_n(0) \cdot Dh^{-1}(0) \\ &= Dg_n(0) \cdot [Dh(0)]^{-1} \\ &= [0 \cdots 0\ 1] \cdot I_n = [0 \cdots 0\ 1]. \end{aligned}$$

Hence $Dk(\mathbf{0}) = I_n$. It follows that k is a diffeomorphism of a neighborhood W_1 of $\mathbf{0}$ in $\mathbb{R}^n$ with an open set W_2 in $\mathbb{R}^n$.

Now let $W_0 = h^{-1}(W_1)$. The diffeomorphisms

$$W_0 \xrightarrow{h} W_1 \xrightarrow{k} W_2$$

are primitive. Furthermore, the composite $k \circ h$ equals $g|W_0$, as we now show.

Given $\mathbf{x} \in W_0$, let $\mathbf{y} = h(\mathbf{x})$. Now

$$(*) \qquad \mathbf{y} = (g_1(\mathbf{x}), \ldots, g_{n-1}(\mathbf{x}), x_n)$$

by definition. Then

$$\begin{aligned} k(\mathbf{y}) &= (y_1, \ldots, y_{n-1}, g_n(h^{-1}(\mathbf{y}))) \quad \text{by definition,} \\ &= (g_1(\mathbf{x}), \ldots, g_{n-1}(\mathbf{x}), g_n(\mathbf{x})) \quad \text{by } (*), \\ &= g(\mathbf{x}). \end{aligned}$$

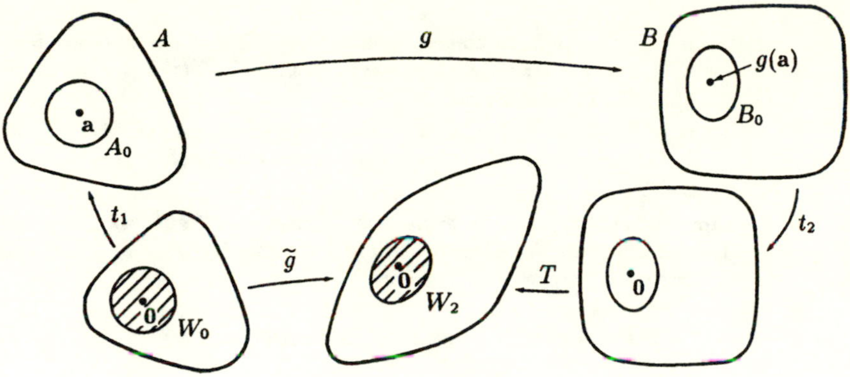

Figure 18.5

Step 4. Now we prove the theorem in the general case.

Given $g : A \to B$, and given $\mathbf{a} \in A$, let C be the matrix $Dg(\mathbf{a})$. Define diffeomorphisms $t_1, t_2, T : \mathbf{R}^n \to \mathbf{R}^n$ by the equations

$$t_1(\mathbf{x}) = \mathbf{x} + \mathbf{a} \quad \text{and} \quad t_2(\mathbf{x}) = \mathbf{x} - g(\mathbf{a}) \quad \text{and} \quad T(\mathbf{x}) = C^{-1} \cdot \mathbf{x}.$$

Let $\tilde{g}$ equal the composite $T \circ t_2 \circ g \circ t_1$. Then $\tilde{g}$ is a diffeomorphism of the open set $t_1^{-1}(A)$ of $\mathbf{R}^n$ with the open set $T(t_2(B))$ of $\mathbf{R}^n$. See Figure 18.5. It has the property that

$$\tilde{g}(\mathbf{0}) = \mathbf{0} \quad \text{and} \quad D\tilde{g}(\mathbf{0}) = I_n;$$

the first equation follows from the definition, while the second follows from the chain rule, since $DT(\mathbf{0}) = C^{-1}$ and $Dt_i = I_n$ for $i = 1, 2$.

By Step 3, there is an open set W_0 about $\mathbf{0}$ contained in $t_1^{-1}(A)$ such that $\tilde{g}|W_0$ factors into a sequence of (two) primitive diffeomorphisms. Let $W_2 = \tilde{g}(W_0)$. Let

$$A_0 = t_1(W_0) \quad \text{and} \quad B_0 = t_2^{-1}T^{-1}(W_2).$$

Then g carries A_0 onto B_0, and $g|A_0$ equals the composite

$$A_0 \xrightarrow{t_1^{-1}} W_0 \xrightarrow{\tilde{g}} W_2 \xrightarrow{T^{-1}} T^{-1}(W_2) \xrightarrow{t_2^{-1}} B_0.$$

By Steps 1 and 2, each of the maps t_1^{-1} and t_2^{-1} and T^{-1} factors into primitive transformations. The theorem follows. □

EXERCISES

1. (a) If $f : \mathbf{R}^2 \to \mathbf{R}^1$ is of class C^1, show that f is not one-to-one. [*Hint:* If $Df(\mathbf{x}) = 0$ for all $\mathbf{x}$, then f is constant. If $Df(\mathbf{x}_0) \neq 0$, apply the implicit function theorem.]
   (b) If $f : \mathbf{R}^1 \to \mathbf{R}^2$ is of class C^1, show that f does not carry $\mathbf{R}^1$ onto $\mathbf{R}^2$. In fact, show that $f(\mathbf{R}^1)$ contains no open set of $\mathbf{R}^2$.

*2. Prove a generalization of Theorem 18.3 in which the statement "h is primitive" is interpreted to mean that h preserves *all but one* coordinate. [*Hint:* First show that if $\mathbf{a}=\mathbf{0}$ and $g(\mathbf{0}) = \mathbf{0}$ and $Dg(\mathbf{0}) = I_n$, then g can be factored locally as $k \circ h$, where

$$h(\mathbf{x}) = (g_1(\mathbf{x}), \ldots, g_{i-1}(\mathbf{x}), x_i, g_{i+1}(\mathbf{x}), \ldots, g_n(\mathbf{x}))$$

and k preserves all but the ith coordinate; and furthermore, $h(\mathbf{0}) = k(\mathbf{0}) = \mathbf{0}$ and $Dh(\mathbf{0}) = Dk(\mathbf{0}) = I_n$. Then proceed inductively.]

3. Let A be open in $\mathbf{R}^m$; let $g : A \to \mathbf{R}^n$. If S is a subset of A, we say that g satisfies the Lipschitz condition on S if the function

$$\lambda(\mathbf{x}, \mathbf{y}) = |g(\mathbf{x}) - g(\mathbf{y})| / |\mathbf{x} - \mathbf{y}|$$

is bounded for $\mathbf{x}, \mathbf{y}$ in S and $\mathbf{x} \neq \mathbf{y}$. We say that g is **locally Lipschitz** if each point of A has a neighborhood on which g satisfies the Lipschitz condition.
   (a) Show that if g is of class C^1, then g is locally Lipschitz.
   (b) Show that if g is locally Lipschitz, then g is continuous.
   (c) Give examples to show that the converses of (a) and (b) do not hold.
   (d) Let g be locally Lipschitz. Show that if C is a compact subset of A, then g satisfies the Lipschitz condition on C. [*Hint:* Show there is a neighborhood V of the diagonal Δ in $C \times C$ such that λ is bounded on $V - \Delta$.]

4. Let A be open in $\mathbf{R}^n$; let $g : A \to \mathbf{R}^n$ be locally Lipschitz. Show that if the subset E of A has measure zero in $\mathbf{R}^n$, then $g(E)$ has measure zero in $\mathbf{R}^n$.

5. Let A and B be open in $\mathbf{R}^n$; let $g : A \to B$ be a one-to-one map carrying A onto B.
   (a) Show that (a) of Theorem 18.2 holds under assumption that g and g^{-1} are continuous.
   (b) Show that (b) of Theorem 18.2 holds under the assumption that g is locally Lipschitz and g^{-1} is continuous.

§19. PROOF OF THE CHANGE OF VARIABLES THEOREM

Now we prove the general change of variables theorem. We prove first the "only if" part of the theorem. It is stated in the following lemma:

Lemma 19.1. *Let $g : A \to B$ be a diffeomorphism of open sets in $\mathbf{R}^n$. Then for every continuous function $f : B \to \mathbf{R}$ that is integrable over B, the function $(f \circ g)|\det Dg|$ is integrable over A, and*

$$\int_B f = \int_A (f \circ g)|\det Dg|.$$

Proof. The proof proceeds in several steps, by which one reduces the proof to successively simpler cases.

Step 1. Let $g : U \to V$ and $h : V \to W$ be diffeomorphisms of open sets in $\mathbf{R}^n$. We show that if the lemma holds for g and for h, then it holds for $h \circ g$.

Suppose $f : W \to \mathbf{R}$ is a continuous function that is integrable over W. It follows from our hypothesis that

$$\int_W f = \int_V (f \circ h)|\det Dh| = \int_U (f \circ h \circ g)|(\det Dh) \circ g|\,|\det Dg|;$$

the second integral exists and equals the first integral because the lemma holds for h; and the third integral exists and equals the second integral because the lemma holds for g. In order to show that the lemma holds for $h \circ g$, it suffices to show that

$$|(\det Dh) \circ g|\,|\det Dg| = |\det D(h \circ g)|.$$

This result follows from the chain rule. We have

$$D(h \circ g)(\mathbf{x}) = Dh(g(\mathbf{x})) \cdot Dg(\mathbf{x}),$$

whence

$$\det D(h \circ g) = [(\det Dh) \circ g] \cdot [\det Dg],$$

as desired.

Step 2. Suppose that for each $\mathbf{x} \in A$, there is a neighborhood U of $\mathbf{x}$ contained in A such that the lemma holds for the diffeomorphism $g : U \to V$ (where $V = g(U)$) and all continuous functions $f : V \to \mathbf{R}$ whose supports are compact subsets of V. Then we show that the lemma holds for g.

Roughly speaking, this statement says that if the lemma holds locally for g and functions f having compact support, then it holds for g and all f.

This is the place in the proof where we use partitions of unity. Write A as the union of a collection of open sets U_α such that if $V_\alpha = g(U_\alpha)$, then the lemma holds for the diffeomorphism $g : U_\alpha \to V_\alpha$ and all continuous functions $f : V_\alpha \to \mathbf{R}$ whose supports are compact subsets of V_α. The union of the open sets V_α equals B. Choose a partition of unity $\{\phi_i\}$ on B, having compact supports, that is dominated by the collection $\{V_\alpha\}$. We show that the collection $\{\phi_i \circ g\}$ is a partition of unity on A, having compact supports. See Figure 19.1.

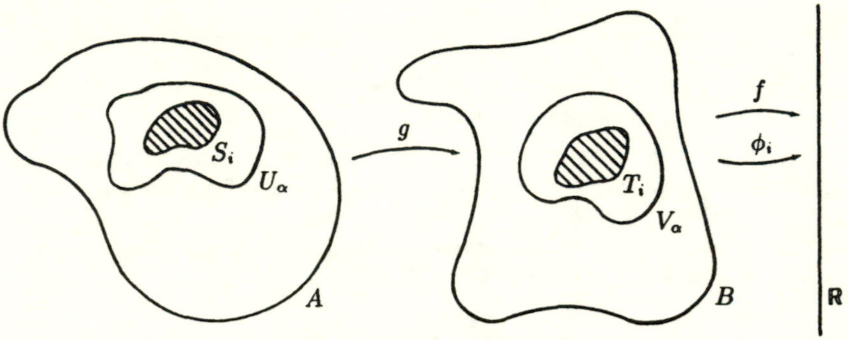

Figure 19.1

First, we note that $\phi_i(g(\mathbf{x})) \geq 0$ for $\mathbf{x} \in A$. Second, we show $\phi_i \circ g$ has compact support. Let $T_i = \text{Support } \phi_i$. The set $g^{-1}(T_i)$ is compact because T_i is compact and g^{-1} is continuous; furthermore, $\phi_i \circ g$ vanishes outside $g^{-1}(T_i)$. The closed set $S_i = \text{Support } (\phi_i \circ g)$ is contained in $g^{-1}(T_i)$, so that S_i is compact. Third, we check the local finiteness condition. Let $\mathbf{x}$ be a point of A. The point $\mathbf{y} = g(\mathbf{x})$ has a neighborhood W that intersects T_i for only finitely many values of i. Then the set $g^{-1}(W)$ is an open set about $\mathbf{x}$ that intersects S_i for at most these same values of i. Fourth, we note that

$$\sum \phi_i(g(\mathbf{x})) = \sum \phi_i(\mathbf{y}) = 1.$$

Thus $\{\phi_i \circ g\}$ is a partition of unity on A.

Now we complete the proof of Step 2. Suppose $f : B \to \mathbf{R}$ is continuous and f is integrable over B. We have

$$\int_B f = \sum_{i=1}^{\infty} [\int_B \phi_i f],$$

by Theorem 16.5. Given i, choose α so that $T_i \subset V_\alpha$. The function $\phi_i f$ is continuous on B and vanishes outside the compact set T_i. Then

$$\int_B \phi_i f = \int_{T_i} \phi_i f = \int_{V_\alpha} \phi_i f,$$

by Lemma 16.4. Our lemma holds by hypothesis for $g : U_\alpha \to V_\alpha$ and the function $\phi_i f$. Therefore

$$\int_{V_\alpha} \phi_i f = \int_{U_\alpha} (\phi_i \circ g)\,(f \circ g)|\det Dg|.$$

Since the integrand on the right vanishes outside the compact set S_i, we can apply Lemma 16.4 again to conclude that

$$\int_B \phi_i f = \int_A (\phi_i \circ g)\,(f \circ g)|\det Dg|.$$

We then sum over i to obtain the equation

$$(*) \qquad \int_B f = \sum_{i=1}^{\infty} [\int_A (\phi_i \circ g)\,(f \circ g)|\det Dg|].$$

Since $|f|$ is integrable over B, equation $(*)$ holds if f is replaced throughout by $|f|$. Since $\{\phi_i \circ g\}$ is a partition of unity on A, it then follows from Theorem 16.5 that $(f \circ g)|\det Dg|$ is integrable over A. We then apply $(*)$ to the function f to conclude that

$$\int_B f = \int_A (f \circ g)|\det Dg|.$$

Step 3. We show that the lemma holds for $n = 1$.

Let $g : A \to B$ be a diffeomorphism of open sets in $\mathbf{R}^1$. Given $x \in A$, let I be a closed interval in A whose interior contains x; and let $J = g(I)$. Now J is an interval in $\mathbf{R}^1$ and g maps Int I onto Int J. (See Theorems 17.1 and 18.2.) Since x is arbitrary, it suffices by Step 2 to prove the lemma holds for the diffeomorphism $g : \text{Int } I \to \text{Int } J$ and any continuous function $f : \text{Int } J \to \mathbf{R}$ whose support is a compact subset of Int J. That is, we wish to verify the equation

$$(**) \qquad \int_{\text{Int } J} f = \int_{\text{Int } I} (f \circ g)|g'|.$$

This is easy. First, we extend f to a continuous function defined on J by letting it vanish on Bd J. Then $(**)$ is equivalent to the equation

$$\int_J f = \int_I (f \circ g)|g'|,$$

in ordinary integrals. But this equation follows from Theorem 17.1.

Step 4. Let $n > 1$. In order to prove the lemma for an arbitrary diffeomorphism $g : A \to B$ of open sets in $\mathbf{R}^n$, we show that it suffices to prove it for a primitive diffeomorphism $h : U \to V$ of open sets in $\mathbf{R}^n$.

Suppose the lemma holds for all primitive diffeomorphisms in $\mathbf{R}^n$. Let $g : A \to B$ be an arbitrary diffeomorphism in $\mathbf{R}^n$. Given $\mathbf{x} \in A$, there exists a neighborhood U_0 of $\mathbf{x}$ and a sequence of primitive diffeomorphisms

$$U_0 \xrightarrow{h_1} U_1 \xrightarrow{h_2} \cdots \xrightarrow{h_k} U_k$$

whose composite equals $g|U_0$. Since the lemma holds for each of the diffeomorphisms h_i, it follows from Step 1 that it holds for $g|U_0$. Then because $\mathbf{x}$ is arbitrary, it follows from Step 2 that it holds for g.

Step 5. We show that if the lemma holds in dimension $n - 1$, it holds in dimension n.

This step completes the proof of the lemma.

In view of Step 4, it suffices to prove the lemma for a primitive diffeomorphism $h : U \to V$ of open sets in $\mathbf{R}^n$. For convenience in notation, we assume that h preserves the last coordinate.

Let $\mathbf{p} \in U$; let $\mathbf{q} = h(\mathbf{p})$. Choose a rectangle Q contained in V whose interior contains $\mathbf{q}$; let $S = h^{-1}(Q)$. By Theorem 18.2, the map h defines a diffeomorphism of Int S with Int Q. Since $\mathbf{p}$ is arbitrary, it suffices by Step 2 to prove that the lemma holds for the diffeomorphism $h : \text{Int } S \to \text{Int } Q$ and any continuous function $f : \text{Int } Q \to \mathbf{R}$ whose support is a compact subset of Int Q. See Figure 19.2.

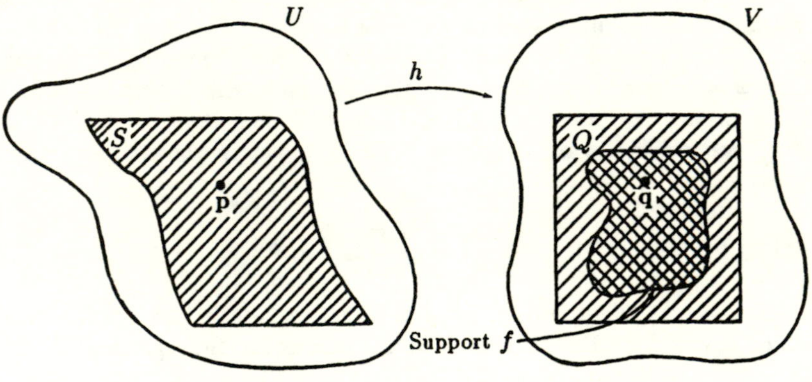

Figure 19.2

Now $(f \circ h)|\det Dh|$ vanishes outside a compact subset of Int S; hence it is integrable over Int S by Lemma 16.4. We need to show that

$$\int_{\text{Int } Q} f = \int_{\text{Int } S} (f \circ h)|\det Dh|.$$

This is an equation involving extended integrals. Since these integrals exist as ordinary integrals, it is by Theorem 15.4 equivalent to the corresponding equation in ordinary integrals.

Let us extend f to $\mathbf{R}^n$ by letting it vanish outside Int Q, and let us define a function $F : \mathbf{R}^n \to \mathbf{R}$ by letting it equal $(f \circ h)|\det Dh|$ on Int S and vanish elsewhere. Then both f and F are continuous, and our desired equation is equivalent to the equation

$$\int_Q f = \int_S F.$$

The rectangle Q has the form $Q = D \times I$, where D is a rectangle in $\mathbf{R}^{n-1}$ and I is a closed interval in $\mathbf{R}$. Since S is compact, its projection on the subspace $\mathbf{R}^{n-1} \times 0$ is compact and thus contained in a set of the form $E \times 0$, where E is a rectangle in $\mathbf{R}^{n-1}$. Because h preserves the last coordinate, the set S is contained in the rectangle $E \times I$. See Figure 19.3.

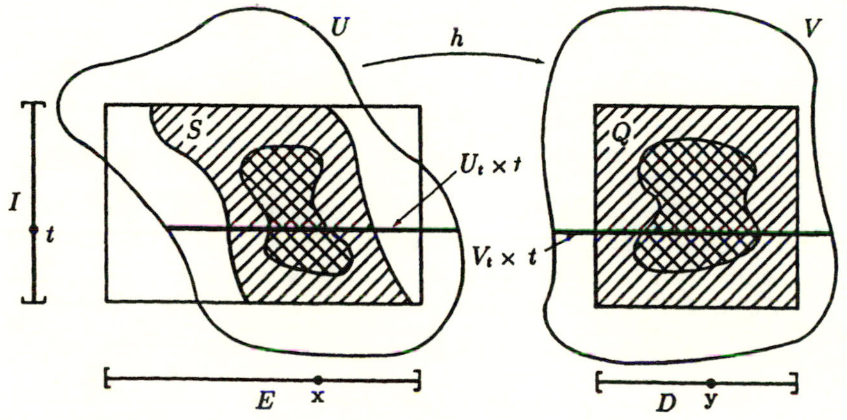

Figure 19.3

Because F vanishes outside S, our desired equation can be written in the form

$$\int_Q f = \int_{E \times I} F,$$

which by the Fubini theorem is equivalent to the equation

$$\int_{t\in I}\int_{\mathbf{y}\in D} f(\mathbf{y},t) = \int_{t\in I}\int_{\mathbf{x}\in E} F(\mathbf{x},t).$$

It suffices to show the inner integrals are equal. This we now do.

The intersections of U and V with $\mathbf{R}^{n-1}\times t$ are sets of the form $U_t\times t$ and $V_t\times t$, respectively, where U_t and V_t are open sets in $\mathbf{R}^{n-1}$. Similarly, the intersection of S with $\mathbf{R}^{n-1}\times t$ has the form $S_t\times t$, where S_t is a compact set in $\mathbf{R}^{n-1}$. Since F vanishes outside S, equality of the "inner integrals" is equivalent to the equation

$$\int_{\mathbf{y}\in D} f(\mathbf{y},t) = \int_{\mathbf{x}\in S_t} F(\mathbf{x},t),$$

and this is in turn equivalent by Lemma 16.4 to the equation

$$\int_{\mathbf{y}\in V_t} f(\mathbf{y},t) = \int_{\mathbf{x}\in U_t} F(\mathbf{x},t).$$

This is an equation in $(n-1)$-dimensional integrals, to which the induction hypothesis applies.

The diffeomorphism $h : U \to V$ has the form

$$h(\mathbf{x},t) = (k(\mathbf{x},t),t)$$

for some C^1 function $k : U \to \mathbf{R}^{n-1}$. The derivative of h has the form

$$Dh = \begin{bmatrix} \partial k/\partial \mathbf{x} & \partial k/\partial t \\ 0\cdots 0 & 1 \end{bmatrix},$$

so that $\det Dh = \det \partial k/\partial\mathbf{x}$. For fixed t, the map $\mathbf{x}\to k(\mathbf{x},t)$ is a C^1 map carrying U_t onto V_t in a one-to-one fashion. Because $\det\partial k/\partial\mathbf{x} = \det Dh \neq 0$, this map is in fact a diffeomorphism of open sets in $\mathbf{R}^{n-1}$.

We apply the induction hypothesis; we have, for fixed t, the equation

$$\int_{\mathbf{y}\in V_t} f(\mathbf{y},t) = \int_{\mathbf{x}\in U_t} f(k(\mathbf{x},t),t)|\det\partial k/\partial\mathbf{x}|.$$

For $\mathbf{x}\in U_t$, the integrand on the right equals

$$f(h(\mathbf{x},t))|\det Dh| = F(\mathbf{x},t).$$

The lemma follows. □

We now prove the "if" part of the change of variables theorem.

Lemma 19.2. *Let $g : A \to B$ be a diffeomorphism of open sets in $\mathbf{R}^n$; let $f : B \to \mathbf{R}$ be continuous. If $(f \circ g)|\det Dg|$ is integrable over A, then f is integrable over B.*

Proof. We apply the lemma just proved to the diffeomorphism $g^{-1} : B \to A$. The function $F = (f \circ g)|\det Dg|$ is continuous on A, and is integrable over A by hypothesis. It follows from Lemma 19.1 that the function

$$(F \circ g^{-1})|\det Dg^{-1}|$$

is integrable over B. But this function equals f. For if $g(\mathbf{x}) = \mathbf{y}$, then

$$(D(g^{-1}))(\mathbf{y}) = [Dg(\mathbf{x})]^{-1}$$

by Theorem 7.4, so that

$$(F \circ g^{-1})(\mathbf{y}) \cdot |(\det D(g^{-1}))(\mathbf{y})| = F(\mathbf{x}) \cdot |1/\det Dg(\mathbf{x})| = f(\mathbf{y}). \quad \square$$

EXAMPLE 1. If it happens that both integrals in the change of variables theorem exist as ordinary integrals, then the theorem implies that these two ordinary integrals are equal. However, it is possible for only one of these integrals, or neither, to exist as an ordinary integral. Consider, for instance, Example 2 of §17. The change of variables theorem, applied to the diffeomorphism $g : (-\pi/2, \pi/2) \to (-1, 1)$ given by $g(x) = \cos x$, implies that

$$\int_{(-1,1)} 1/(1-y^2)^{1/2} = \int_{(-\pi/2,\pi/2)} 1.$$

Here the integral on the right exists as an ordinary integral, but the integral on the left does not.

EXERCISES

1. Let A be the region in $\mathbf{R}^2$ bounded by the curve $x^2 - xy + 2y^2 = 1$. Express the integral $\int_A xy$ as an integral over the unit ball in $\mathbf{R}^2$ centered at $\mathbf{0}$. [*Hint:* Complete the square.]

2. (a) Express the volume of the solid in $\mathbf{R}^3$ bounded below by the surface $z = x^2 + 2y^2$, and above by the plane $z = 2x + 6y + 1$, as the integral of a suitable function over the unit ball in $\mathbf{R}^2$ centered at $\mathbf{0}$.

   (b) Find this volume.

3. Let $\pi_k : \mathbf{R}^n \to \mathbf{R}$ be the kth projection function, defined by the equation $\pi_k(\mathbf{x}) = x_k$. Let S be a rectifiable set in $\mathbf{R}^n$ with non-zero volume.

The centroid of S is defined to be the point $\mathbf{c}(S)$ of $\mathbf{R}^n$ whose k^{th} coordinate, for each k, is given by the equation

$$c_k(S) = [1/v(S)] \int_S \pi_k.$$

We say that S is symmetric with respect to the subspace $x_k = 0$ of $\mathbf{R}^n$ if the transformation

$$h(\mathbf{x}) = (x_1, \ldots, x_{k-1}, -x_k, x_{k+1}, \ldots, x_n)$$

carries S onto itself. In this case, show that $c_k(S) = 0$.

4. Find the centroid of the upper half-ball of radius a in $\mathbf{R}^3$. (See Exercise 2 of §17.)

5. Let A be an open rectifiable set in $\mathbf{R}^{n-1}$. Given the point $\mathbf{p}$ in $\mathbf{R}^n$ with $p_n > 0$, let S be the subset of $\mathbf{R}^n$ defined by the equation

$$S = \{\mathbf{x} \mid \mathbf{x} = (1-t)\mathbf{a} + t\mathbf{p}, \quad \text{where} \quad \mathbf{a} \in A \times 0 \quad \text{and} \quad 0 < t < 1\}.$$

Then S is the union of all open line segments in $\mathbf{R}^n$ joining $\mathbf{p}$ to points of $A \times 0$; its closure is called the cone with base $\bar{A} \times 0$ and vertex $\mathbf{p}$. Figure 19.4 illustrates the case $n = 3$.

(a) Define a diffeomorphism g of $A \times (0,1)$ with S.

(b) Find $v(S)$ in terms of $v(A)$.

*(c) Show that the centroid $\mathbf{c}(S)$ of S lies on the line segment joining $\mathbf{c}(A)$ and $\mathbf{p}$; express it in terms of $\mathbf{c}(A)$ and $\mathbf{p}$.

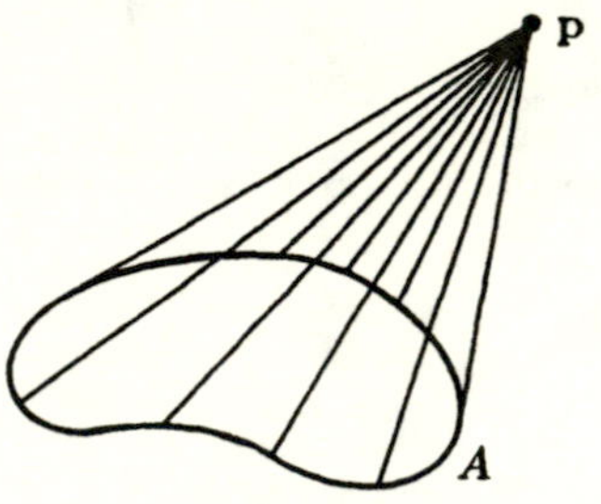

Figure 19.4

*6. Let $B^n(a)$ denote the closed ball of radius a in $\mathbf{R}^n$, centered at $\mathbf{0}$.

(a) Show that

$$v(B^n(a)) = \lambda_n a^n$$

for some constant λ_n. Then $\lambda_n = v(B^n(1))$.

(b) Compute λ_1 and λ_2.

(c) Compute λ_n in terms of λ_{n-2}.

(d) Obtain a formula for λ_n. [*Hint:* Consider two cases, according as n is even or odd.]

*7. (a) Find the centroid of the upper half-ball

$$B_+^n(a) = \{\mathbf{x} \mid \mathbf{x} \in B^n(a) \quad \text{and} \quad x_n \geq 0\}$$

in terms of λ_n and λ_{n-1} and a, where $\lambda_n = v(B^n(1))$.

(b) Express $\mathbf{c}(B_+^n(a))$ in terms of $\mathbf{c}(B_+^{n-2}(a))$.

§20. APPLICATIONS OF CHANGE OF VARIABLES

The meaning of the determinant

We now give a geometric interpretation of the determinant function.

Theorem 20.1. *Let A be an n by n matrix. Let $h : \mathbf{R}^n \to \mathbf{R}^n$ be the linear transformation $h(\mathbf{x}) = A \cdot \mathbf{x}$. Let S be a rectifiable set in $\mathbf{R}^n$, and let $T = h(S)$. Then*

$$v(T) = |\det A| \cdot v(S).$$

Proof. Consider first the case where A is non-singular. Then h is a diffeomorphism of $\mathbf{R}^n$ with itself; h carries Int S onto Int T; and T is rectifiable. We have

$$v(T) = v(\text{Int } T) = \int_{\text{Int } T} 1 = \int_{\text{Int } S} |\det Dh|$$

by the change of variables theorem. Hence

$$v(T) = \int_{\text{Int } S} |\det A| = |\det A| \cdot v(S).$$

Consider now the case where A is singular; then $\det A = 0$. We show that $v(T) = 0$. Since S is bounded, so is T. The transformation h carries $\mathbf{R}^n$ onto a linear subspace V of $\mathbf{R}^n$ of dimension p less than n, which has measure zero in $\mathbf{R}^n$, as you can check. Then $\overline{T}$ is closed and bounded and has measure zero in $\mathbb{R}^n$. The function 1_T is continuous and vanishes outside $\overline{T}$; hence the integral $\int_T 1$ exists and equals 0. □

This theorem gives one interpretation of the number $|\det A|$; it is the factor by which the linear transformation $h(\mathbf{x}) = A \cdot \mathbf{x}$ multiplies volumes. Here is another interpretation.

Definition. Let $\mathbf{a}_1, \dots, \mathbf{a}_k$ be independent vectors in $\mathbb{R}^n$. We define the k-dimensional **parallelopiped** $\mathcal{P} = \mathcal{P}(\mathbf{a}_1, \dots, \mathbf{a}_k)$ to be the set of all $\mathbf{x}$ in $\mathbb{R}^n$ such that

$$\mathbf{x} = c_1\mathbf{a}_1 + \cdots + c_k\mathbf{a}_k$$

for scalars c_i with $0 \leq c_i \leq 1$. The vectors $\mathbf{a}_1, \dots, \mathbf{a}_k$ are called the **edges** of $\mathcal{P}$.

A few sketches will convince you that a 2-dimensional parallelopiped is what we usually call a "parallelogram," and a 3-dimensional one is what we usually call a "parallelopiped." See Figure 20.1, which pictures parallelograms in $\mathbb{R}^2$ and $\mathbb{R}^3$ and a 3-dimensional parallelopiped in $\mathbb{R}^3$.

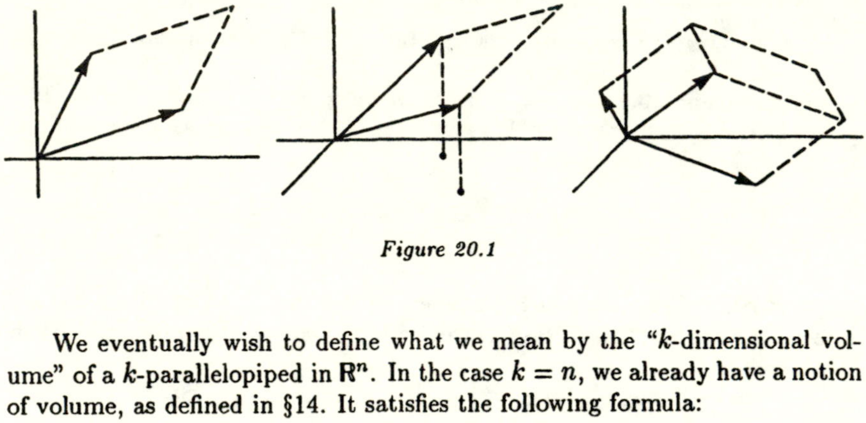

Figure 20.1

We eventually wish to define what we mean by the "k-dimensional volume" of a k-parallelopiped in $\mathbb{R}^n$. In the case $k = n$, we already have a notion of volume, as defined in §14. It satisfies the following formula:

Theorem 20.2. *Let $\mathbf{a}_1, \dots, \mathbf{a}_n$ be n independent vectors in $\mathbb{R}^n$. Let $A = [\mathbf{a}_1 \ \dots \ \mathbf{a}_n]$ be the n by n matrix with columns $\mathbf{a}_1, \dots, \mathbf{a}_n$. Then*

$$v(\mathcal{P}(\mathbf{a}_1, \dots, \mathbf{a}_n)) = |\det A|.$$

Proof. Consider the linear transformation $h : \mathbb{R}^n \to \mathbb{R}^n$ given by $h(\mathbf{x}) = A \cdot \mathbf{x}$. Then h carries the unit basis vectors $\mathbf{e}_1, \dots, \mathbf{e}_n$ to the vectors $\mathbf{a}_1, \dots, \mathbf{a}_n$, since $A \cdot \mathbf{e}_j = \mathbf{a}_j$ by direct computation. Furthermore, h

carries the unit cube $I^n = [0,1]^n$ onto the parallelopiped $\mathcal{P}(\mathbf{a}_1, \ldots, \mathbf{a}_n)$. By the preceding theorem,

$$v(\mathcal{P}(\mathbf{a}_1, \ldots, \mathbf{a}_n)) = |\det A| \cdot v(I^n) = |\det A|. \quad \square$$

EXAMPLE 1. In calculus, one studies the 3-dimensional version of this formula. One learns that the volume of the parallelopiped with edges $\mathbf{a}$, $\mathbf{b}$, $\mathbf{c}$ is given (up to sign) by the "triple scalar product"

$$\mathbf{a} \cdot (\mathbf{b} \times \mathbf{c}) = \det [\mathbf{a}\ \mathbf{b}\ \mathbf{c}].$$

(We write $\mathbf{a}$, $\mathbf{b}$, and $\mathbf{c}$ as column matrices here, as usual.) One learns also that the *sign* of the triple scalar product depends on whether the triple $\mathbf{a}$, $\mathbf{b}$, $\mathbf{c}$ is "right-handed" or "left-handed." We now generalize this second notion to $\mathbf{R}^n$, and indeed, to an arbitrary finite-dimensional vector space V.

Definition. Let V be an n-dimensional vector space. An n-tuple $(\mathbf{a}_1, \ldots, \mathbf{a}_n)$ of independent vectors in V is called an **n-frame** in V. In $\mathbf{R}^n$, we call such a frame **right-handed** if

$$\det [\mathbf{a}_1 \cdots \mathbf{a}_n] > 0;$$

we call it **left-handed** otherwise. The collection of all right-handed frames in $\mathbf{R}^n$ is called an **orientation** of $\mathbf{R}^n$; and so is the collection of all left-handed frames. More generally, choose a linear isomorphism $T : \mathbf{R}^n \to V$, and define one **orientation** of V to consist of all frames of the form $(T(\mathbf{a}_1), \ldots, T(\mathbf{a}_n))$ for which $(\mathbf{a}_1, \ldots, \mathbf{a}_n)$ is a right-handed frame in $\mathbf{R}^n$, and the other orientation of V to consist of all such frames for which $(\mathbf{a}_1, \ldots, \mathbf{a}_n)$ is left-handed. Thus V has two orientations; each is called the **reverse**, or the **opposite**, of the other.

It is easy to see that this notion is well-defined (independent of the choice of T). Note that in an arbitrary n-dimensional vector space, there is no well-defined notion of "right-handed," although there is a well-defined notion of orientation.

EXAMPLE 2. In $\mathbf{R}^1$, a frame consists of a single non-zero number; it is right-handed if it is positive, and left-handed if it is negative. In $\mathbf{R}^2$, a frame $(\mathbf{a}_1, \mathbf{a}_2)$ is right-handed if one must rotate $\mathbf{a}_1$ in a counterclockwise direction through an angle less than π to make it point in the same direction as $\mathbf{a}_2$. (See the exercises.) In $\mathbf{R}^3$, a frame $(\mathbf{a}_1, \mathbf{a}_2, \mathbf{a}_3)$ is right-handed if curling the fingers of one's right hand in the direction from $\mathbf{a}_1$ to $\mathbf{a}_2$ makes one's thumb point in the direction of $\mathbf{a}_3$. See Figure 20.2.

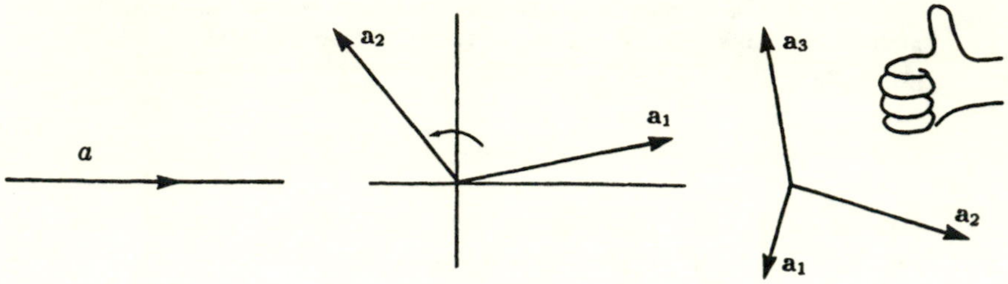

Figure 20.2

One way to justify this statement is to note that if one has a frame $(\mathbf{a}_1(t), \mathbf{a}_2(t), \mathbf{a}_3(t))$ that varies continuously as a function of t for $0 \leq t \leq 1$, and if the frame is right-handed when $t = 0$, then it remains right-handed for all t. For the function $\det [\mathbf{a}_1\ \mathbf{a}_2\ \mathbf{a}_3]$ cannot change sign, by the intermediate-value theorem. Then since the frame $(\mathbf{e}_1, \mathbf{e}_2, \mathbf{e}_3)$ satisfies the "curled right-hand rule" as well as the condition $\det [\mathbf{e}_1\, \mathbf{e}_2\, \mathbf{e}_3] > 0$, so does the frame corresponding to any other position of the "curled right hand" in 3-dimensional space.

We now obtain another interpretation of the sign of the determinant.

Theorem 20.3. *Let C be a non-singular n by n matrix. Let $h : \mathbf{R}^n \to \mathbf{R}^n$ be the linear transformation $h(\mathbf{x}) = C \cdot \mathbf{x}$. Let $(\mathbf{a}_1, \ldots, \mathbf{a}_n)$ be a frame in $\mathbf{R}^n$. If $\det C > 0$, the the frames*

$$(\mathbf{a}_1, \ldots, \mathbf{a}_n) \quad \textit{and} \quad (h(\mathbf{a}_1), \ldots, h(\mathbf{a}_n))$$

belong to the same orientation of $\mathbf{R}^n$; if $\det C < 0$, they belong to opposite orientations of $\mathbf{R}^n$.

If $\det C > 0$, we say h is **orientation-preserving**; if $\det C < 0$, we say h is **orientation-reversing**.

Proof. Let $\mathbf{b}_i = h(\mathbf{a}_i)$ for each i. Then

$$C \cdot [\mathbf{a}_1 \cdots \mathbf{a}_n] = [\mathbf{b}_1 \cdots \mathbf{b}_n],$$

so that

$$(\det C) \cdot \det [\mathbf{a}_1 \cdots \mathbf{a}_n] = \det [\mathbf{b}_1 \cdots \mathbf{b}_n].$$

If $\det C > 0$, then $\det [\mathbf{a}_1 \cdots \mathbf{a}_n]$ and $\det [\mathbf{b}_1 \cdots \mathbf{b}_n]$ have the same sign; if $\det C < 0$, they have opposite signs. □

Invariance of volume under isometries

Definition. The vectors $\mathbf{a}_1, \ldots, \mathbf{a}_k$ of $\mathbf{R}^n$ are said to form an **orthogonal set** if $\langle \mathbf{a}_i, \mathbf{a}_j \rangle = 0$ for $i \neq j$. They form an **orthonormal set** if they satisfy the additional condition $\langle \mathbf{a}_i, \mathbf{a}_i \rangle = 1$ for all i. If the vectors $\mathbf{a}_1, \ldots, \mathbf{a}_k$ form an orthogonal set and are non-zero, then the vectors $\mathbf{a}_1/\|\mathbf{a}_1\|, \ldots, \mathbf{a}_k/\|\mathbf{a}_k\|$ form an orthonormal set.

An orthogonal set of non-zero vectors $\mathbf{a}_1, \ldots, \mathbf{a}_k$ is always independent. For, given the equation

$$d_1\mathbf{a}_1 + \cdots + d_k\mathbf{a}_k = \mathbf{0},$$

one takes the dot product of both sides with $\mathbf{a}_i$ to obtain the equation $d_i\langle \mathbf{a}_i, \mathbf{a}_i \rangle = 0$, which implies (since $\mathbf{a}_i \neq \mathbf{0}$) that $d_i = 0$.

An orthogonal set of non-zero vectors in $\mathbf{R}^n$ that consists of n vectors is thus a basis for $\mathbf{R}^n$. The set $\mathbf{e}_1, \ldots, \mathbf{e}_n$ is one such basis for $\mathbf{R}^n$, but there are many others.

Definition. An n by n matrix A is called an **orthogonal matrix** if the columns of A form an orthonormal set. This condition is equivalent to the matrix equation

$$A^{\mathrm{tr}} \cdot A = I_n,$$

as you can check.

If A is orthogonal, then A is square and A^{tr} is a left inverse for A; it follows that A^{tr} is also a right inverse for A. Thus A is an orthogonal matrix if and only if A is non-singular and $A^{\mathrm{tr}} = A^{-1}$.

Note that if A is orthogonal, then $\det A = \pm 1$. For

$$(\det A)^2 = (\det A^{\mathrm{tr}})(\det A) = \det(A^{\mathrm{tr}} \cdot A) = \det I_n = 1.$$

The set of orthogonal matrices forms what is called, in modern algebra, a *group*. That is the substance of the following theorem:

Theorem 20.4. *Let A, B, C be orthogonal n by n matrices. Then:*

(a) *$A \cdot B$ is orthogonal.*

(b) *$A \cdot (B \cdot C) = (A \cdot B) \cdot C$.*

(c) *There is an orthogonal matrix I_n such that $A \cdot I_n = I_n \cdot A = A$ for all orthogonal A.*

(d) *Given A, there is an orthogonal matrix A^{-1} such that $A \cdot A^{-1} = A^{-1} \cdot A = I_n$.*

Proof. To check (a), we compute

$$(A \cdot B)^{tr} \cdot (A \cdot B) = (B^{tr} \cdot A^{tr}) \cdot (A \cdot B)$$
$$= B^{tr} \cdot B = I_n.$$

Condition (b) is immediate and (c) follows from the fact that I_n is orthogonal. To check (d), we note that since A^{tr} equals A^{-1},

$$I_n = A \cdot A^{tr} = (A^{tr})^{tr} \cdot A^{tr} = (A^{-1})^{tr} \cdot A^{-1}.$$

Thus A^{-1} is orthogonal, as desired. □

Definition. The linear transformation $h : \mathbf{R}^n \to \mathbf{R}^n$ given by

$$h(\mathbf{x}) = A \cdot \mathbf{x}$$

is called an **orthogonal transformation** if A is an orthogonal matrix. This condition is equivalent to the requirement that h carry the basis $\mathbf{e}_1, \dots, \mathbf{e}_n$ for $\mathbf{R}^n$ to an orthonormal basis for $\mathbf{R}^n$.

Definition. Let $h : \mathbf{R}^n \to \mathbf{R}^n$. We say that h is a (euclidean) **isometry** if

$$\|h(\mathbf{x}) - h(\mathbf{y})\| = \|\mathbf{x} - \mathbf{y}\|$$

for all $\mathbf{x}, \mathbf{y} \in \mathbf{R}^n$. Thus an isometry is a map that preserves euclidean distances.

Theorem 20.5. *Let $h : \mathbf{R}^n \to \mathbf{R}^n$ be a map such that $h(\mathbf{0}) = \mathbf{0}$.*

(a) *The map h is an isometry if and only if it preserves dot products.*

(b) *The map h is an isometry if and only if it is an orthogonal transformation.*

Proof. (a) Given $\mathbf{x}$ and $\mathbf{y}$, we compute:

$$(1) \qquad \|h(\mathbf{x}) - h(\mathbf{y})\|^2 = \langle h(\mathbf{x}), h(\mathbf{x})\rangle - 2\langle h(\mathbf{x}), h(\mathbf{y})\rangle + \langle h(\mathbf{y}), h(\mathbf{y})\rangle$$

$$(2) \qquad \|\mathbf{x} - \mathbf{y}\|^2 = \langle \mathbf{x}, \mathbf{x}\rangle - 2\langle \mathbf{x}, \mathbf{y}\rangle + \langle \mathbf{y}, \mathbf{y}\rangle.$$

If h preserves dot products, then the right sides of (1) and (2) are equal; thus h preserves euclidean distances as well. Conversely, suppose h preserves euclidean distances. Then in particular, for all $\mathbf{x}$,

$$\|h(\mathbf{x}) - h(\mathbf{0})\| = \|\mathbf{x} - \mathbf{0}\|,$$

so that $\|h(\mathbf{x})\| = \|\mathbf{x}\|$. Then the first and last terms on the right side of (1) are equal to the corresponding terms on the right side of (2). Furthermore, the left sides of (1) and (2) are equal by hypothesis. It follows that

$$\langle h(\mathbf{x}), h(\mathbf{y})\rangle = \langle \mathbf{x}, \mathbf{y}\rangle,$$

as desired.

(b) Let $h(\mathbf{x}) = A \cdot \mathbf{x}$, where A is orthogonal; we show h is an isometry. By (a), it suffices to show h preserves dot products. Now the dot product of $h(\mathbf{x})$ and $h(\mathbf{y})$ can be expressed as the matrix product

$$h(\mathbf{x})^{\text{tr}} \cdot h(\mathbf{y})$$

if $h(\mathbf{x})$ and $h(\mathbf{y})$ are written as column matrices (as usual). We compute

$$\begin{aligned} h(\mathbf{x})^{\text{tr}} \cdot h(\mathbf{y}) &= (A \cdot \mathbf{x})^{\text{tr}} \cdot (A \cdot \mathbf{y}) \\ &= \mathbf{x}^{\text{tr}} \cdot A^{\text{tr}} \cdot A \cdot \mathbf{y} = \mathbf{x}^{\text{tr}} \cdot \mathbf{y}. \end{aligned}$$

Thus h preserves dot products, so it is an isometry.

Conversely, let h be an isometry with $h(\mathbf{0}) = \mathbf{0}$. Let $\mathbf{a}_i$ be the vector $\mathbf{a}_i = h(\mathbf{e}_i)$ for all i; let A be the matrix $A = [\mathbf{a}_1 \cdots \mathbf{a}_n]$. Since h preserves dot products by (a), the vectors $\mathbf{a}_1, \dots, \mathbf{a}_n$ are orthonormal; thus A is an orthogonal matrix. We show that $h(\mathbf{x}) = A \cdot \mathbf{x}$ for all $\mathbf{x}$; then the proof is complete.

Since the vectors $\mathbf{a}_i$ form a basis for $\mathbf{R}^n$, for each $\mathbf{x}$ the vector $h(\mathbf{x})$ can be written uniquely in the form

$$h(\mathbf{x}) = \sum_{i=1}^{n} \alpha_i(\mathbf{x}) \mathbf{a}_i,$$

for certain real-valued functions $\alpha_i(\mathbf{x})$ of $\mathbf{x}$. Because the $\mathbf{a}_i$ are orthonormal,

$$\langle h(\mathbf{x}), \mathbf{a}_j\rangle = \alpha_j(\mathbf{x})$$

for each j. Because h preserves dot products,

$$\langle h(\mathbf{x}), \mathbf{a}_j\rangle = \langle h(\mathbf{x}), h(\mathbf{e}_j)\rangle = \langle \mathbf{x}, \mathbf{e}_j\rangle = x_j$$

for all j. Thus $\alpha_j(\mathbf{x}) = x_j$, so that

$$h(\mathbf{x}) = \sum_{i=1}^{n} x_i \mathbf{a}_i = [\mathbf{a}_1 \cdots \mathbf{a}_n] \cdot \begin{bmatrix} x_1 \\ \vdots \\ x_n \end{bmatrix} = A \cdot \mathbf{x}. \quad \square$$

Theorem 20.6. *Let $h : \mathbf{R}^n \to \mathbf{R}^n$. Then h is an isometry if and only if it equals an orthogonal transformation followed by a translation, that is, if and only if h has the form*

$$h(\mathbf{x}) = A \cdot \mathbf{x} + \mathbf{p},$$

where A is an orthogonal matrix.

Proof. Given h, let $\mathbf{p} = h(\mathbf{0})$, and define $k(\mathbf{x}) = h(\mathbf{x}) - \mathbf{p}$. Then

$$\|k(\mathbf{x}) - k(\mathbf{y})\| = \|h(\mathbf{x}) - h(\mathbf{y})\|,$$

by direct computation. Thus k is an isometry if and only if h is an isometry.

Since $k(\mathbf{0}) = \mathbf{0}$, the map k is an isometry if and only if $k(\mathbf{x}) = A \cdot \mathbf{x}$, where A is orthogonal. This in turn occurs if and only if $h(\mathbf{x}) = A \cdot \mathbf{x} + \mathbf{p}$. $\square$

Theorem 20.7. *Let $h : \mathbf{R}^n \to \mathbf{R}^n$ be an isometry. If S is a rectifiable set in $\mathbf{R}^n$, then the set $T = h(S)$ is rectifiable, and $v(T) = v(S)$.*

Proof. The map h is of the form $h(\mathbf{x}) = A \cdot \mathbf{x} + \mathbf{p}$, where A is orthogonal. Then $Dh(\mathbf{x}) = A$, and it follows from the change of variables theorem that

$$v(T) = |\det A| \cdot v(S) = v(S). \quad \square$$

EXERCISES

1. Show that if h is an orthogonal transformation, then h carries *every* orthonormal set to an orthonormal set.

2. Find a linear transformation $h : \mathbf{R}^n \to \mathbf{R}^n$ that preserves volumes but is not an isometry.

3. Let V be an arbitrary n-dimensional vector space. Show that the two orientations of V are well-defined.

4. Consider the vectors $\mathbf{a}_i$ in $\mathbf{R}^3$ such that

$$[\mathbf{a}_1\ \mathbf{a}_2\ \mathbf{a}_3\ \mathbf{a}_4] = \begin{bmatrix} 1 & 0 & 1 & 1 \\ 1 & 0 & 1 & 1 \\ 1 & 1 & 2 & 0 \end{bmatrix}.$$

   Let V be the subspace of $\mathbf{R}^3$ spanned by $\mathbf{a}_1$ and $\mathbf{a}_2$. Show that $\mathbf{a}_3$ and $\mathbf{a}_4$ also span V, and that the frames $(\mathbf{a}_1, \mathbf{a}_2)$ and $(\mathbf{a}_3, \mathbf{a}_4)$ belong to opposite orientations of V.

5. Given θ and ϕ, let

$$\mathbf{a}_1 = (\cos\theta, \sin\theta) \quad \text{and} \quad \mathbf{a}_2 = \big(\cos(\theta+\phi), \sin(\theta+\phi)\big).$$

   Show that $(\mathbf{a}_1, \mathbf{a}_2)$ is right-handed if $0 < \phi < \pi$, and left-handed if $-\pi < \phi < 0$. What happens if ϕ equals 0 or π?

5

Manifolds

We have studied the notion of volume for bounded subsets of euclidean space; if A is a bounded rectifiable set in $\mathbf{R}^k$, its volume is defined by the equation

$$v(A) = \int_A 1.$$

When $k = 1$, it is common to call $v(A)$ the length of A; when $k = 2$, it is common to call $v(A)$ the area of A.

Now in calculus one studies the notion of length not only for subsets of $\mathbf{R}^1$, but also for smooth *curves* in $\mathbf{R}^2$ and $\mathbf{R}^3$. And one studies the notion of area not only for subsets of $\mathbf{R}^2$, but also for smooth *surfaces* in $\mathbf{R}^3$. In this chapter, we introduce the k-dimensional analogues of curves and surfaces; they are called *k-manifolds* in $\mathbf{R}^n$. And we define a notion of k-dimensional volume for such objects. We also define what we mean by the integral of a scalar function over a k-manifold with respect to k-volume, generalizing notions defined in calculus for curves and surfaces.

§21. THE VOLUME OF A PARALLELOPIPED

We begin by studying parallelopipeds. Let $\mathcal{P}$ be a k-dimensional parallelopiped in $\mathbf{R}^n$, with $k < n$. We wish to define a notion of k-dimensional volume for $\mathcal{P}$. (Its n-dimensional volume is of course zero, since it lies in a k-dimensional subspace of $\mathbf{R}^n$, which has measure zero in $\mathbf{R}^n$.) How shall we proceed? There are two conditions that it is reasonable that such a volume function should satisfy. We know that an orthogonal transformation of $\mathbf{R}^n$ preserves n-dimensional volume; it is reasonable to require that it preserve k-dimensional volume as well. Second, if the parallelopiped happens to lie in the subspace $\mathbf{R}^k \times 0$ of $\mathbf{R}^n$, then it is reasonable to require that its k-dimensional volume agree with the usual notion of volume for a k-dimensional parallelopiped in $\mathbf{R}^k$. These two "reasonable" conditions determine k-dimensional volume completely, as we shall see.

We begin with a result from linear algebra which may already be familiar to you.

Lemma 21.1. *Let W be a linear subspace of $\mathbf{R}^n$ of dimension k. Then there is an orthonormal basis for $\mathbf{R}^n$ whose first k elements form a basis for W.*

Proof. By Theorem 1.2, there is a basis $\mathbf{a}_1, \ldots, \mathbf{a}_n$ for $\mathbf{R}^n$ whose first k elements form a basis for W. There is a standard procedure for forming from these vectors an orthogonal set of vectors $\mathbf{b}_1, \ldots, \mathbf{b}_n$ such that for each i, the vectors $\mathbf{b}_1, \ldots, \mathbf{b}_i$ span the same space as the vectors $\mathbf{a}_1, \ldots, \mathbf{a}_i$. It is called the *Gram-Schmidt process*; we recall it here.

Given $\mathbf{a}_1, \ldots, \mathbf{a}_n$, we set

$$\mathbf{b}_1 = \mathbf{a}_1,$$

$$\mathbf{b}_2 = \mathbf{a}_2 - \lambda_{21}\mathbf{b}_1,$$

and for general i,

$$\mathbf{b}_i = \mathbf{a}_i - \lambda_{i1}\mathbf{b}_1 - \lambda_{i2}\mathbf{b}_2 - \cdots - \lambda_{i,i-1}\mathbf{b}_{i-1},$$

where the λ_{ij} are scalars yet to be specified. No matter what these scalars are, however, we note that for each j the vector $\mathbf{a}_j$ equals a linear combination of the vectors $\mathbf{b}_1, \ldots, \mathbf{b}_j$. Furthermore, for each j the vector $\mathbf{b}_j$ can be written as a linear combination of the vectors $\mathbf{a}_1, \ldots, \mathbf{a}_j$. (The proof proceeds by induction.) These two facts imply that, for each i, $\mathbf{a}_1, \ldots, \mathbf{a}_i$ and $\mathbf{b}_1, \ldots, \mathbf{b}_i$ span the same subspace of $\mathbf{R}^n$. It also follows that the vectors $\mathbf{b}_1, \ldots, \mathbf{b}_n$ are independent, for there are n of them, and they span $\mathbf{R}^n$ as we have just noted. In particular, none of the $\mathbf{b}_i$ can equal 0.

Now we note that the scalars λ_{ij} may in fact be chosen so that the vectors $\mathbf{b}_i$ are mutually orthogonal. One proceeds by induction. If the vectors $\mathbf{b}_1, \ldots, \mathbf{b}_{i-1}$ are mutually orthogonal, one simply takes the dot product of both sides of the equation for $\mathbf{b}_i$ with each of the vectors $\mathbf{b}_j$ for $j = 1, \ldots, i-1$ to obtain the equation

$$\langle \mathbf{b}_i, \mathbf{b}_j \rangle = \langle \mathbf{a}_i, \mathbf{b}_j \rangle - \lambda_{ij} \langle \mathbf{b}_j, \mathbf{b}_j \rangle.$$

Since $\mathbf{b}_j \neq \mathbf{0}$, there is a (unique) value of λ_{ij} that makes the right side of this equation vanish. With this choice of the scalars λ_{ij}, the vector $\mathbf{b}_i$ is orthogonal to each of the vectors $\mathbf{b}_1, \ldots, \mathbf{b}_{i-1}$.

Once we have the non-zero orthogonal vectors $\mathbf{b}_i$, we merely divide each by its norm $\|\mathbf{b}_i\|$ to find the desired orthonormal basis for $\mathbf{R}^n$. □

Theorem 21.2. *Let W be a k-dimensional linear subspace of $\mathbf{R}^n$. There is an orthogonal transformation $h : \mathbf{R}^n \to \mathbf{R}^n$ that carries W onto the subspace $\mathbf{R}^k \times \mathbf{0}$ of $\mathbf{R}^n$.*

Proof. Choose an orthonormal basis $\mathbf{b}_1, \ldots, \mathbf{b}_n$ for $\mathbf{R}^n$ such that the first k basis elements $\mathbf{b}_1, \ldots, \mathbf{b}_k$ form a basis for W. Let $g : \mathbf{R}^n \to \mathbf{R}^n$ be the linear transformation $g(\mathbf{x}) = B \cdot \mathbf{x}$, where B is the matrix with successive columns $\mathbf{b}_1, \ldots, \mathbf{b}_n$. Then g is an orthogonal transformation, and $g(\mathbf{e}_i) = \mathbf{b}_i$ for all i. In particular, g carries $\mathbf{R}^k \times \mathbf{0}$, which has basis $\mathbf{e}_1, \ldots, \mathbf{e}_k$, onto W. The inverse of g is the transformation we seek. □

Now we obtain our notion of k-dimensional volume.

Theorem 21.3. *There is a unique function V that assigns, to each k-tuple $(\mathbf{x}_1, \ldots, \mathbf{x}_k)$ of elements of $\mathbf{R}^n$, a non-negative number such that:*

(1) *If $h : \mathbf{R}^n \to \mathbf{R}^n$ is an orthogonal transformation, then*

$$V(h(\mathbf{x}_1), \ldots, h(\mathbf{x}_k)) = V(\mathbf{x}_1, \ldots, \mathbf{x}_k).$$

(2) *If $\mathbf{y}_1, \ldots, \mathbf{y}_k$ belong to the subspace $\mathbf{R}^k \times \mathbf{0}$ of $\mathbf{R}^n$, so that*

$$\mathbf{y}_i = \begin{bmatrix} \mathbf{z}_i \\ \mathbf{0} \end{bmatrix}$$

for $\mathbf{z}_i \in \mathbf{R}^k$, then

$$V(\mathbf{y}_1, \ldots, \mathbf{y}_k) = |\det [\mathbf{z}_1 \cdots \mathbf{z}_k]|.$$

The function V vanishes if and only if the vectors $\mathbf{x}_1, \dots, \mathbf{x}_k$ are dependent. It satisfies the equation

$$V(\mathbf{x}_1, \dots, \mathbf{x}_k) = [\det(X^{\text{tr}} \cdot X)]^{1/2},$$

where X is the n by k matrix $X = [\mathbf{x}_1 \cdots \mathbf{x}_k]$.

We often denote $V(\mathbf{x}_1, \dots, \mathbf{x}_k)$ simply by $V(X)$.

Proof. Given $X = [\mathbf{x}_1 \cdots \mathbf{x}_k]$, define

$$F(X) = \det(X^{\text{tr}} \cdot X).$$

Step 1. If $h : \mathbf{R}^n \to \mathbf{R}^n$ is an orthogonal transformation, given by the equation $h(\mathbf{x}) = A \cdot \mathbf{x}$, where A is an orthogonal matrix, then

$$F(A \cdot X) = \det((A \cdot X)^{\text{tr}} \cdot (A \cdot X))$$

$$= \det(X^{\text{tr}} \cdot X) = F(X).$$

Furthermore, if Z is a k by k matrix, and if Y is the n by k matrix

$$Y = \begin{bmatrix} Z \\ 0 \end{bmatrix},$$

then

$$F(Y) = \det([Z^{\text{tr}}\ 0] \cdot \begin{bmatrix} Z \\ 0 \end{bmatrix})$$

$$= \det(Z^{\text{tr}} \cdot Z) = \det^2 Z.$$

Step 2. It follows that F is non-negative. For given $\mathbf{x}_1, \dots, \mathbf{x}_k$ in $\mathbf{R}^n$, let W be a k-dimensional subspace of $\mathbf{R}^n$ containing them. (If the $\mathbf{x}_i$ are independent, W is unique.) Let $h(\mathbf{x}) = A \cdot \mathbf{x}$ be an orthogonal transformation of $\mathbf{R}^n$ carrying W onto the subspace $\mathbf{R}^k \times \mathbf{0}$. Then $A \cdot X$ has the form

$$A \cdot X = \begin{bmatrix} Z \\ 0 \end{bmatrix},$$

so that $F(X) = F(A \cdot X) = \det^2 Z \geq 0$. Note that $F(X) = 0$ if and only if the columns of Z are dependent, and this occurs if and only if the vectors $\mathbf{x}_1, \dots, \mathbf{x}_k$ are dependent.

Step 3. Now we define $V(X) = \left(F(X)\right)^{1/2}$. It follows from the computations of Step 1 that V satisfies conditions (1) and (2). And it follows from the computation of Step 2 that V is uniquely characterized by these two conditions. □

Definition. If $\mathbf{x}_1, \dots, \mathbf{x}_k$ are independent vectors in $\mathbf{R}^n$, we define the k-dimensional **volume** of the parallelopiped $\mathcal{P} = \mathcal{P}(\mathbf{x}_1, \dots, \mathbf{x}_k)$ to be the number $V(\mathbf{x}_1, \dots, \mathbf{x}_k)$, which is positive.

EXAMPLE 1. Consider two independent vectors $\mathbf{a}$ and $\mathbf{b}$ in $\mathbf{R}^3$; let X be the matrix $X = [\mathbf{a}\ \mathbf{b}]$. Then $V(X)$ is the area of the parallelogram with edges $\mathbf{a}$ and $\mathbf{b}$. Let θ be the angle between $\mathbf{a}$ and $\mathbf{b}$, defined by the equation $\langle \mathbf{a}, \mathbf{b} \rangle = \|\mathbf{a}\|\ \|\mathbf{b}\| \cos\theta$. Then

$$\begin{aligned}
\left(V(X)\right)^2 &= \det(X^{tr} \cdot X) \\
&= \det \begin{bmatrix} \|\mathbf{a}\|^2 & \langle \mathbf{a}, \mathbf{b} \rangle \\ \langle \mathbf{b}, \mathbf{a} \rangle & \|\mathbf{b}\|^2 \end{bmatrix} \\
&= \|\mathbf{a}\|^2 \|\mathbf{b}\|^2 (1 - \cos^2\theta) = \|\mathbf{a}\|^2 \|\mathbf{b}\|^2 \sin^2\theta.
\end{aligned}$$

Figure 21.1 shows why this number is interpreted in calculus as the square of the area of the parallelogram with edges $\mathbf{a}$ and $\mathbf{b}$.

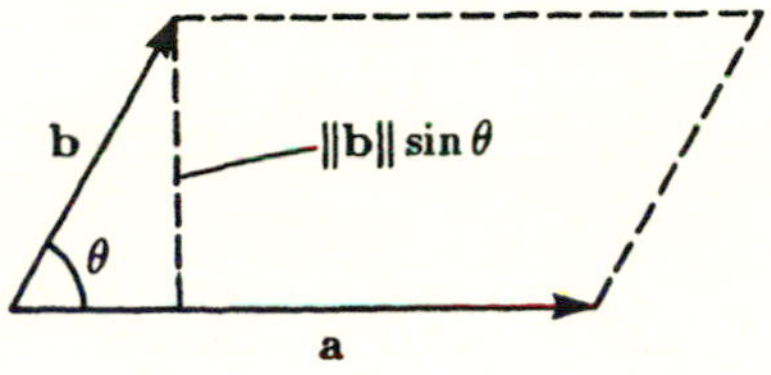

Figure 21.1

In calculus one studies another formula for the area of the parallelogram with edges $\mathbf{a}$ and $\mathbf{b}$. If $\mathbf{a} \times \mathbf{b}$ is the **cross product** of $\mathbf{a}$ and $\mathbf{b}$, defined by the equation

$$\mathbf{a} \times \mathbf{b} = \det \begin{bmatrix} a_2 & b_2 \\ a_3 & b_3 \end{bmatrix} \mathbf{e}_1 - \det \begin{bmatrix} a_1 & b_1 \\ a_3 & b_3 \end{bmatrix} \mathbf{e}_2 + \det \begin{bmatrix} a_1 & b_1 \\ a_2 & b_2 \end{bmatrix} \mathbf{e}_3,$$

then one learns in calculus that the number $\|\mathbf{a} \times \mathbf{b}\|$ equals the area of $\mathcal{P}(\mathbf{a}, \mathbf{b})$. This is justified by verifying directly that

$$\|\mathbf{a}\|^2\ \|\mathbf{b}\|^2 - \langle \mathbf{a}, \mathbf{b} \rangle^2 = \|\mathbf{a} \times \mathbf{b}\|^2.$$

Often this verification is left as an "exercise for the reader." Some exercise!

Just as there are for a parallelogram in $\mathbf{R}^3$, there are for a k-parallelopiped in $\mathbf{R}^n$ two different formulas for its k-dimensional volume. The first is the formula given in the preceding theorem. It is very convenient for theoretical purposes, but sometimes not very pleasant for computational purposes. The second, which is a generalization of the cross-product formula just discussed, is often more convenient to use in practice. We derive it now; it will be used in some of the examples and exercises.

Definition. Let $\mathbf{x}_1, \ldots, \mathbf{x}_k$ be vectors in $\mathbf{R}^n$ with $k \leq n$. Let X be the matrix $X = [\mathbf{x}_1 \cdots \mathbf{x}_k]$. If $I = (i_1, \ldots, i_k)$ is a k-tuple of integers such that $1 \leq i_1 < i_2 < \cdots < i_k \leq n$, we call I an **ascending** k**-tuple** from the set $\{1, \ldots, n\}$, and we denote by

$$X_I \quad \text{or by} \quad X(i_1, \ldots, i_k)$$

the k by k submatrix of X consisting of rows $i_1, \ldots, i_k$ of X.

More generally, if I is *any* k-tuple of integers from the set $\{1, \ldots, n\}$, not necessarily distinct nor arranged in any particular order, we use this same notation to denote the k by k matrix whose successive rows are rows $i_1, \ldots, i_k$ of X. It need not be a submatrix of X in this case, of course.

***Theorem 21.4.** *Let X be an n by k matrix with $k \leq n$. Then*

$$V(X) = [\sum_{[I]} \det^2 X_I]^{1/2},$$

where the symbol $[I]$ indicates that the summation extends over all ascending k-tuples from the set $\{1, \ldots, n\}$.

This theorem may be thought of as a Pythagorean theorem for k-volume. It states that the square of the volume of a k-parallelopiped $\mathcal{P}$ in $\mathbf{R}^n$ is equal to the sum of the squares of the volumes of the k-parallelopipeds obtained by projecting $\mathcal{P}$ onto the various coordinate k-planes of $\mathbf{R}^n$.

Proof. Let X have size n by k. Let

$$F(X) = \det(X^{\text{tr}} \cdot X) \quad \text{and} \quad G(X) = \sum_{[I]} \det^2 X_I.$$

Proving the theorem is equivalent to showing that $F(X) = G(X)$ for all X.

Step 1. The theorem holds when $k = 1$ or $k = n$. If $k = 1$, then X is a column matrix with entries $\lambda_1, \ldots, \lambda_n$, say. Then

$$F(X) = \sum (\lambda_i)^2 = G(X).$$

If $k = n$, the summation in the definition of G has only one term, and

$$F(X) = \det^2 X = G(X).$$

Step 2. If $X = [\mathbf{x}_1 \cdots \mathbf{x}_k]$ and the $\mathbf{x}_i$ are orthogonal, then

$$F(X) = \|\mathbf{x}_1\|^2 \|\mathbf{x}_2\|^2 \cdots \|\mathbf{x}_k\|^2.$$

The general entry of $X^{\text{tr}} \cdot X$ is $\mathbf{x}_i^{\text{tr}} \cdot \mathbf{x}_j$, which is the dot product of $\mathbf{x}_i$ and $\mathbf{x}_j$. Thus if the $\mathbf{x}_i$ are orthogonal, $X^{\text{tr}} \cdot X$ is a diagonal matrix with diagonal entries $\|\mathbf{x}_i\|^2$.

Step 3. Consider the following two elementary column operations, where $j \neq \ell$:

(1) Exchange columns j and ℓ.

(2) Replace column j by itself plus c times column ℓ.

We show that applying either of these operations to X does not change the values of F or G.

Given an elementary *row* operation, with corresponding elementary matrix E, then $E \cdot X$ equals the matrix obtained by applying this elementary row operation to X. One can compute the effect of applying the corresponding elementary *column* operation to X by transposing X, premultiplying by E, and then transposing back. Thus the matrix obtained by applying an elementary column operation to X is the matrix

$$(E \cdot X^{\text{tr}})^{\text{tr}} = X \cdot E^{\text{tr}}.$$

It follows that these two operations do not change the value of F. For

$$\begin{aligned} F(X \cdot E^{\text{tr}}) &= \det(E \cdot X^{\text{tr}} \cdot X \cdot E^{\text{tr}}) \\ &= (\det E)\big(\det(X^{\text{tr}} \cdot X)\big)(\det E^{\text{tr}}) \\ &= F(X), \end{aligned}$$

since $\det E = \pm 1$ for these two elementary operations.

Nor do these operations change the value of G. Note that if one applies one of these elementary column operations to X and then deletes all rows but $i_1, \ldots, i_k$, the result is the same as if one had first deleted all rows but $i_1, \ldots, i_k$ and then applied the elementary column operation. This means that

$$(X \cdot E^{\text{tr}})_I = X_I \cdot E^{\text{tr}}.$$

We then compute

$$\begin{aligned} G(X \cdot E^{\text{tr}}) &= \sum_{[I]} \det{}^2 (X \cdot E^{\text{tr}})_I \\ &= \sum_{[I]} \det{}^2 (X_I \cdot E^{\text{tr}}) \\ &= \sum_{[I]} (\det{}^2 X_I)(\det{}^2 E^{\text{tr}}) \\ &= G(X). \end{aligned}$$

Step 4. In order to prove the theorem for all matrices of a given size, we show that it suffices to prove it in the special case where all the entries of the bottom row are zero except possibly for the last entry, and the columns form an orthogonal set.

Given X, if the last row of X has a non-zero entry, we may by elementary operations of the specified types bring the matrix to the form

$$D = \begin{bmatrix} & * & \\ 0 \cdots & 0 & \lambda \end{bmatrix},$$

where $\lambda \neq 0$. If the last row of X has no non-zero entry, it is already of this form, with $\lambda = 0$. One now applies the Gram-Schmidt process to the columns of this matrix. The first column is left as is. At the general step, the jth column is replaced by itself minus scalar multiples of the earlier columns. The Gram-Schmidt process thus involves only elementary column operations of type (2). And the zeros in the last row remain unchanged during the process. At the end of the process, the columns are orthogonal, and the matrix still has the form of D.

Step 5. We prove the theorem, by induction on n.

If $n = 1$, then $k = 1$ and Step 1 applies. If $n = 2$, then $k = 1$ or $k = 2$, and Step 1 applies. Now suppose the theorem holds for matrices having fewer than n rows. We prove it for matrices of size n by k. In view of Step 1, we need only consider the case $1 < k < n$. In view of Step 4, we may assume that all entries in the bottom row of X, except possibly for the last, are zero, and that the columns of X are orthogonal. Then X has the form

$$X = \begin{bmatrix} \mathbf{b}_1 & \cdots & \mathbf{b}_{k-1} & \mathbf{b}_k \\ 0 & \cdots & 0 & \lambda \end{bmatrix};$$

the vectors $\mathbf{b}_i$ of $\mathbf{R}^{n-1}$ are orthogonal because the columns of X are orthogonal vectors in $\mathbf{R}^n$. For convenience in notation, let B and C denote the matrices

$$B = [\mathbf{b}_1 \cdots \mathbf{b}_k] \quad \text{and} \quad C = [\mathbf{b}_1 \cdots \mathbf{b}_{k-1}].$$

We compute $F(X)$ in terms of B and C as follows:

$$F(X) = \|\mathbf{b}_1\|^2 \cdots \|\mathbf{b}_{k-1}\|^2 \left(\|\mathbf{b}_k\|^2 + \lambda^2 \right) \quad \text{by Step 2,}$$

$$= F(B) + \lambda^2 F(C).$$

To compute $G(X)$, we break the summation in the definition of $G(X)$ into two parts, according to the value of i_k. We have

$$(*) \qquad G(X) = \sum_{i_k < n} \det{}^2 X_I + \sum_{i_k = n} \det{}^2 X_I.$$

Now if $I = (i_1, \ldots, i_k)$ is an ascending k-tuple with $i_k < n$, then $X_I = B_I$. Hence the first summation in $(*)$ equals $G(B)$. On the other hand, if $i_k = n$, one computes

$$\det X(i_1, \ldots, i_{k-1}, n) = \pm\lambda \det C(i_1, \ldots, i_{k-1}).$$

It follows that the second summation in $(*)$ equals $\lambda^2 G(C)$. Then

$$G(X) = G(B) + \lambda^2 G(C).$$

The induction hypothesis tells us that $F(B) = G(B)$ and $F(C) = G(C)$. It follows that $F(X) = G(X)$. $\square$

EXERCISES

1. Let

$$X = \begin{bmatrix} 1 & 0 & 0 \\ 0 & 1 & 0 \\ 0 & 0 & 1 \\ a & b & c \end{bmatrix}.$$

   (a) Find $X^{tr} \cdot X$.

   (b) Find $V(X)$.

2. Let $\mathbf{x}_1, \ldots, \mathbf{x}_k$ be vectors in $\mathbf{R}^n$. Show that

$$V(\mathbf{x}_1, \ldots, \lambda\mathbf{x}_i, \ldots, \mathbf{x}_k) = |\lambda| V(\mathbf{x}_1, \ldots, \mathbf{x}_k).$$

3. Let $h : \mathbf{R}^n \to \mathbf{R}^n$ be the function $h(\mathbf{x}) = \lambda\mathbf{x}$. If $\mathcal{P}$ is a k-dimensional parallelopiped in $\mathbf{R}^n$, find the volume of $h(\mathcal{P})$ in terms of the volume of $\mathcal{P}$.

4. (a) Use Theorem 21.4 to verify the last equation stated in Example 1.

   (b) Verify this equation by direct computation(!).

5. Prove the following:

   Theorem. *Let W be an n-dimensional vector space with an inner product. Then there exists a unique real-valued function $V(\mathbf{x}_1, \ldots, \mathbf{x}_k)$ of k-tuples of vectors of W such that:*

   (i) *Exchanging $\mathbf{x}_i$ with $\mathbf{x}_j$ does not change the value of V.*

   (ii) *Replacing $\mathbf{x}_i$ by $\mathbf{x}_i + c\mathbf{x}_j$ (for $j \neq i$) does not change the value of V.*

   (iii) *Replacing $\mathbf{x}_i$ by $\lambda\mathbf{x}_i$ multiplies the value of V by $|\lambda|$.*

   (iv) *If the $\mathbf{x}_i$ are orthonormal, then $V(\mathbf{x}_1, \ldots, \mathbf{x}_k) = 1$.*

   Proof. (a) Prove uniqueness. [*Hint:* Use the Gram-Schmidt process.]

   (b) Prove existence. [*Hint:* If $f : W \to \mathbf{R}^n$ is a linear transformation that carries an orthonormal basis to an orthonormal basis, then f carries the inner product on W to the dot product on $\mathbf{R}^n$.]

§22. THE VOLUME OF A PARAMETRIZED-MANIFOLD

Now we define what we mean by a parametrized-manifold in $\mathbf{R}^n$, and we define the volume of such an object. This definition generalizes the definitions given in calculus for the length of a parametrized-curve, and the area of a parametrized-surface, in $\mathbf{R}^3$.

Definition. Let $k \leq n$. Let A be open in $\mathbf{R}^k$, and let $\alpha : A \to \mathbf{R}^n$ be a map of class $C^r (r \geq 1)$. The set $Y = \alpha(A)$, together with the map α, constitute what is called **parametrized-manifold,** of dimension k. We denote this parametrized-manifold by Y_α; and we define the (k-dimensional) **volume** of Y_α by the equation

$$v(Y_\alpha) = \int_A V(D\alpha),$$

provided the integral exists.

Let us give a plausibility argument to justify this definition of volume. Suppose A is the interior of a rectangle Q in $\mathbf{R}^k$, and suppose $\alpha : A \to \mathbf{R}^n$ can be extended to be of class C^r in a neighborhood of Q. Let $Y = \alpha(A)$.

Let P be a partition of Q. Consider one of the subrectangles

$$R = [a_1, a_1 + h_1] \times \cdots \times [a_k, a_k + h_k]$$

determined by P. Now R is mapped by α onto a "curved rectangle" contained in Y. The edge of R having endpoints $\mathbf{a}$ and $\mathbf{a} + h_i\mathbf{e}_i$ is mapped by α into a curve in $\mathbf{R}^n$; the vector joining the initial point of this curve to the final point is the vector

$$\alpha(\mathbf{a} + h_i\mathbf{e}_i) - \alpha(\mathbf{a}).$$

A first-order approximation to this vector is, as we know, the vector

$$\mathbf{v}_i = D\alpha(\mathbf{a}) \cdot h_i\mathbf{e}_i = (\partial\alpha/\partial x_i) \cdot h_i.$$

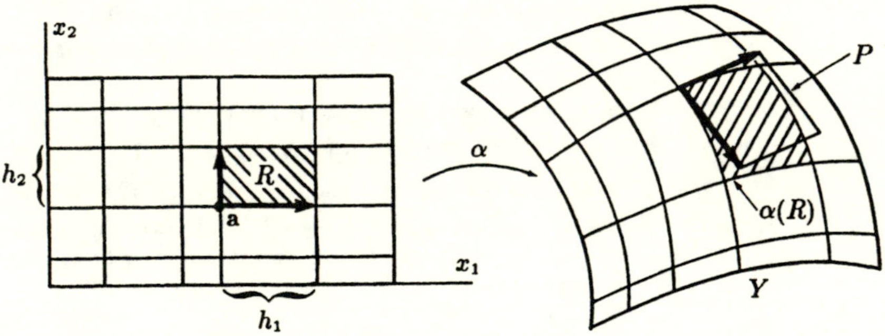

Figure 22.1

It is *plausible* therefore to consider the k-dimensional parallelopiped $\mathcal{P}$ whose edges are the vectors $\mathbf{v}_i$ to be in some sense a first-order approximation to the "curved rectangle" $\alpha(R)$. See Figure 22.1. The k-dimensional volume of $\mathcal{P}$ is the number

$$\begin{aligned} V(\mathbf{v}_1, \ldots, \mathbf{v}_k) &= V(\partial\alpha/\partial x_1, \ldots, \partial\alpha/\partial x_k) \cdot (h_1 \cdots h_k) \\ &= V\big(D\alpha(\mathbf{a})\big) \cdot v(R). \end{aligned}$$

When we sum this expression over all subrectangles R, we obtain a number which lies between the lower and upper sums for the function $V(D\alpha)$ relative to the partition P. Hence this sum is an approximation to the integral

$$\int_A V(D\alpha);$$

the approximation may be made as close as we wish by choosing an appropriate partition P.

We now define the integral of a scalar function over a parametrized-manifold.

Definition. Let A be open in $\mathbf{R}^k$; let $\alpha : A \to \mathbf{R}^n$ be of class C^r; let $Y = \alpha(A)$. Let f be a real-valued continuous function defined at each point of Y. We define the **integral of f over Y_α, with respect to volume,** by the equation

$$\int_{Y_\alpha} f \, \mathrm{d}V = \int_A (f \circ \alpha) V(D\alpha),$$

provided this integral exists.

Here we are reverting to "calculus notation" in using the meaningless symbol $\mathbf{d}V$ to denote the "integral with respect to volume." Note that in this notation,

$$v(Y_\alpha) = \int_{Y_\alpha} \mathrm{d}V.$$

We show that this integral is "invariant under reparametrization."

Theorem 22.1. *Let $g : A \to B$ be a diffeomorphism of open sets in $\mathbf{R}^k$. Let $\beta : B \to \mathbf{R}^n$ be a map of class C^r; let $Y = \beta(B)$. Let $\alpha = \beta \circ g$; then $\alpha : A \to \mathbf{R}^n$ and $Y = \alpha(A)$. If $f : Y \to \mathbf{R}$ is a continuous function, then f is integrable over Y_β if and only if it is integrable over Y_α; in this case*

$$\int_{Y_\alpha} f \, \mathrm{d}V = \int_{Y_\beta} f \, \mathrm{d}V.$$

In particular, $v(Y_\alpha) = v(Y_\beta)$.

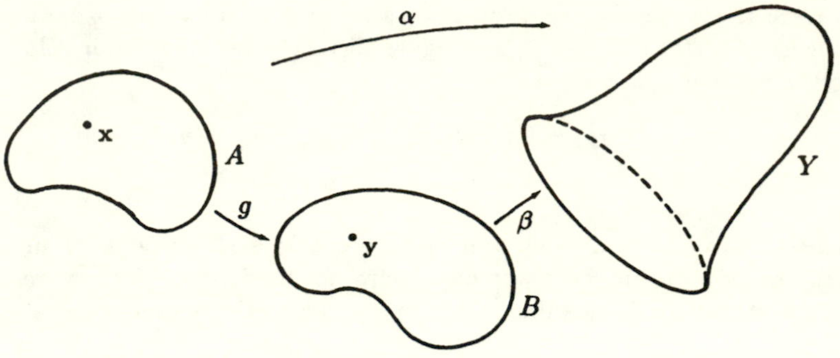

Figure 22.2

Proof. We must show that

$$\int_B (f \circ \beta) V(D\beta) = \int_A (f \circ \alpha) V(D\alpha),$$

where one integral exists if the other does. See Figure 22.2.

The change of variables theorem tells us that

$$\int_B (f \circ \beta) V(D\beta) = \int_A ((f \circ \beta) \circ g)(V(D\beta) \circ g) |\det Dg|.$$

We show that

$$(V(D\beta) \circ g) |\det Dg| = V(D\alpha),$$

and the proof is complete. Let $\mathbf{x}$ denote the general point of A; let $\mathbf{y} = g(\mathbf{x})$. By the chain rule,

$$D\alpha(\mathbf{x}) = D\beta(\mathbf{y}) \cdot Dg(\mathbf{x}).$$

Then

$$\begin{aligned}[V(D\alpha(\mathbf{x}))]^2 &= \det(Dg(\mathbf{x})^{tr} \cdot D\beta(\mathbf{y})^{tr} \cdot D\beta(\mathbf{y}) \cdot Dg(\mathbf{x})) \\ &= \det(Dg(\mathbf{x}))^2 [V(D\beta(\mathbf{y}))]^2.\end{aligned}$$

Our desired equation follows. □

A remark on notation. In this book, we shall use the symbol $\mathbf{d}V$ when dealing with the integral with respect to volume, to avoid confusion with the differential operator d and the notation $\int_A d\omega$, which we shall introduce in succeeding chapters. The integrals $\int_A \mathbf{d}V$ and $\int_A d\omega$ are quite different notions. It is however common in the literature to use the same symbol d in both situations, and the reader must determine from the context which meaning is intended.

EXAMPLE 1. Let A be an open interval in $\mathbf{R}^1$, and let $\alpha : A \to \mathbf{R}^n$ be a map of class C^r. Let $Y = \alpha(A)$. Then Y_α is called a **parametrized-curve** in $\mathbf{R}^n$, and its 1-dimensional volume is often called its **length**. This length is given by the formula

$$v(Y_\alpha) = \int_A V(D\alpha) = \int_A \left[\left(\frac{d\alpha_1}{dt}\right)^2 + \cdots + \left(\frac{d\alpha_n}{dt}\right)^2 \right]^{1/2}$$

since $D\alpha$ is the column matrix whose entries are the functions $d\alpha_i/dt$. This formula may be familiar to you from calculus, in the case $n = 3$, as the formula for computing the arc length of a parametrized-curve.

EXAMPLE 2. Consider the parametrized-curve

$$\alpha(t) = (a\cos t, a\sin t) \quad \text{for} \quad 0 < t < 3\pi.$$

Using the formula of Example 1, we compute its length as

$$\int_0^{3\pi} [a^2 \sin^2 t + a^2 \cos^2 t]^{1/2} = 3\pi a.$$

See Figure 22.3. Since α is not one-to-one, what this number measures is not the actual length of the image set (which is the circle of radius a) but rather the distance travelled by a particle whose equation of motion is $\mathbf{x} = \alpha(t)$ for $0 < t < 3\pi$. We shall later restrict ourselves to parametrizations that are one-to-one, to avoid this situation.

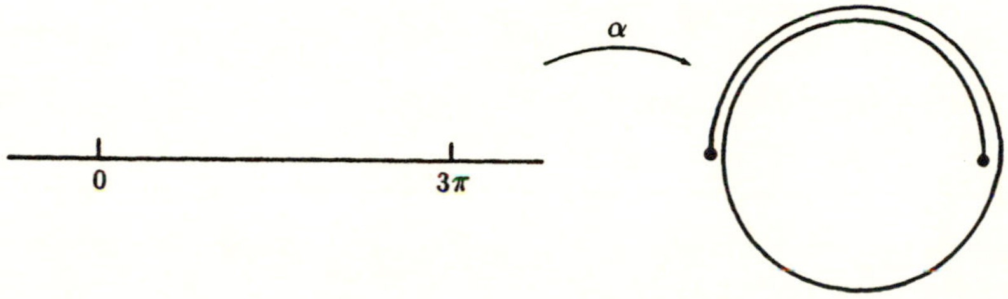

Figure 22.3

EXAMPLE 3. Let A be open in $\mathbf{R}^2$; let $\alpha : A \to \mathbf{R}^n$ be of class C^r; let $Y = \alpha(A)$. Then Y_α is called a **parametrized-surface** in $\mathbf{R}^n$, and its 2-dimensional volume is often called its **area**.

Let us consider the case $n = 3$. If we use (x, y) as the general point of $\mathbf{R}^2$, then $D\alpha = [\partial\alpha/\partial x \quad \partial\alpha/\partial y]$, and

$$v(Y_\alpha) = \int_A V(D\alpha) = \int_A \left\| \frac{\partial\alpha}{\partial x} \times \frac{\partial\alpha}{\partial y} \right\|.$$

(See Example 1 of the preceding section.) In particular, if α has the form

$$\alpha(x,y) = (x,y,f(x,y)),$$

where $f : A \to \mathbf{R}$ is a C^r function, then Y is simply the graph of f, and we have

$$D\alpha = \begin{bmatrix} 1 & 0 \\ 0 & 1 \\ \partial f/\partial x & \partial f/\partial y \end{bmatrix},$$

so that

$$v(Y_\alpha) = \int_A [1 + (\partial f/\partial x)^2 + (\partial f/\partial y)^2]^{1/2}.$$

You may recognize these as formulas for surface area given in calculus.

EXAMPLE 4. Suppose A is the open disc $x^2 + y^2 < a^2$ in $\mathbf{R}^2$, and f is the function

$$f(x,y) = [a^2 - x^2 - y^2]^{1/2}.$$

The graph of f is called a hemisphere of radius a. See Figure 22.4.

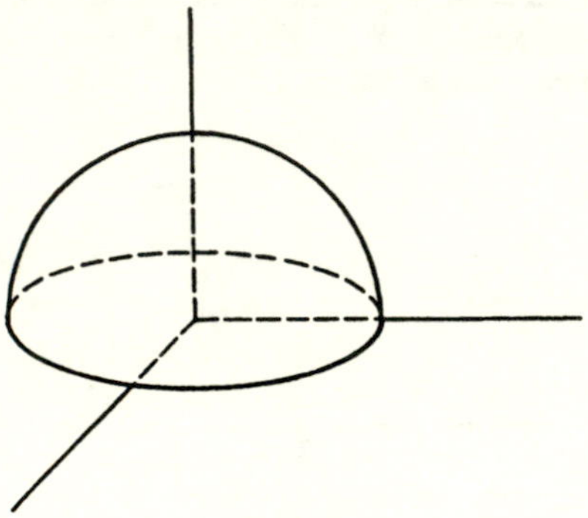

Figure 22.4

Let $\alpha(x,y) = (x,y,f(x,y))$. You can check that

$$V(D\alpha) = a/(a^2 - x^2 - y^2)^{1/2},$$

so that (using polar coordinates)

$$v(Y_\alpha) = \int_B ar/(a^2 - r^2)^{1/2},$$

where B is the open set $(0, a) \times (0, 2\pi)$ in the (r, θ)-plane. This is an improper integral, so we cannot use the Fubini theorem, which was proved only for the ordinary integral. Instead, we integrate over the set $(0, a_n) \times (0, 2\pi)$ using the Fubini theorem, where $0 < a_n < a$, and then we let $a_n \to a$. We have

$$v(Y_\alpha) = \lim_{n\to\infty} (-2\pi a)[(a^2 - a_n^2)^{1/2} - a] = 2\pi a^2.$$

A different method for computing this area, one that avoids improper integrals, is given in §25.

EXERCISES

1. Let A be open in $\mathbf{R}^k$; let $\alpha : A \to \mathbf{R}^n$ be of class C^r; let $Y = \alpha(A)$. Suppose $h : \mathbf{R}^n \to \mathbf{R}^n$ is an isometry; let $Z = h(Y)$ and let $\beta = h \circ \alpha$. Show that Y_α and Z_β have the same volume.

2. Let A be open in $\mathbf{R}^k$; let $f : A \to \mathbf{R}$ be of class C^r; let Y be the graph of f in $\mathbf{R}^{k+1}$, parametrized by the function $\alpha : A \to \mathbf{R}^{k+1}$ given by $\alpha(\mathbf{x}) = (\mathbf{x}, f(\mathbf{x}))$. Express $v(Y_\alpha)$ as an integral.

3. Let A be open in $\mathbf{R}^k$; let $\alpha : A \to \mathbf{R}^n$ be of class C^r; let $Y = \alpha(A)$. The **centroid** $\mathbf{c}(Y_\alpha)$ of the parametrized-manifold Y_α is the point of $\mathbf{R}^n$ whose i^{th} coordinate is given by the equation

$$c_i(Y_\alpha) = [1/v(Y_\alpha)] \int_{Y_\alpha} \pi_i \, dV,$$

where $\pi_i : \mathbf{R}^n \to \mathbf{R}$ is the ith projection function.

(a) Find the centroid of the parametrized-curve

$$\alpha(t) = (a \cos t, a \sin t) \quad \text{with} \quad 0 < t < \pi.$$

(b) Find the centroid of the hemisphere of radius a in $\mathbf{R}^3$. (See Example 4.)

*4. The following exercise gives a strong plausibility argument justifying our definition of volume. We consider only the case of a surface in $\mathbf{R}^3$, but a similar result holds in general.

Given three points $\mathbf{a}, \mathbf{b}, \mathbf{c}$ in $\mathbf{R}^3$, let C be the matrix with columns $\mathbf{b}-\mathbf{a}$ and $\mathbf{c}-\mathbf{a}$. The transformation $h : \mathbf{R}^2 \to \mathbf{R}^3$ given by $h(\mathbf{x}) = C \cdot \mathbf{x} + \mathbf{a}$ carries $\mathbf{0}, \mathbf{e}_1, \mathbf{e}_2$ to $\mathbf{a}$, $\mathbf{b}$, $\mathbf{c}$, respectively. The image Y under h of the set

$$A = \{(x, y) \mid x > 0 \quad \text{and} \quad y > 0 \quad \text{and} \quad x + y < 1\}$$

is called the (open) **triangle** in $\mathbf{R}^3$ with vertices $\mathbf{a}$, $\mathbf{b}$, $\mathbf{c}$. See Figure 22.5. The area of the parametrized-surface Y_h is one-half the area of the parallelopiped with edges $\mathbf{b} - \mathbf{a}$ and $\mathbf{c} - \mathbf{a}$, as you can check.

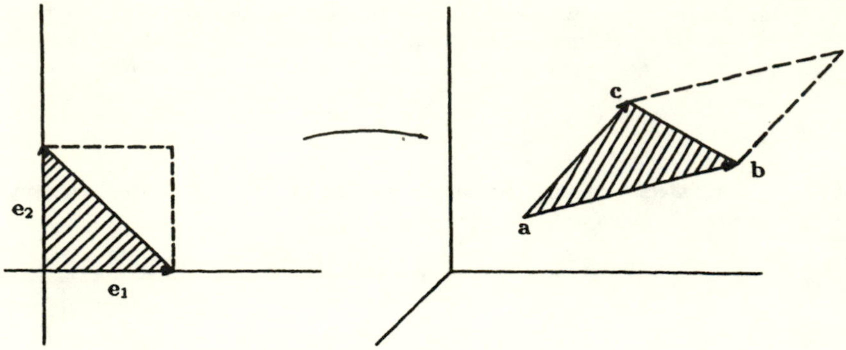

Figure 22.5

Now let Q be a rectangle in $\mathbf{R}^2$ and let $\alpha : Q \to \mathbf{R}^3$; suppose α extends to a map of class C^r defined in an open set containing Q. Let P be a partition of Q. Let R be a subrectangle determined by P, say

$$R = [a, a+h] \times [b, b+k].$$

Consider the triangle $\Delta_1(R)$ having vertices

$$\alpha(a,b), \quad \alpha(a+h,b), \quad \text{and} \quad \alpha(a+h,b+k)$$

and the triangle $\Delta_2(R)$ having vertices

$$\alpha(a,b), \quad \alpha(a,b+k), \quad \text{and} \quad \alpha(a+h,b+k).$$

We consider these two triangles to be an approximation to the "curved rectangle" $\alpha(R)$. See Figure 22.6. We then define

$$A(P) = \sum_R [v(\Delta_1(R)) + v(\Delta_2(R))],$$

where the sum extends over all subrectangles R determined by P. This number is the area of a polyhedral surface that approximates $\alpha(Q)$. Prove the following:

Theorem. *Let Q be a rectangle in $\mathbf{R}^2$ and let $\alpha : A \to \mathbf{R}^3$ be a map of class C^r defined in an open set containing Q. Given $\epsilon > 0$, there is a $\delta > 0$ such that for every partition P of Q of mesh less than δ,*

$$\left| A(P) - \int_Q V(D\alpha) \right| < \epsilon.$$

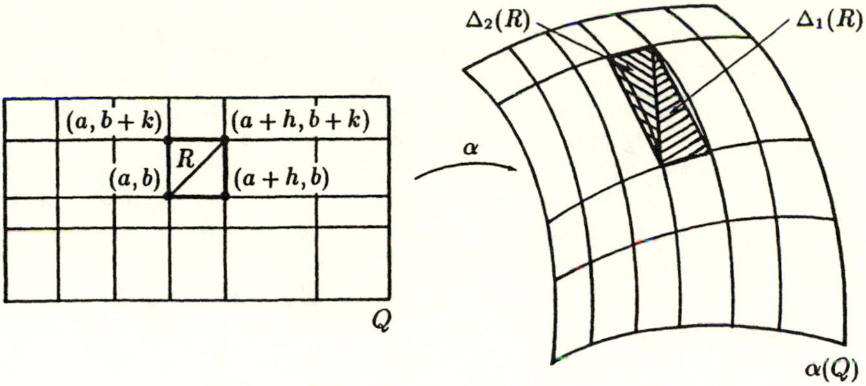

Figure 22.6

Proof. (a) Given points $\mathbf{x}_1, \dots, \mathbf{x}_6$ of Q, let

$$\mathcal{D}\alpha(\mathbf{x}_1, \dots, \mathbf{x}_6) = \begin{bmatrix} D_1\alpha_1(\mathbf{x}_1) & D_2\alpha_1(\mathbf{x}_4) \\ D_1\alpha_2(\mathbf{x}_2) & D_2\alpha_2(\mathbf{x}_5) \\ D_1\alpha_3(\mathbf{x}_3) & D_2\alpha_3(\mathbf{x}_6) \end{bmatrix}.$$

Then $\mathcal{D}\alpha$ is just the matrix $D\alpha$ with its entries evaluated at different points of Q. Show that if R is a subrectangle determined by P, then there are points $\mathbf{x}_1, \dots, \mathbf{x}_6$ of R such that

$$v(\Delta_1(R)) = \frac{1}{2} V(\mathcal{D}\alpha(\mathbf{x}_1, \dots, \mathbf{x}_6)) \cdot v(R).$$

Prove a similar result for $v(\Delta_2(R))$.

(b) Given $\epsilon > 0$, show one can choose $\delta > 0$ so that if $\mathbf{x}_i, \mathbf{y}_i \in Q$ with $|\mathbf{x}_i - \mathbf{y}_i| < \delta$ for $i = 1, \dots, 6$, then

$$|V(\mathcal{D}\alpha(\mathbf{x}_1, \dots, \mathbf{x}_6)) - V(\mathcal{D}\alpha(\mathbf{y}_1, \dots, \mathbf{y}_6))| < \epsilon.$$

(c) Prove the theorem.

§23. MANIFOLDS IN $\mathbb{R}^n$

Manifolds form one of the most important classes of spaces in mathematics. They are useful in such diverse fields as differential geometry, theoretical physics, and algebraic topology. We shall restrict ourselves in this book to manifolds that are submanifolds of euclidean space $\mathbb{R}^n$. In a final chapter, we define abstract manifolds and discuss how our results generalize to that case.

We begin by defining a particular kind of manifold.

Definition. Let $k > 0$. Suppose that M is a subspace of $\mathbb{R}^n$ having the following property: For each $\mathbf{p} \in M$, there is a set V containing $\mathbf{p}$ that is open in M, a set U that is open in $\mathbb{R}^k$, and a continuous map $\alpha : U \to V$ carrying U onto V in a one-to-one fashion, such that:

(1) α is of class C^r.

(2) $\alpha^{-1} : V \to U$ is continuous.

(3) $D\alpha(\mathbf{x})$ has rank k for each $\mathbf{x} \in U$.

Then M is called a **k-manifold without boundary** in $\mathbb{R}^n$, of class C^r. The map α is called a **coordinate patch** on M about $\mathbf{p}$.

Let us explore the geometric meaning of the various conditions in this definition.

EXAMPLE 1. Consider the case $k = 1$. If α is a coordinate patch on M, the condition that $D\alpha$ have rank 1 means merely that $D\alpha \neq 0$. This condition rules out the possibility that M could have "cusps" and "corners." For example, let $\alpha : \mathbb{R} \to \mathbb{R}^2$ be given by the equation $\alpha(t) = (t^3, t^2)$, and let M be the image set of α. Then M has a cusp at the origin. (See Figure 23.1.) Here α is of class C^∞ and α^{-1} is continuous, but $D\alpha$ does not have rank 1 at $t = 0$.

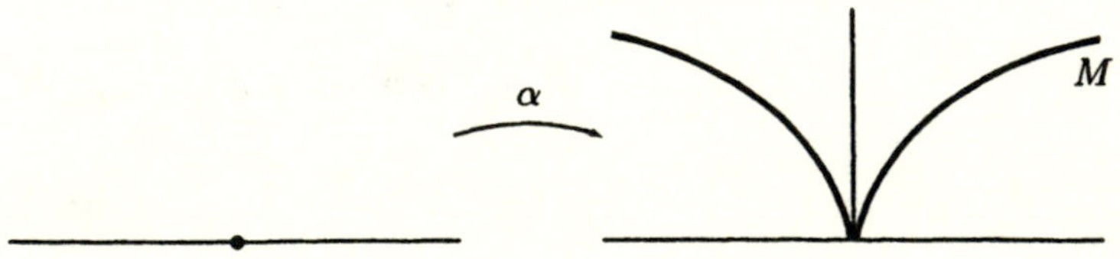

Figure 23.1

Similarly, let $\beta : \mathbb{R} \to \mathbb{R}^2$ be given by $\beta(t) = (t^3, |t^3|)$, and let N be the image set of β. Then N has a corner at the origin. (See Figure 23.2.) Here

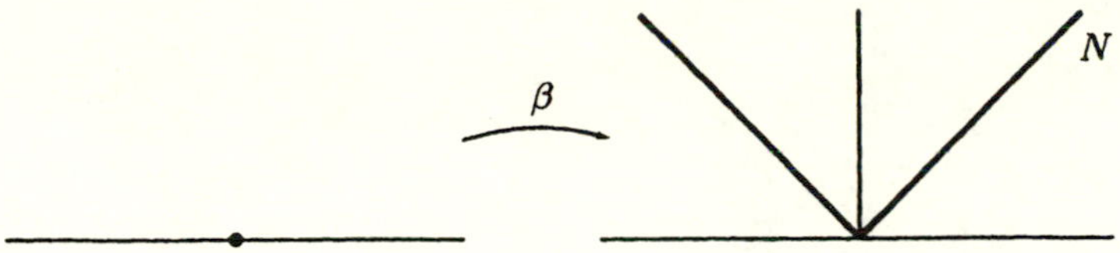

Figure 23.2

β is of class C^2 (as you can check) and β^{-1} is continuous, but $D\beta$ does not have rank 1 at $t = 0$.

EXAMPLE 2. Consider the case $k = 2$. The condition that $D\alpha(\mathbf{a})$ have rank 2 means that the columns $\partial\alpha/\partial x_1$ and $\partial\alpha/\partial x_2$ of $D\alpha$ are independent at $\mathbf{a}$. Note that $\partial\alpha/\partial x_j$ is the velocity vector of the curve $f(t) = \alpha(\mathbf{a} + t\mathbf{e}_j)$ and is thus tangent to the surface M. Then $\partial\alpha/\partial x_1$ and $\partial\alpha/\partial x_2$ span a 2-dimensional "tangent plane" to M. See Figure 23.3.

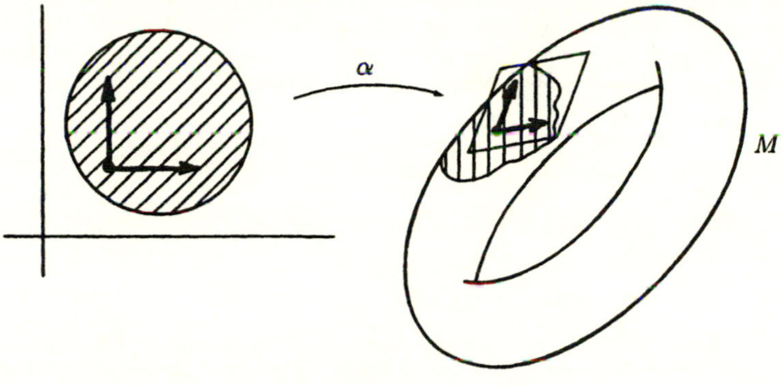

Figure 23.3

As an example of what can happen when this condition fails, consider the function $\alpha : \mathbf{R}^2 \to \mathbf{R}^3$ given by the equation

$$\alpha(x, y) = (x(x^2 + y^2),\ y(x^2 + y^2),\ x^2 + y^2),$$

and let M be the image set of α. Then M fails to have a tangent plane at the origin. See Figure 23.4. The map α is of class C^∞ and α^{-1} is continuous, but $D\alpha$ does not have rank 2 at $\mathbf{0}$.

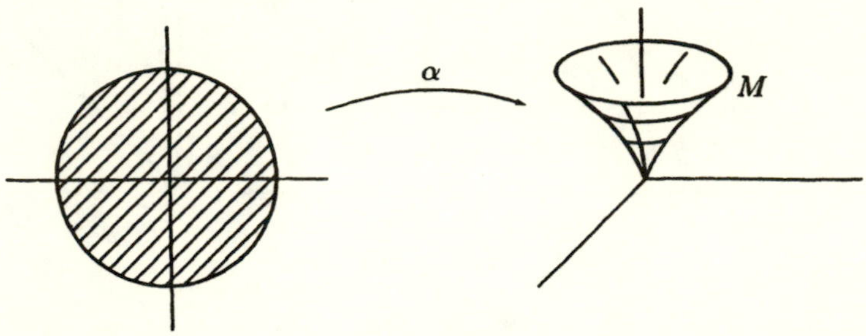

Figure 23.4

EXAMPLE 3. The condition that α^{-1} be continuous also rules out various sorts of "pathological behavior." For instance, let α be the map

$$\alpha(t) = (\sin 2t)(|\cos t|, \sin t) \quad \text{for} \quad 0 < t < \pi,$$

and let M be the image set of α. Then M is a "figure eight" in the plane. The map α is of class C^1 with $D\alpha$ of rank 1, and α maps the interval $(0, \pi)$ in a one-to-one fashion onto M. But the function α^{-1} is not continuous. For continuity of α^{-1} means that α carries any set U_0 that is open in U onto a set that is open in M. In this case, the image of the smaller interval U_0 pictured in Figure 23.5 is *not* open in M. Another way of seeing that α^{-1} is not continuous is to note that points near $\mathbf{0}$ in M need not map under α^{-1} to points near $\pi/2$.

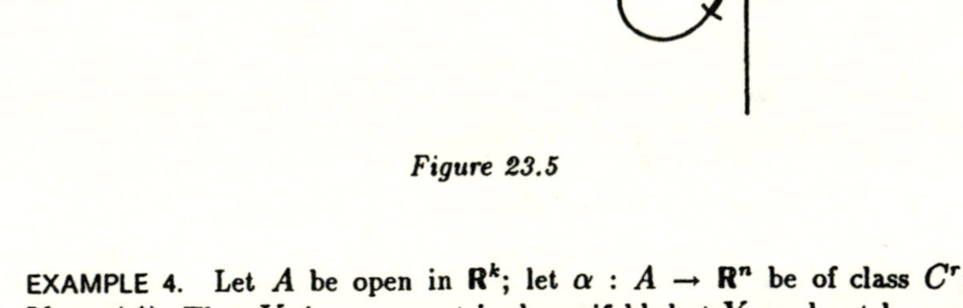

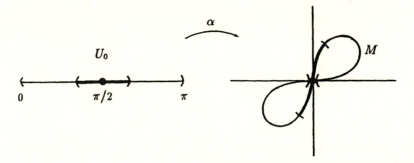

Figure 23.5

EXAMPLE 4. Let A be open in $\mathbf{R}^k$; let $\alpha : A \to \mathbf{R}^n$ be of class C^r; let $Y = \alpha(A)$. Then Y_α is a parametrized-manifold; but Y need not be a manifold. However, if α is one-to-one and α^{-1} is continuous and $D\alpha$ has rank k,

then Y is a manifold without boundary, and in fact Y is covered by the single coordinate patch α.

Now we define what we mean by a manifold in general. We must first generalize our notion of differentiability to functions that are defined on arbitrary subsets of $\mathbf{R}^k$.

Definition. Let S be a subset of $\mathbf{R}^k$; let $f : S \to \mathbf{R}^n$. We say that f is **of class C^r on S** if f may be extended to a function $g : U \to \mathbf{R}^n$ that is of class C^r on an open set U of $\mathbf{R}^k$ containing S.

It follows from this definition that a composite of C^r functions is of class C^r. Suppose $S \subset \mathbf{R}^k$ and $f_1 : S \to \mathbf{R}^n$ is of class C^r. Next, suppose that $T \subset \mathbf{R}^n$ and $f_1(S) \subset T$ and $f_2 : T \to \mathbf{R}^p$ is of class C^r. Then $f_2 \circ f_1 : S \to \mathbf{R}^p$ is of class C^r. For if g_1 is a C^r extension of f_1 to an open set U in $\mathbf{R}^k$, and if g_2 is a C^r extension of f_2 to an open set V in $\mathbf{R}^n$, then $g_2 \circ g_1$ is a C^r extension of $f_2 \circ f_1$ that is defined on the open set $g_1^{-1}(V)$ of $\mathbf{R}^k$ containing S.

The following lemma shows that f is of class C^r if it is locally of class C^r:

Lemma 23.1. *Let S be a subset of $\mathbf{R}^k$; let $f : S \to \mathbf{R}^n$. If for each $\mathbf{x} \in S$, there is a neighborhood $U_{\mathbf{x}}$ of $\mathbf{x}$ and a function $g_{\mathbf{x}} : U_{\mathbf{x}} \to \mathbf{R}^n$ of class C^r that agrees with f on $U_{\mathbf{x}} \cap S$, then f is of class C^r on S.*

Proof. The lemma was given as an exercise in §16; we provide a proof here. Cover S by the neighborhoods $U_{\mathbf{x}}$; let A be the union of these neighborhoods; let $\{\phi_i\}$ be a partition of unity on A of class C^r dominated by the collection $\{U_{\mathbf{x}}\}$. For each i, choose one of the neighborhoods $U_{\mathbf{x}}$ containing the support of ϕ_i, and let g_i denote the C^r function $g_{\mathbf{x}} : U_{\mathbf{x}} \to \mathbf{R}^n$. The C^r function $\phi_i g_i : U_{\mathbf{x}} \to \mathbf{R}^n$ vanishes outside a closed subset of $U_{\mathbf{x}}$; we extend it to a C^r function h_i on all of A by letting it vanish outside $U_{\mathbf{x}}$. Then we define

$$g(\mathbf{x}) = \sum_{i=1}^{\infty} h_i(\mathbf{x})$$

for each $\mathbf{x} \in A$. Each point of A has a neighborhood on which g equals a finite sum of functions h_i; thus g is of class C^r on this neighborhood and hence on all of A. Furthermore, if $\mathbf{x} \in S$, then

$$h_i(\mathbf{x}) = \phi_i(\mathbf{x}) g_i(\mathbf{x}) = \phi_i(\mathbf{x}) f(\mathbf{x})$$

for each i for which $\phi_i(\mathbf{x}) \neq 0$. Hence if $\mathbf{x} \in S$,

$$g(\mathbf{x}) = \sum_{i=1}^{\infty} \phi_i(\mathbf{x}) f(\mathbf{x}) = f(\mathbf{x}). \quad \square$$

Definition. Let $\mathbf{H}^k$ denote **upper half-space** in $\mathbf{R}^k$, consisting of those $\mathbf{x} \in \mathbf{R}^k$ for which $x_k \geq 0$. Let $\mathbf{H}^k_+$ denote the **open upper half-space**, consisting of those $\mathbf{x}$ for which $x_k > 0$.

We shall be particularly interested in functions defined on sets that are open in $\mathbf{H}^k$ but not open in $\mathbf{R}^k$. In this situation, we have the following useful result:

Lemma 23.2. *Let U be open in $\mathbf{H}^k$ but not in $\mathbf{R}^k$; let $\alpha : U \to \mathbf{R}^n$ be of class C^r. Let $\beta : U' \to \mathbf{R}^n$ be a C^r extension of α defined on an open set U' of $\mathbf{R}^k$. Then for $\mathbf{x} \in U$, the derivative $D\beta(\mathbf{x})$ depends only on the function α and is independent of the extension β. It follows that we may denote this derivative by $D\alpha(\mathbf{x})$ without ambiguity.*

Proof. Note that to calculate the partial derivative $\partial\beta/\partial x_j$ at $\mathbf{x}$, we form the difference quotient

$$[\beta(\mathbf{x} + h\mathbf{e}_j) - \beta(\mathbf{x})]/h$$

and take the limit as h approaches 0. For calculation purposes, it suffices to let h approach 0 through positive values. In that case, if $\mathbf{x}$ is in $\mathbf{H}^k$ then so is $\mathbf{x} + h\mathbf{e}_j$. Since the functions β and α agree at points of $\mathbf{H}^k$, the value of $D\beta(\mathbf{x})$ depends only on α. See Figure 23.6. □

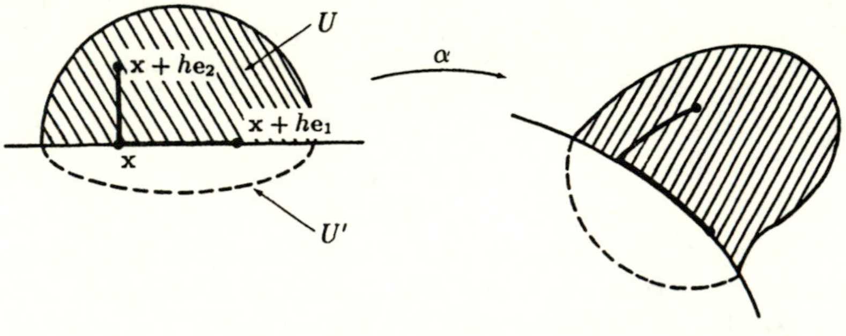

Figure 23.6

Now we define what we mean by a manifold.

Definition. Let $k > 0$. A **k-manifold in $\mathbf{R}^n$ of class C^r** is a subspace M of $\mathbf{R}^n$ having the following property: For each $\mathbf{p} \in M$, there is an open set V of M containing $\mathbf{p}$, a set U that is open in either $\mathbb{R}^k$ or $\mathbb{H}^k$, and a continuous map $\alpha : U \to V$ carrying U onto V in a one-to-one fashion, such that:

(1) α is of class C^r.

(2) $\alpha^{-1} : V \to U$ is continuous.

(3) $D\alpha(\mathbf{x})$ has rank k for each $\mathbf{x} \in U$.

The map α is called a **coordinate patch** on M about $\mathbf{p}$.

We extend the definition to the case $k = 0$ by declaring a discrete collection of points in $\mathbb{R}^n$ to be a **0-manifold** in $\mathbb{R}^n$.

Note that a manifold without boundary is simply the special case of a manifold where all the coordinate patches have domains that are open in $\mathbb{R}^k$.

Figure 23.7 illustrates a 2-manifold in $\mathbb{R}^3$. Indicated are two coordinate patches on M, one whose domain is open in $\mathbb{R}^2$ and the other whose domain is open in $\mathbb{H}^2$ but not in $\mathbb{R}^2$.

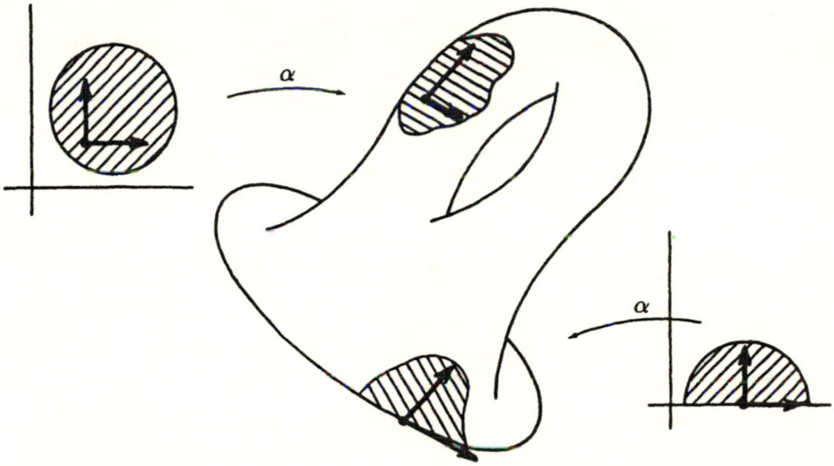

Figure 23.7

It seems clear from this figure that in a k-manifold, there are two kinds of points, those that have neighborhoods that look like open k-balls, and those that do not but instead have neighborhoods that look like open half-balls of dimension k. The latter points constitute what we shall call the *boundary* of M. Making this definition precise, however, requires a certain amount of effort. We shall deal with this question in the next section.

We close this section with the following elementary result:

Lemma 23.3. *Let M be a manifold in $\mathbb{R}^n$, and let $\alpha : U \to V$ be a coordinate patch on M. If U_0 is a subset of U that is open in U, then the restriction of α to U_0 is also a coordinate patch on M.*

Proof. The fact that U_0 is open in U and α^{-1} is continuous implies that the set $V_0 = \alpha(U_0)$ is open in V. Then U_0 is open in $\mathbb{R}^k$ or $\mathbb{H}^k$ (according as U is open in $\mathbb{R}^k$ or $\mathbb{H}^k$), and V_0 is open in M. Then the map $\alpha|U_0$ is a coordinate patch on M: it carries U_0 onto V_0 in a one-to-one fashion; it is of class C^r because α is; its inverse is continuous being simply a restriction of α^{-1}; and its derivative has rank k because $D\alpha$ does. □

Note that this result would not hold if we had not required α^{-1} to be continuous. The map α of Example 3 satisfies all the other conditions for a coordinate patch, but the restricted map $\alpha|U_0$ is not a coordinate patch on M, because its image is not open in M.

EXERCISES

1. Let $\alpha : \mathbb{R} \to \mathbb{R}^2$ be the map $\alpha(x) = (x, x^2)$; let M be the image set of α. Show that M is a 1-manifold in $\mathbb{R}^2$ covered by the single coordinate patch α.
2. Let $\beta : \mathbb{H}^1 \to \mathbb{R}^2$ be the map $\beta(x) = (x, x^2)$; let N be the image set of β. Show that N is a 1-manifold in $\mathbb{R}^2$.
3. (a) Show that the unit circle S^1 is a 1-manifold in $\mathbb{R}^2$.

   (b) Show that the function $\alpha : [0,1) \to S^1$ given by
$$\alpha(t) = (\cos 2\pi t, \sin 2\pi t)$$
   is not a coordinate patch on S^1.
4. Let A be open in $\mathbb{R}^k$; let $f : A \to \mathbb{R}$ be of class C^r. Show that the graph of f is a k-manifold in $\mathbb{R}^{k+1}$.
5. Show that if M is a k-manifold without boundary in $\mathbb{R}^m$, and if N is an ℓ-manifold in $\mathbb{R}^n$, then $M \times N$ is a $k + \ell$ manifold in $\mathbb{R}^{m+n}$.
6. (a) Show that $I = [0,1]$ is a 1-manifold in $\mathbb{R}^1$.

   (b) Is $I \times I$ a 2-manifold in $\mathbb{R}^2$? Justify your answer.

§24. THE BOUNDARY OF A MANIFOLD

In this section, we make precise what we mean by the boundary of a manifold; and we prove a theorem that is useful in practice for constructing manifolds.

To begin, we derive an important property of coordinate patches, namely, the fact that they "overlap differentiably." We make this statement more precise as follows:

Theorem 24.1. *Let M be a k-manifold in $\mathbf{R}^n$, of class C^r. Let $\alpha_0 : U_0 \to V_0$ and $\alpha_1 : U_1 \to V_1$ be coordinate patches on M, with $W = V_0 \cap V_1$ non-empty. Let $W_i = \alpha_i^{-1}(W)$. Then the map*

$$\alpha_1^{-1} \circ \alpha_0 : W_0 \to W_1$$

is of class C^r, and its derivative is non-singular.

Typical cases are pictured in Figure 24.1. We often call $\alpha_1^{-1} \circ \alpha_0$ the **transition function** between the coordinate patches α_0 and α_1.

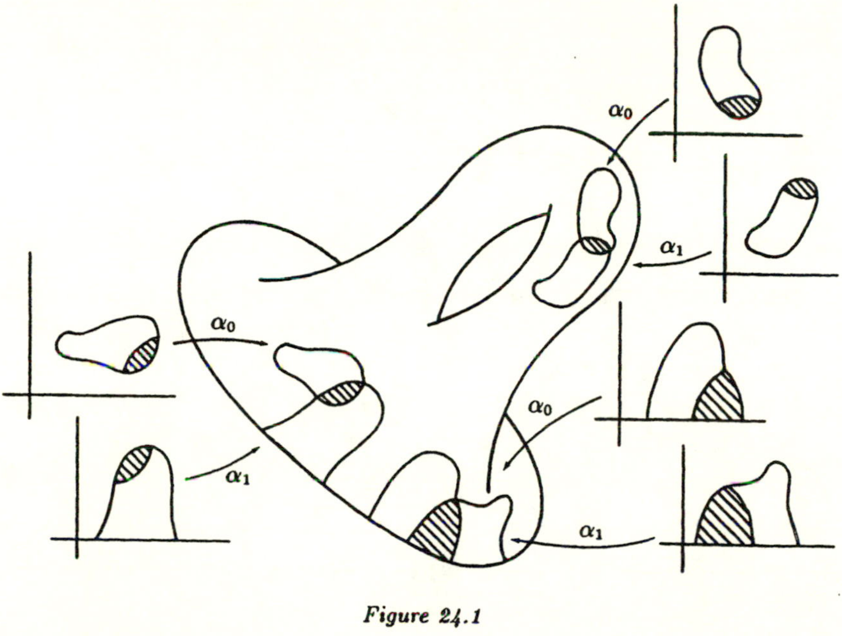

Figure 24.1

Proof. It suffices to show that if $\alpha : U \to V$ is a coordinate patch on M, then $\alpha^{-1} : V \to \mathbf{R}^k$ is of class C^r, as a map of the subset V of $\mathbf{R}^n$ into $\mathbf{R}^k$. For

then it follows that, since α_0 and α_1^{-1} are of class C^r, so is their composite $\alpha_1^{-1} \circ \alpha_0$. The same argument applies to show $\alpha_0^{-1} \circ \alpha_1$ is of class C^r; then the chain rule implies that both these transition functions have non-singular derivatives.

To prove that α^{-1} is of class C^r, it suffices (by Lemma 23.1) to show that it is locally of class C^r. Let $\mathbf{p}_0$ be a point of V; let $\alpha^{-1}(\mathbf{p}_0) = \mathbf{x}_0$. We show α^{-1} extends to a C^r function defined in a neighborhood of $\mathbf{p}_0$ in $\mathbf{R}^n$.

Let us first consider the case where U is open in $\mathbf{H}^k$ but not in $\mathbf{R}^k$. By assumption, we can extend α to a C^r map β of an open set U' of $\mathbf{R}^k$ into $\mathbf{R}^n$. Now $D\alpha(\mathbf{x}_0)$ has rank k, so some k rows of this matrix are independent; assume for convenience the first k rows are independent. Let $\pi : \mathbf{R}^n \to \mathbf{R}^k$ project $\mathbf{R}^n$ onto its first k coordinates. Then the map $g = \pi \circ \beta$ maps U' into $\mathbf{R}^k$, and $Dg(\mathbf{x}_0)$ is non-singular. By the inverse function theorem, g is a C^r diffeomorphism of an open set W of $\mathbf{R}^k$ about $\mathbf{x}_0$ with an open set in $\mathbf{R}^k$. See Figure 24.2.

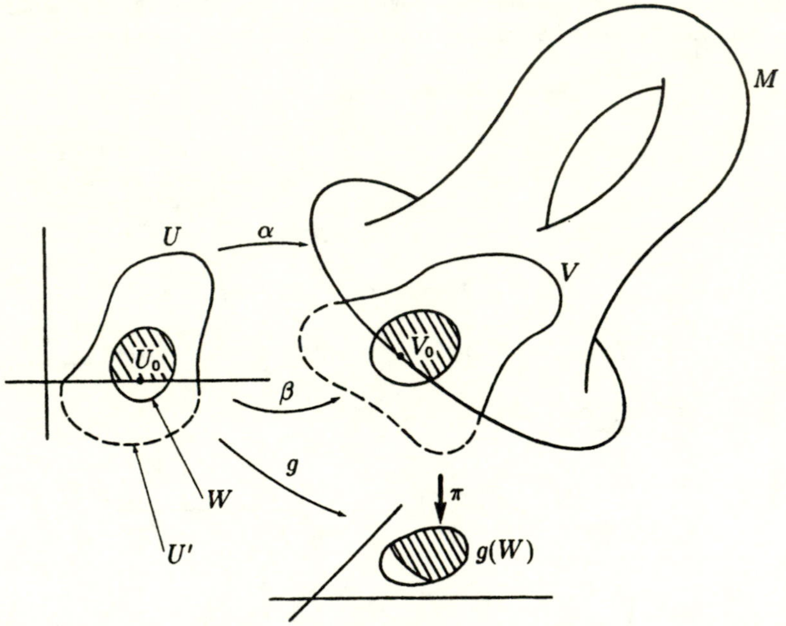

Figure 24.2

We show that the map $h = g^{-1} \circ \pi$, which is of class C^r, is the desired extension of α^{-1} to a neighborhood A of $\mathbf{p}_0$. To begin, note that the set $U_0 = W \cap U$ is open in U, so that the set $V_0 = \alpha(U_0)$ is open in V; this means there is an open set A of $\mathbf{R}^n$ such that $A \cap V = V_0$. We can choose A so it is contained in the domain of h (by intersecting with $\pi^{-1}(g(W))$ if necessary). Then $h : A \to \mathbf{R}^k$ is of class C^r; and if $\mathbf{p} \in A \cap V = V_0$, then we let $\mathbf{x} = \alpha^{-1}(\mathbf{p})$ and compute

$$h(\mathbf{p}) = h(\alpha(\mathbf{x})) = g^{-1}(\pi(\alpha(\mathbf{x}))) = g^{-1}(g(\mathbf{x})) = \mathbf{x} = \alpha^{-1}(\mathbf{p}),$$

as desired.

A similar argument holds if U is open in $\mathbf{R}^k$. In this case, we set $U' = U$ and $\beta = \alpha$, and the preceding argument proceeds unchanged. □

Now we define the boundary of a manifold.

Definition. Let M be a k-manifold in $\mathbf{R}^n$; let $\mathbf{p} \in M$. If there is a coordinate patch $\alpha : U \to V$ on M about $\mathbf{p}$ such that U is open in $\mathbf{R}^k$, we say $\mathbf{p}$ is an **interior point** of M. Otherwise, we say $\mathbf{p}$ is a **boundary point** of M. We denote the set of boundary points of M by ∂M, and call this set the **boundary** of M.

Note that our use here of the terms "interior" and "boundary" has nothing to do with the way these terms are used in general topology. *Any* subset S of $\mathbf{R}^n$ has an interior and a boundary and an exterior in the topological sense, which we denote by Int S and Bd S and Ext S, respectively. For a manifold M, we denote its boundary by ∂M and its interior by $M - \partial M$.

Given M, one can readily identify the boundary points of M by use of the following criterion:

Lemma 24.2. *Let M be a k-manifold in $\mathbf{R}^n$; let $\alpha : U \to V$ be a coordinate patch about the point $\mathbf{p}$ of M.*

(a) *If U is open in $\mathbf{R}^k$, then $\mathbf{p}$ is an interior point of M.*

(b) *If U is open in $\mathbf{H}^k$ and if $\mathbf{p} = \alpha(\mathbf{x}_0)$ for $\mathbf{x}_0 \in \mathbf{H}^k_+$, then $\mathbf{p}$ is an interior point of M.*

(c) *If U is open in $\mathbf{H}^k$ and $p = \alpha(\mathbf{x}_0)$ for $\mathbf{x}_0 \in \mathbf{R}^{k-1} \times 0$, then $\mathbf{p}$ is a boundary point of M.*

Proof. (a) is immediate from the definition. (b) is almost as easy. Given $\alpha : U \to V$ as in (b), let $U_0 = U \cap \mathbf{H}^k_+$ and let $V_0 = \alpha(U_0)$. Then $\alpha|U_0$, mapping U_0 onto V_0, is a coordinate patch about $\mathbf{p}$, with U_0 open in $\mathbf{R}^k$.

We prove (c). Let $\alpha_0 : U_0 \to V_0$ be a coordinate patch about $\mathbf{p}$, with U_0 open in $\mathbf{H}^k$ and $\mathbf{p} = \alpha_0(\mathbf{x}_0)$ for $\mathbf{x}_0 \in \mathbf{R}^{k-1} \times 0$. We assume there is a coordinate patch $\alpha_1 : U_1 \to V_1$ about $\mathbf{p}$ with U_1 open in $\mathbf{R}^k$, and derive a contradiction.

Since V_0 and V_1 are open in M, the set $W = V_0 \cap V_1$ is also open in M. Let $W_i = \alpha_i^{-1}(W)$ for $i = 0, 1$; then W_0 is open in $\mathbf{H}^k$ and contains $\mathbf{x}_0$, and W_1 is open in $\mathbf{R}^k$. The preceding theorem tells us that the transition function

$$\alpha_0^{-1} \circ \alpha_1 : W_1 \to W_0$$

is a map of class C^r carrying W_1 onto W_0 in a one-to-one fashion, with non-singular derivative. Then Theorem 8.2 tells us that the image set of this map is open in $\mathbf{R}^k$. But W_0 is contained in $\mathbf{H}^k$ and contains the point $\mathbf{x}_0$ of $\mathbf{R}^{k-1} \times 0$, so it is *not* open in $\mathbf{R}^k$! See Figure 24.3. □

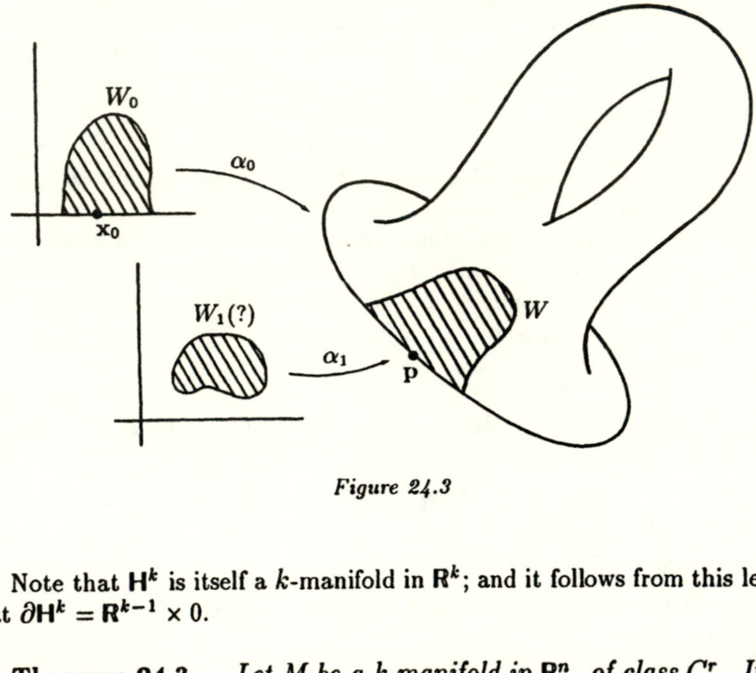

Figure 24.3

Note that $\mathbf{H}^k$ is itself a k-manifold in $\mathbf{R}^k$; and it follows from this lemma that $\partial\mathbf{H}^k = \mathbf{R}^{k-1} \times 0$.

Theorem 24.3. *Let M be a k-manifold in $\mathbf{R}^n$, of class C^r. If ∂M is non-empty, then ∂M is a $k-1$ manifold without boundary in $\mathbf{R}^n$ of class C^r.*

Proof. Let $\mathbf{p} \in \partial M$. Let $\alpha : U \to V$ be a coordinate patch on M about $\mathbf{p}$. Then U is open in $\mathbf{H}^k$ and $\mathbf{p} = \alpha(\mathbf{x}_0)$ for some $\mathbf{x}_0 \in \partial\mathbf{H}^k$. By the preceding lemma, each point of $U \cap \mathbf{H}^k_+$ is mapped by α to an interior point of M, and each point of $U \cap (\partial\mathbf{H}^k)$ is mapped to a point of ∂M. Thus the restriction of α to $U \cap (\partial \mathbf{H}^k)$ carries this set in a one-to-one fashion onto the open set $V_0 = V \cap \partial M$ of ∂M. Let U_0 be the open set of $\mathbf{R}^{k-1}$ such that $U_0 \times 0 = U \cap \partial \mathbf{H}^k$; if $\mathbf{x} \in U_0$, define $\alpha_0(\mathbf{x}) = \alpha(\mathbf{x}, 0)$. Then $\alpha_0 : U_0 \to V_0$ is a coordinate patch on ∂M. It is of class C^r because α is, and its derivative has rank $k-1$ because $D\alpha_0(\mathbf{x})$ consists simply of the first $k-1$ columns of the matrix $D\alpha(\mathbf{x}, 0)$. The inverse α_0^{-1} is continuous because it equals the restriction to V_0 of the continuous function α^{-1}, followed by projection of $\mathbf{R}^k$ onto its first $k-1$ coordinates. □

The coordinate patch α_0 on ∂M constructed in the proof of this theorem is said to be obtained by **restricting** the coordinate patch α on M.

Finally, we prove a theorem that is useful in practice for constructing manifolds.

Theorem 24.4. *Let $\mathcal{O}$ be open in $\mathbf{R}^n$; let $f : \mathcal{O} \to \mathbf{R}$ be of class C^r. Let M be the set of points $\mathbf{x}$ for which $f(\mathbf{x}) = 0$; let N be the set of points for which $f(\mathbf{x}) \geq 0$. Suppose M is non-empty and $Df(\mathbf{x})$ has rank 1 at each point of M. Then N is an n-manifold in $\mathbf{R}^n$ and $\partial N = M$.*

Proof. Suppose first that $\mathbf{p}$ is a point of N such that $f(\mathbf{p}) > 0$. Let U be the open set in $\mathbf{R}^n$ consisting of all points $\mathbf{x}$ for which $f(\mathbf{x}) > 0$; let $\alpha : U \to U$ be the identity map. Then α is (trivially) a coordinate patch on N about $\mathbf{p}$ whose domain is open in $\mathbf{R}^n$.

Now suppose that $f(\mathbf{p}) = 0$. Since $Df(\mathbf{p})$ is non-zero, at least one of the partial derivatives $D_i f(\mathbf{p})$ is non-zero. Suppose $D_n f(\mathbf{p}) \neq 0$. Define $F : \mathcal{O} \to \mathbf{R}^n$ by the equation $F(\mathbf{x}) = (x_1, \dots, x_{n-1}, f(\mathbf{x}))$. Then

$$DF = \begin{bmatrix} I_{n-1} & 0 \\ * & D_n f \end{bmatrix},$$

so that $DF(\mathbf{p})$ is non-singular. It follows that F is a diffeomorphism of a neighborhood A of $\mathbf{p}$ in $\mathbf{R}^n$ with an open set B of $\mathbf{R}^n$. Furthermore, F carries the open set $A \cap N$ of N onto the open set $B \cap \mathbf{H}^n$ of $\mathbf{H}^n$, since $\mathbf{x} \in N$ if and only if $f(\mathbf{x}) \geq 0$. It also carries $A \cap M$ onto $B \cap \partial \mathbf{H}^n$, since $\mathbf{x} \in M$ if and only if $f(\mathbf{x}) = 0$. Then $F^{-1} : B \cap \mathbf{H}^n \to A \cap N$ is the required coordinate patch on N. See Figure 24.4. □

Definition. Let $B^n(a)$ consist of all points $\mathbf{x}$ of $\mathbf{R}^n$ for which $\|\mathbf{x}\| \leq a$, and let $S^{n-1}(a)$ consist of all $\mathbf{x}$ for which $\|\mathbf{x}\| = a$. We call them the ***n*-ball** and the **$n-1$ sphere**, respectively, of radius a.

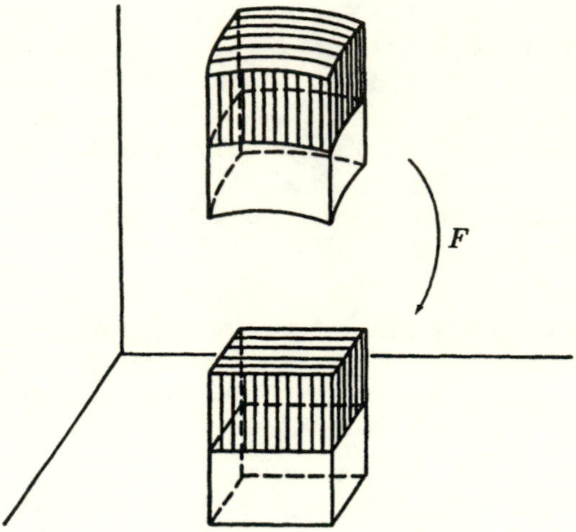

Figure 24.4

Corollary 24.5. *The n-ball $B^n(a)$ is an n-manifold in $\mathbb{R}^n$ of class C^∞, and $S^{n-1}(a) = \partial B^n(a)$.*

Proof. We apply the preceding theorem to the function $f(\mathbf{x}) = a^2 - \|\mathbf{x}\|^2$. Then

$$Df(\mathbf{x}) = [(-2x_1) \quad \cdots \quad (-2x_n)],$$

which is non-zero at each point of $S^{n-1}(a)$. □

EXERCISES

1. Show that the solid torus is a 3-manifold, and its boundary is the torus T. (See the exercises of §17.) [*Hint:* Write the equation for T in cartesian coordinates and apply Theorem 24.4.]
2. Prove the following:

   Theorem. *Let $f : \mathbb{R}^{n+k} \to \mathbb{R}^n$ be of class C^r. Let M be the set of all $\mathbf{x}$ such that $f(\mathbf{x}) = \mathbf{0}$. Assume that M is non-empty and that $Df(\mathbf{x})$ has rank n for $\mathbf{x} \in M$. Then M is a k-manifold without boundary in $\mathbb{R}^{n+k}$. Furthermore, if N is the set of all $\mathbf{x}$ for which*

   $$f_1(\mathbf{x}) = \cdots = f_{n-1}(\mathbf{x}) = 0 \quad \text{and} \quad f_n(\mathbf{x}) \geq 0,$$

   and if the matrix

   $$\partial(f_1, \ldots, f_{n-1})/\partial\mathbf{x}$$

has rank $n-1$ at each point of N, then N is a $k+1$ manifold, and $\partial N = M$.

[*Hint:* Examine the proof of the implicit function theorem.]

3. Let $f, g : \mathbf{R}^3 \to \mathbf{R}$ be of class C^r. Under what conditions can you be sure that the solution set of the system of equations $f(x,y,z) = 0$, $g(x,y,z) = 0$ is a smooth curve without singularities (i.e., a 1-manifold without boundary)?

4. Show that the upper hemisphere of $S^{n-1}(a)$, defined by the equation

$$E_+^{n-1}(a) = S^{n-1}(a) \cap \mathbf{H}^n,$$

is an $n-1$ manifold. What is its boundary?

5. Let $\mathcal{O}(3)$ denote the set of all orthogonal 3 by 3 matrices, considered as a subspace of $\mathbf{R}^9$.
   (a) Define a C^∞ function $f : \mathbf{R}^9 \to \mathbf{R}^6$ such that $\mathcal{O}(3)$ is the solution set of the equation $f(\mathbf{x}) = \mathbf{0}$.
   (b) Show that $\mathcal{O}(3)$ is a compact 3-manifold in $\mathbf{R}^9$ without boundary. [*Hint:* Show the rows of $Df(\mathbf{x})$ are independent if $\mathbf{x} \in \mathcal{O}(3)$.]

6. Let $\mathcal{O}(n)$ denote the set of all orthogonal n by n matrices, considered as a subspace of $\mathbf{R}^N$, where $N = n^2$. Show $\mathcal{O}(n)$ is a compact manifold without boundary. What is its dimension?

   The manifold $\mathcal{O}(n)$ is a particular example of what is called a **Lie group** (pronounced "lee group"). It is a group under the operation of matrix multiplication; it is a C^∞ manifold; and the product operation and the map $A \to A^{-1}$ are C^∞ maps. Lie groups are of increasing importance in theoretical physics, as well as in mathematics.

§25. INTEGRATING A SCALAR FUNCTION OVER A MANIFOLD

Now we define what we mean by the integral of a continuous scalar function f over a manifold M in $\mathbf{R}^n$. For simplicity, we shall restrict ourselves to the case where M is compact. The extension to the general case can be carried out by methods analogous to those used in §16 in treating the extended integral.

First we define the integral in the case where the support of f can be covered by a single coordinate patch.

Definition. Let M be a compact k-manifold in $\mathbf{R}^n$, of class C^r. Let $f : M \to \mathbf{R}^n$ be a continuous function. Let C = Support f; then C is compact. Suppose there is a coordinate patch $\alpha : U \to V$ on M such that $C \subset V$. Now $\alpha^{-1}(C)$ is compact. Therefore, by replacing U by a smaller open set if necessary, we can assume that U is bounded. We define the **integral of f over M** by the equation

$$\int_M f\, dV = \int_{\text{Int } U} (f \circ \alpha) V(D\alpha).$$

Here Int $U = U$ if U is open in $\mathbf{R}^k$, and Int $U = U \cap \mathbf{H}^k_+$ if U is open in $\mathbf{H}^k$ but not in $\mathbf{R}^k$.

It is easy to see this integral exists as an ordinary integral, and hence as an extended integral: The function $F = (f \circ \alpha)V(D\alpha)$ is continuous on U and vanishes outside the compact set $\alpha^{-1}(C)$; hence F is bounded. If U is open in $\mathbf{R}^k$, then F vanishes near each point $\mathbf{x}_0$ of Bd U. If U is not open in $\mathbf{R}^k$, then F vanishes near each point of Bd U not in $\partial\mathbf{H}^k$, a set that has measure zero in $\mathbf{R}^k$. In either case, F is integrable over U and hence over Int U. See Figure 25.1.

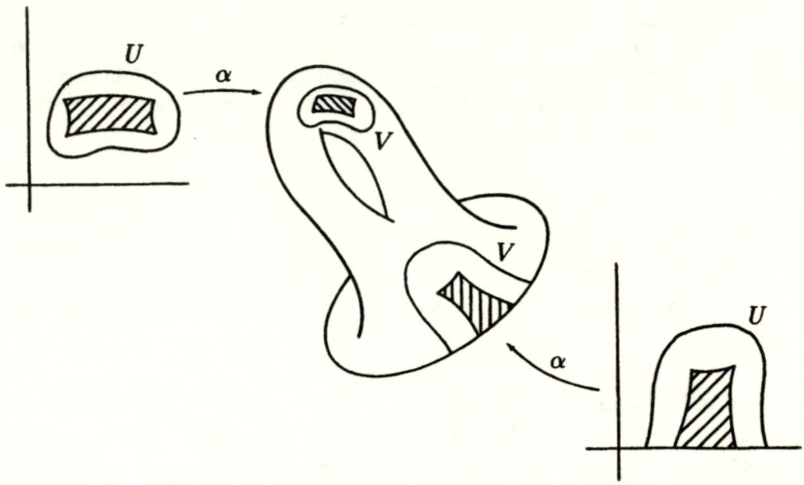

Figure 25.1

Lemma 25.1. *If the support of f can be covered by a single coordinate patch, the integral $\int_M f\, dV$ is well-defined, independent of the choice of coordinate patch.*

Proof. We prove a preliminary result. Let $\alpha : U \to V$ be a coordinate patch containing the support of f. Let W be an open set in U such that $\alpha(W)$ also contains the support of f. Then

$$\int_{\text{Int } W} (f \circ \alpha)V(D\alpha) = \int_{\text{Int } U} (f \circ \alpha)V(D\alpha);$$

the (ordinary) integrals over W and U are equal because the integrand vanishes outside W; then one applies Theorem 13.6.

Let $\alpha_i : U_i \to V_i$ for $i = 0, 1$ be coordinate patches on M such that both V_0 and V_1 contain the support of f. We wish to show that

$$\int_{\text{Int } U_0} (f \circ \alpha_0)V(D\alpha_0) = \int_{\text{Int } U_1} (f \circ \alpha_1)V(D\alpha_1).$$

Let $W = V_0 \cap V_1$ and let $W_i = \alpha_i^{-1}(W)$. In view of the result of the preceding paragraph, it suffices to show that this equation holds with U_i replaced by W_i, for $i = 0, 1$. Since $\alpha_1^{-1} \circ \alpha_0 : \text{Int } W_0 \to \text{Int } W_1$ is a diffeomorphism, this result follows at once from Theorem 22.1. □

To define $\int_M f \, dV$ in general, we use a partition of unity on M.

Lemma 25.2. *Let M be a compact k-manifold in $\mathbf{R}^n$, of class C^r. Given a covering of M by coordinate patches, there exists a finite collection of C^∞ functions $\phi_1, \ldots, \phi_\ell$ mapping $\mathbf{R}^n$ into $\mathbf{R}$ such that:*

(1) $\phi_i(\mathbf{x}) \geq 0$ *for all* $\mathbf{x}$.

(2) *Given i, the support of ϕ_i is compact and there is a coordinate patch $\alpha_i : U_i \to V_i$ belonging to the given covering such that*

$$((\text{Support } \phi_i) \cap M) \subset V_i.$$

(3) $\sum \phi_i(\mathbf{x}) = 1$ *for* $\mathbf{x} \in M$.

We call $\{\phi_1, \ldots, \phi_\ell\}$ a **partition of unity** on M dominated by the given collection of coordinate patches.

Proof. For each coordinate patch $\alpha : U \to V$ belonging to the given collection, choose an open set A_V of $\mathbf{R}^n$ such that $A_V \cap M = V$. Let A be the union of the sets A_V. Choose a partition of unity on A that is dominated by this open covering of A. Local finiteness guarantees that all but finitely many of the functions in the partition of unity vanish identically on M. Let $\phi_1, \ldots, \phi_\ell$ be those that do not. □

Definition. Let M be a compact k-manifold in $\mathbf{R}^n$, of class C^r. Let $f : M \to \mathbf{R}$ be a continuous function. Choose a partition of unity $\phi_1, \ldots, \phi_\ell$

on M that is dominated by the collection of all coordinate patches on M. We define the **integral of f over M** by the equation

$$\int_M f\,dV = \sum_{i=1}^{\ell}[\int_M (\phi_i f)\,dV].$$

Then we define the **(k-dimensional) volume of M** by the equation

$$v(M) = \int_M 1\,dV.$$

If the support of f happens to lie in a single coordinate patch $\alpha : U \to V$, this definition agrees with the preceding definition. For in that case, letting $A = \text{Int } U$, we have

$$\begin{aligned}
\sum_{i=1}^{\ell}[\int_M (\phi_i f)\,dV] &= \sum_{i=1}^{\ell}[\int_A (\phi_i \circ \alpha)(f \circ \alpha)V(D\alpha)] && \text{by definition,} \\
&= \int_A [\sum_{i=1}^{\ell} (\phi_i \circ \alpha)(f \circ \alpha)V(D\alpha)] && \text{by linearity,} \\
&= \int_A (f \circ \alpha)V(D\alpha) && \text{because } \sum_{i=1}^{\ell}(\phi_i \circ \alpha) = 1 \text{ on } A, \\
&= \int_M f\,dV && \text{by definition.}
\end{aligned}$$

We note also that this definition is independent of the choice of the partition of unity. Let $\psi_1, \ldots, \psi_m$ be another choice for the partition of unity. Because the support of $\psi_j f$ lies in a single coordinate patch, we can apply the computation just given (replacing f by $\psi_j f$) to conclude that

$$\sum_{i=1}^{\ell}[\int_M (\phi_i \psi_j f)\,dV] = \int_M (\psi_j f)\,dV.$$

Summing over j, we have

$$\sum_{j=1}^{m}\sum_{i=1}^{\ell}[\int_M (\phi_i \psi_j f)\,dV] = \sum_{j=1}^{m}[\int_M (\psi_j f)\,dV].$$

Symmetry shows that this double summation also equals

$$\sum_{i=1}^{\ell}[\int_M (\phi_i f)\,dV],$$

as desired.

Linearity of the integral follows at once. We state it formally as a theorem:

Theorem 25.3. *Let M be a compact k-manifold in $\mathbf{R}^n$, of class C^r. Let $f, g : M \to \mathbf{R}$ be continuous. Then*

$$\int_M (af + bg)\ \mathrm{d}V \quad = \quad a\int_M f\ \mathrm{d}V \quad + \quad b\int_M g\ \mathrm{d}V. \ \square$$

This definition of the integral $\int_M f\ \mathrm{d}V$ is satisfactory for theoretical purposes, but not for practical purposes. If one wishes actually to integrate a function over the $n-1$ sphere S^{n-1}, for example, what one does is to break S^{n-1} into suitable "pieces," integrate over each piece separately, and add the results together. We now prove a theorem that makes this procedure more precise. We shall use this result in some examples and exercises.

Definition. Let M be a compact k-manifold in $\mathbf{R}^n$, of class C^r. A subset D of M is said to have **measure zero in** M if it can be covered by countably many coordinate patches $\alpha_i : U_i \to V_i$ such that the set

$$D_i = \alpha_i^{-1}(D \cap V_i)$$

has measure zero in $\mathbf{R}^k$ for each i.

An equivalent definition is to require that for *any* coordinate patch $\alpha : U \to V$ on M, the set $\alpha^{-1}(D \cap V)$ have measure zero in $\mathbf{R}^k$. To verify this fact, it suffices to show that $\alpha^{-1}(D \cap V \cap V_i)$ has measure zero for each i. And this follows from the fact that the set $\alpha_i^{-1}(D \cap V \cap V_i)$ has measure zero because it is a subset of D_i, and that $\alpha^{-1} \circ \alpha_i$ is of class C^r.

***Theorem 25.4.** *Let M be a compact k-manifold in $\mathbf{R}^n$, of class C^r. Let $f : M \to \mathbf{R}$ be a continuous function. Suppose that $\alpha_i : A_i \to M_i$, for $i = 1, \ldots, N$, is a coordinate patch on M, such that A_i is open in $\mathbf{R}^k$ and M is the disjoint union of the open sets $M_1, \ldots, M_N$ of M and a set K of measure zero in M. Then*

$$(*) \qquad \int_M f\ \mathrm{d}V = \sum_{i=1}^{N} [\int_{A_i} (f \circ \alpha_i) V(D\alpha_i)].$$

This theorem says that $\int_M f\ \mathrm{d}V$ can be evaluated by breaking M up into pieces that are parametrized-manifolds and integrating f over each piece separately.

Proof. Since both sides of $(*)$ are linear in f, it suffices to prove the theorem in the case where the set C = Support f is covered by a single coordinate patch $\alpha : U \to V$. We can assume that U is bounded. Then

$$\int_M f\ \mathrm{d}V = \int_{\text{Int } U} (f \circ \alpha) V(D\alpha),$$

by definition.

Step 1. Let $W_i = \alpha^{-1}(M_i \cap V)$ and let $L = \alpha^{-1}(K \cap V)$. Then W_i is open in $\mathbf{R}^k$, and L has measure zero in $\mathbf{R}^k$; and U is the disjoint union of L and the sets W_i. See Figures 25.2 and 25.3. We show first that

$$\int_M f \, dV = \sum_i \left[\int_{W_i} (f \circ \alpha) V(D\alpha)\right].$$

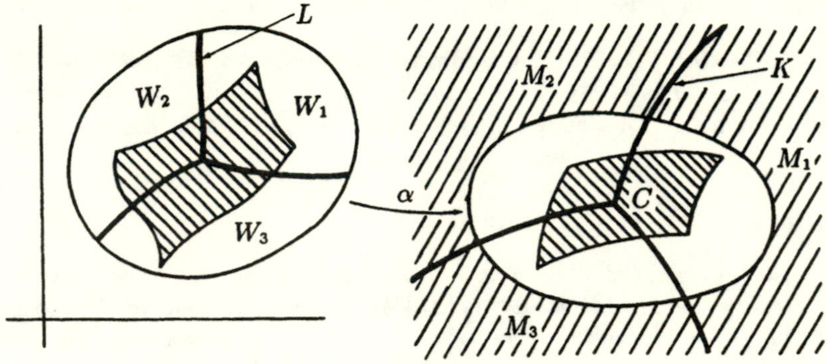

Figure 25.2

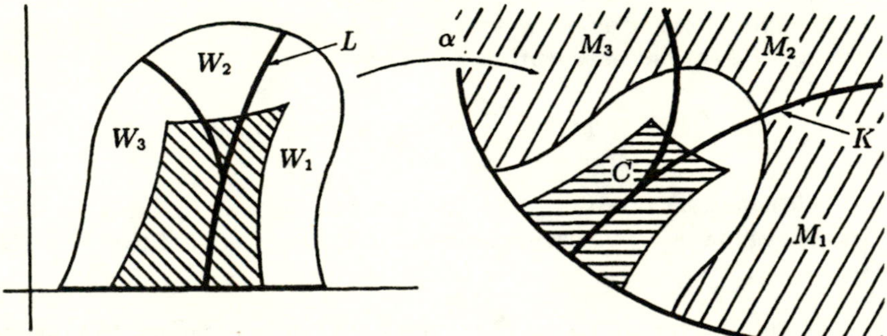

Figure 25.3

Note that these integrals over W_i exist as ordinary integrals. For the function $F = (f \circ \alpha)V(D\alpha)$ is bounded, and F vanishes near each point of

Bd W_i not in L. Then we note that

$$\sum_i [\int_{W_i} F] = \int_{(\text{Int } U)-L} F \qquad \text{by additivity,}$$

$$= \int_{\text{Int } U} F \qquad \text{since } L \text{ has measure zero,}$$

$$= \int_M f \, dV \qquad \text{by definition.}$$

Step 2. We complete the proof by showing that

$$\int_{W_i} F = \int_{A_i} F_i,$$

where $F_i = (f \circ \alpha_i)V(D\alpha_i)$. See Figure 25.4.

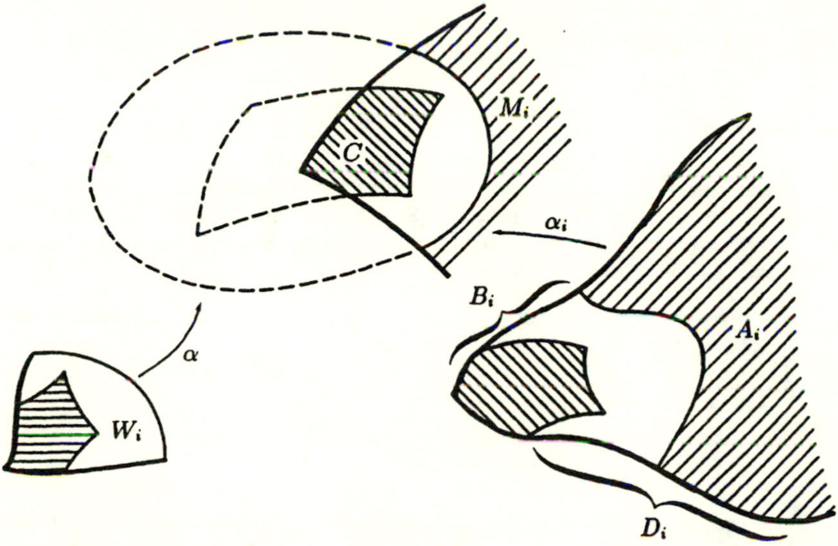

Figure 25.4

The map $\alpha_i^{-1} \circ \alpha$ is a diffeomorphism carrying W_i onto the open set

$$B_i = \alpha_i^{-1}(M_i \cap V)$$

of $\mathbf{R}^k$. It follows from the change of variables theorem that

$$\int_{W_i} F = \int_{B_i} F_i,$$

just as in Theorem 22.1. To complete the proof, we show that

$$\int_{B_i} F_i = \int_{A_i} F_i.$$

These integrals may not be ordinary integrals, so some care is required.

Since $C = \text{Support } f$ is closed in M, the set $\alpha_i^{-1}(C)$ is closed in A_i and its complement

$$D_i = A_i - \alpha_i^{-1}(C)$$

is open in A_i and thus in $\mathbf{R}^k$. The function F_i vanishes on D_i. We apply additivity of the extended integral to conclude that

$$\int_{A_i} F_i = \int_{B_i} F_i + \int_{D_i} F_i - \int_{B_i \cap D_i} F_i.$$

The last two integrals vanish. □

EXAMPLE 1. Consider the 2-sphere $S^2(a)$ of radius a in $\mathbf{R}^3$. We computed the area of its open upper hemisphere as $2\pi a^2$. (See Example 4 of §22.). Since the reflection map $(x, y, z) \to (x, y, -z)$ is an isometry of $\mathbf{R}^3$, the open lower hemisphere also has area $2\pi a^2$. (See the exercises of §22.) Since the upper and lower hemispheres constitute all of the sphere except for a set of measure zero in the sphere, it follows that $S^2(a)$ has area $4\pi a^2$.

EXAMPLE 2. Here is an alternate method for computing the area of the 2-sphere; it involves no improper integrals.

Given $z_0 \in \mathbf{R}$ with $|z_0| < a$, the intersection of $S^2(a)$ with the plane $z = z_0$ is the circle

$$z = z_0; \quad x^2 + y^2 = a^2 - (z_0)^2.$$

This fact suggests that we parametrize $S^2(a)$ by the function $\alpha : A \to \mathbf{R}^3$ given by the equation

$$\alpha(t, z) = \big((a^2 - z^2)^{1/2} \cos t,\ (a^2 - z^2)^{1/2} \sin t,\ z\big),$$

where A is the set of all (t, z) for which $0 < t < 2\pi$ and $|z| < a$. It is easy to check that α is a coordinate patch that covers all of $S^2(a)$ except for a great-circle arc, which has measure zero in the sphere. See Figure 25.5. By the preceding theorem, we may use this coordinate patch to compute the area of $S^2(a)$. We have

$$D\alpha = \begin{bmatrix} -(a^2 - z^2)^{1/2} \sin t & (-z \cos t)/(a^2 - z^2)^{1/2} \\ (a^2 - z^2)^{1/2} \cos t & (-z \sin t)/(a^2 - z^2)^{1/2} \\ 0 & 1 \end{bmatrix},$$

whence $V(D\alpha) = a$, as you can check. Then $v(S^2(a)) = \int_A a = 4\pi a^2$.

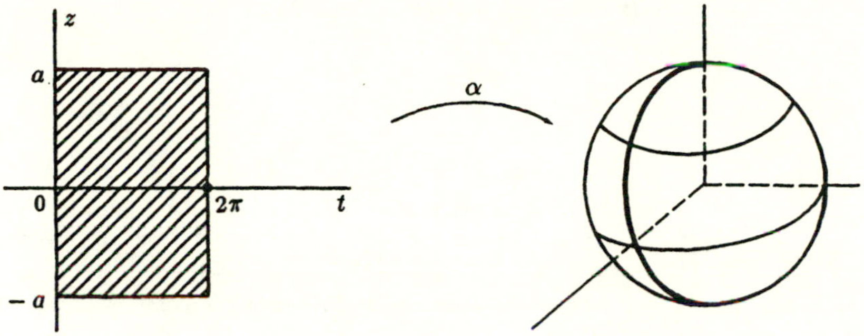

Figure 25.5

EXERCISES

1. Check the computations made in Example 2.
2. Let $\alpha(t), \beta(t), f(t)$ be real-valued functions of class C^1 on $[0,1]$, with $f(t) > 0$. Suppose M is a 2-manifold in $\mathbf{R}^3$ whose intersection with the plane $z = t$ is the circle

$$(x - \alpha(t))^2 + (y - \beta(t))^2 = (f(t))^2; \quad z = t$$

if $0 \leq t \leq 1$, and is empty otherwise.

(a) Set up an integral for the area of M. [*Hint:* Proceed as in Example 2.]

(b) Evaluate when α and β are constant and $f(t) = (1 + t)^{1/2}$.

(c) What form does the integral take when f is constant and $\alpha(t) = 0$ and $\beta(t) = at$? (This integral cannot be evaluated in terms of the elementary functions.)

3. Consider the torus T of Exercise 7 of §17.

(a) Find the area of this torus. [*Hint:* The cylindrical coordinate transformation carries a cylinder onto T. Parametrize the cylinder using the fact that its cross-section are circles.]

(b) Find the area of that portion of T satisfying the condition $x^2 + y^2 \geq b^2$.

4. Let M be a compact k-manifold in $\mathbf{R}^n$. Let $h : \mathbf{R}^n \to \mathbf{R}^n$ be an isometry; let $N = h(M)$. Let $f : N \to \mathbf{R}$ be a continuous function. Show that N is a k-manifold in $\mathbf{R}^n$, and

$$\int_N f \, dV = \int_M (f \circ h) \, dV.$$

Conclude that M and N have the same volume.

5. (a) Express the volume of $S^n(a)$ in terms of the volume of $B^{n-1}(a)$. [*Hint:* Follow the pattern of Example 2.]

(b) Show that for $t > 0$,

$$v(S^n(t)) = D\,v(B^{n+1}(t)).$$

[*Hint:* Use the result of Exercise 6 of §19.]

6. The centroid of a compact manifold M in $\mathbf{R}^n$ is defined by a formula like that given in Exercise 3 of §22. Show that if M is symmetric with respect to the subspace $x_i = 0$ of $\mathbf{R}^n$, then $c_i(M) = 0$.

*7. Let $E^n_+(a)$ denote the intersection of $S^n(a)$ with upper half-space $\mathbf{H}^{n+1}$. Let $\lambda_n = v(B^n(1))$.

(a) Find the centroid of $E^n_+(a)$ in terms of λ_n and λ_{n-1}.

(b) Find the centroid of $E^n_+(a)$ in terms of the centroid of $B^{n-1}_+(a)$. (See the exercises of §19.)

8. Let M and N be compact manifolds without boundary in $\mathbf{R}^m$ and $\mathbf{R}^n$, respectively.

(a) Let $f : M \to \mathbf{R}$ and $g : N \to \mathbf{R}$ be continuous. Show that

$$\int_{M \times N} f \cdot g \, dV = [\int_M f \, dV]\,[\int_N g \, dV].$$

[*Hint:* Consider the case where the supports of f and g are contained in coordinate patches.]

(b) Show that $v(M \times N) = v(M) \cdot v(N)$.

(c) Find the area of the 2-manifold $S^1 \times S^1$ in $\mathbf{R}^4$.

6

Differential Forms

We have treated, with considerable generality, two of the major topics of multivariable calculus—differentiation and integration. We now turn to the third topic. It is commonly called "vector integral calculus," and its major theorems bear the names of Green, Gauss, and Stokes. In calculus, one limits oneself to curves and surfaces in $\mathbf{R}^3$. We shall deal more generally with k-manifolds in $\mathbf{R}^n$. In dealing with this general situation, one finds that the concepts of linear algebra and vector calculus are no longer adequate. One needs to introduce concepts that are more sophisticated; they constitute a subject called multilinear algebra that is a sequel to linear algebra.

In the first three sections of this chapter, we introduce this subject; in these sections we use only the material on linear algebra treated in Chapter 1. In the remainder of the chapter, we combine the notions of multilinear algebra with results about differentiation from Chapter 2 to define and study differential forms in $\mathbf{R}^n$. Differential forms and their operators are what are used to replace vector and scalar fields and their operators—grad, curl, and div—when one passes from $\mathbf{R}^3$ to $\mathbf{R}^n$.

In the succeeding chapter, additional topics, including integration, manifolds, and the change of variables theorem, will be brought into the picture, in order to treat the generalized version of Stokes' theorem in $\mathbf{R}^n$.

§26. MULTILINEAR ALGEBRA

Tensors

Definition. Let V be a vector space. Let $V^k = V \times \cdots \times V$ denote the set of all k-tuples $(\mathbf{v}_1, \ldots, \mathbf{v}_k)$ of vectors of V. A function $f : V^k \to \mathbf{R}$ is said to be **linear in the ith variable** if, given fixed vectors $\mathbf{v}_j$ for $j \neq i$, the function $T : V \to \mathbf{R}$ defined by

$$T(\mathbf{v}) = f(\mathbf{v}_1, \ldots, \mathbf{v}_{i-1}, \mathbf{v}, \mathbf{v}_{i+1}, \ldots, \mathbf{v}_k)$$

is linear. The function f is said to be **multilinear** if it is linear in the ith variable for each i. Such a function f is also called a **k-tensor,** or a **tensor of order k,** on V. We denote the set of all k-tensors on V by the symbol $\mathcal{L}^k(V)$. If $k = 1$, then $\mathcal{L}^1(V)$ is just the set of all linear transformations $f : V \to \mathbf{R}$. It is sometimes called the **dual space** of V and denoted by V^*.

How this notion of tensor relates to the tensors used by physicists and geometers remains to be seen.

Theorem 26.1. *The set of all k-tensors on V constitutes a vector space if we define*

$$(f + g)(\mathbf{v}_1, \ldots, \mathbf{v}_k) = f(\mathbf{v}_1, \ldots, \mathbf{v}_k) + g(\mathbf{v}_1, \ldots, \mathbf{v}_k),$$
$$(cf)(\mathbf{v}_1, \ldots, \mathbf{v}_k) = c(f(\mathbf{v}_1, \ldots, \mathbf{v}_k)).$$

Proof. The proof is left as an exercise. The zero tensor is the function whose value is zero on every k-tuple of vectors. □

Just as is the case with linear transformations, a multilinear transformation is entirely determined once one knows its values on basis elements. That we now prove.

Lemma 26.2. *Let $\mathbf{a}_1, \ldots, \mathbf{a}_n$ be a basis for V. If $f, g : V^k \to \mathbf{R}$ are k-tensors on V, and if*

$$f(\mathbf{a}_{i_1}, \ldots, \mathbf{a}_{i_k}) = g(\mathbf{a}_{i_1}, \ldots, \mathbf{a}_{i_k})$$

for every k-tuple $I = (i_1, \ldots, i_k)$ of integers from the set $\{1, \ldots, n\}$, then $f = g$.

Note that there is no requirement here that the integers $i_1, \ldots, i_k$ be distinct or arranged in any particular order.

Proof. Given an arbitrary k-tuple $(\mathbf{v}_1, \ldots, \mathbf{v}_k)$ of vectors of V, let us express each $\mathbf{v}_i$ in terms of the given basis, writing

$$\mathbf{v}_i = \sum_{j=1}^{n} c_{ij}\mathbf{a}_j.$$

Then we compute

$$\begin{aligned} f(\mathbf{v}_1, \ldots, \mathbf{v}_k) &= \sum_{j_1=1}^{n} c_{1j_1} f(\mathbf{a}_{j_1}, \mathbf{v}_2, \ldots, \mathbf{v}_k) \\ &= \sum_{j_1=1}^{n} \sum_{j_2=1}^{n} c_{1j_1} c_{2j_2} f(\mathbf{a}_{j_1}, \mathbf{a}_{j_2}, \mathbf{v}_3, \ldots, \mathbf{v}_k), \end{aligned}$$

and so on. Eventually we obtain the equation

$$f(\mathbf{v}_1, \ldots, \mathbf{v}_k) = \sum_{1 \leq j_1, \ldots, j_k \leq n} c_{1j_1} c_{2j_2} \cdots c_{kj_k} f(\mathbf{a}_{j_1}, \ldots, \mathbf{a}_{j_k}).$$

The same computation holds for g. It follows that f and g agree on all k-tuples of vectors if they agree on all k-tuples of basis elements. □

Just as a linear transformation from V to W can be defined by specifying its values arbitrarily on basis elements for V, a k-tensor on V can be defined by specifying its values arbitrarily on k-tuples of basis elements. That fact is a consequence of the next theorem.

Theorem 26.3. *Let V be a vector space with basis $\mathbf{a}_1, \ldots, \mathbf{a}_n$. Let $I = (i_1, \ldots, i_k)$ be a k-tuple of integers from the set $\{1, \ldots, n\}$. There is a unique k-tensor ϕ_I on V such that, for every k-tuple $J = (j_1, \ldots, j_k)$ from the set $\{1, \ldots, n\}$,*

$$(*) \qquad \phi_I(\mathbf{a}_{j_1}, \cdots, \mathbf{a}_{j_k}) = \begin{cases} 0 & \text{if } I \neq J, \\ 1 & \text{if } I = J. \end{cases}$$

The tensors ϕ_I form a basis for $\mathcal{L}^k(V)$.

The tensors ϕ_I are called the **elementary k-tensors** on V corresponding to the basis $\mathbf{a}_1, \ldots, \mathbf{a}_n$ for V. Since they form a basis for $\mathcal{L}^k(V)$ and since there are n^k distinct k-tuples from the set $\{1, \ldots, n\}$, the space $\mathcal{L}^k(V)$ must have dimension n^k. When $k = 1$, the basis for V^* formed by the elementary tensors $\phi_1, \ldots, \phi_n$ is called the basis for V^* dual to the given basis for V.

Proof. Uniqueness follows from the preceding lemma. We prove existence as follows: First, consider the case $k = 1$. We know that we can determine a linear transformation $\phi_i : V \to \mathbf{R}$ by specifying its values arbitrarily on basis elements. So we can define ϕ_i by the equation

$$\phi_i(\mathbf{a}_j) = \begin{cases} 0 & \text{if} \quad i \neq j, \\ 1 & \text{if} \quad i = j. \end{cases}$$

These then are the desired 1-tensors. In the case $k > 1$, we define ϕ_I by the equation

$$\phi_I(\mathbf{v}_1, \ldots, \mathbf{v}_k) = [\phi_{i_1}(\mathbf{v}_1)] \cdot [\phi_{i_2}(\mathbf{v}_2)] \cdots [\phi_{i_k}(\mathbf{v}_k)].$$

It follows, from the facts that (1) each ϕ_i is linear and (2) multiplication is distributive, that ϕ_I is multilinear. One checks readily that it has the required value on $(\mathbf{a}_{j_1}, \cdots, \mathbf{a}_{j_k})$.

We show that the tensors ϕ_I form a basis for $\mathcal{L}^k(V)$. Given a k-tensor f on V, we show that it can be written uniquely as a linear combination of the tensors ϕ_I. For each k-tuple $I = (i_1, \ldots, i_k)$, let d_I be the scalar defined by the equation

$$d_I = f(\mathbf{a}_{i_1}, \ldots, \mathbf{a}_{i_k}).$$

Then consider the k-tensor

$$g = \sum_J d_J \phi_J,$$

where the summation extends over all k-tuples J of integers from the set $\{1, \ldots, n\}$. The value of g on the k-tuple $(\mathbf{a}_{i_1}, \ldots, \mathbf{a}_{i_k})$ equals d_I, by $(*)$, and the value of f on this k-tuple equals the same thing by definition. Then the preceding lemma implies that $f = g$. Uniqueness of this representation of f follows from the preceding lemma. $\square$

It follows from this theorem that given scalars d_I for all I, there is exactly one k-tensor f such that $f(\mathbf{a}_{i_1}, \ldots, \mathbf{a}_{i_k}) = d_I$ for all I. Thus a k-tensor may be defined by specifying its values arbitrarily on k-tuples of basis elements.

EXAMPLE 1. Consider the case $V = \mathbf{R}^n$. Let $\mathbf{e}_1, \dots, \mathbf{e}_n$ be the usual basis for $\mathbf{R}^n$; let $\phi_1, \dots, \phi_n$ be the dual basis for $\mathcal{L}^1(V)$. Then if $\mathbf{x}$ has components $x_1, \dots, x_n$, we have

$$\phi_i(\mathbf{x}) = \phi_i(x_1\mathbf{e}_1 + \cdots + x_n\mathbf{e}_n) = x_i.$$

Thus $\phi_i : \mathbf{R}^n \to \mathbf{R}$ equals projection onto the ith coordinate.

More generally, given $I = (i_1, \dots, i_k)$, the elementary tensor ϕ_I satisfies the equation

$$\phi_I(\mathbf{x}_1, \dots, \mathbf{x}_k) = \phi_{i_1}(\mathbf{x}_1) \cdots \phi_{i_k}(\mathbf{x}_k).$$

Let us write $X = [\mathbf{x}_1 \cdots \mathbf{x}_k]$, and let x_{ij} denote the entry of X in row i and column j. Then $\mathbf{x}_j$ is the vector having components $x_{1j}, \dots, x_{nj}$. In this notation,

$$\phi_I(\mathbf{x}_1, \dots, \mathbf{x}_k) = x_{i_1 1}\, x_{i_2 2} \cdots x_{i_k k}.$$

Thus ϕ_I is just a *monomial* in the components of the vectors $\mathbf{x}_1, \dots, \mathbf{x}_k$; and the general k-tensor on $\mathbf{R}^n$ is a linear combination of such monomials.

It follows that the general 1-tensor on $\mathbf{R}^n$ is a function of the form

$$f(\mathbf{x}) = d_1 x_1 + \cdots + d_n x_n,$$

for some scalars d_i. The general 2-tensor on $\mathbf{R}^n$ has the form

$$g(\mathbf{x}, \mathbf{y}) = \sum_{i,j=1}^{n} d_{ij} x_i y_j,$$

for some scalars d_{ij}. And so on.

The tensor product

Now we introduce a product operation into the set of all tensors on V. The product of a k-tensor and an ℓ-tensor will be a $k + \ell$ tensor.

Definition. Let f be a k-tensor on V and let g be an ℓ-tensor on V. We define a $k + \ell$ tensor $f \otimes g$ on V by the equation

$$(f \otimes g)(\mathbf{v}_1, \dots, \mathbf{v}_{k+\ell}) = f(\mathbf{v}_1, \dots, \mathbf{v}_k) \cdot g(\mathbf{v}_{k+1}, \dots, \mathbf{v}_{k+\ell}).$$

It is easy to check that the function $f \otimes g$ is multilinear; it is called the **tensor product** of f and g.

We list some of the properties of this product operation:

Theorem 26.4. *Let f, g, h be tensors on V. Then the following properties hold:*

(1) *(Associativity).* $f \otimes (g \otimes h) = (f \otimes g) \otimes h$.

(2) *(Homogeneity).* $(cf) \otimes g = c(f \otimes g) = f \otimes (cg)$.

(3) *(Distributivity). Suppose f and g have the same order. Then:*

$$(f + g) \otimes h = f \otimes h + g \otimes h,$$
$$h \otimes (f + g) = h \otimes f + h \otimes g.$$

(4) *Given a basis $\mathbf{a}_1, \ldots, \mathbf{a}_n$ for V, the corresponding elementary tensors ϕ_I satisfy the equation*

$$\phi_I = \phi_{i_1} \otimes \phi_{i_2} \otimes \cdots \otimes \phi_{i_k},$$

where $I = (i_1, \ldots, i_k)$.

Note that no parentheses are needed in the expression for ϕ_I given in (4), since $\otimes$ is associative. Note also that nothing is said here about commutativity. The reason is obvious; it almost never holds.

Proof. The proofs are straightforward. Associativity is proved, for instance, by noting that (if f, g, h have orders k, ℓ, m, respectively)

$$\begin{aligned}&(f \otimes (g \otimes h))\,(\mathbf{v}_1, \ldots, \mathbf{v}_{k+\ell+m}) \\ &\qquad = f(\mathbf{v}_1, \ldots, \mathbf{v}_k) \cdot g(\mathbf{v}_{k+1}, \ldots, \mathbf{v}_{k+\ell}) \cdot h(\mathbf{v}_{k+\ell+1}, \ldots, \mathbf{v}_{k+\ell+m}).\end{aligned}$$

The value of $(f \otimes g) \otimes h$ on the given tuple is the same. □

The action of a linear transformation

Finally, we examine how tensors behave with respect to linear transformation of the underlying vector spaces.

Definition. Let $T : V \to W$ be a linear transformation. We define the **dual transformation**

$$T^* : \mathcal{L}^k(W) \to \mathcal{L}^k(V),$$

(which goes in the opposite direction) as follows: If f is in $\mathcal{L}^k(W)$, and if $\mathbf{v}_1, \ldots, \mathbf{v}_k$ are vectors in V, then

$$(T^* f)(\mathbf{v}_1, \ldots, \mathbf{v}_k) = f(T(\mathbf{v}_1), \ldots, T(\mathbf{v}_k)).$$

The transformation T^* is the composite of the transformation $T \times \cdots \times T$ and the transformation f, as indicated in the following diagram:

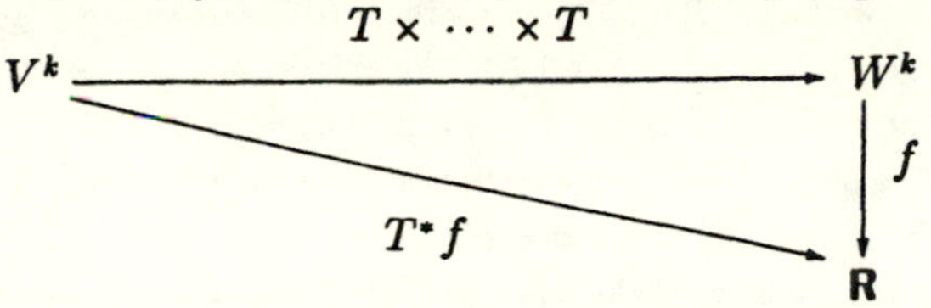

It is immediate from the definition that $T^* f$ is multilinear, since T is linear and f is multilinear. It is also true that T^* *itself* is linear, as a map of tensors, as we now show.

Theorem 26.5. *Let $T : V \to W$ be a linear transformation; let*

$$T^* : \mathcal{L}^k(W) \to \mathcal{L}^k(V)$$

be the dual transformation. Then:

(1) *T^* is linear.*

(2) *$T^*(f \otimes g) = T^* f \otimes T^* g$.*

(3) *If $S : W \to X$ is a linear transformation, then $(S \circ T)^* f = T^*(S^* f)$.*

Proof. The proofs are straightforward. One verifies (1), for instance, as follows:

$$\begin{aligned}(T^*(af + bg))\,(\mathbf{v}_1, \ldots, \mathbf{v}_k) &= (af + bg)\,(T(\mathbf{v}_1), \ldots, T(\mathbf{v}_k)) \\ &= a\,f(T(\mathbf{v}_1), \ldots, T(\mathbf{v}_k)) + bg(T(\mathbf{v}_1), \ldots, T(\mathbf{v}_k)) \\ &= aT^* f(\mathbf{v}_1, \ldots, \mathbf{v}_k) + bT^* g(\mathbf{v}_1, \ldots, \mathbf{v}_k),\end{aligned}$$

whence $T^*(af + bg) = aT^* f + bT^* g$. $\square$

The following diagrams illustrate property (3):

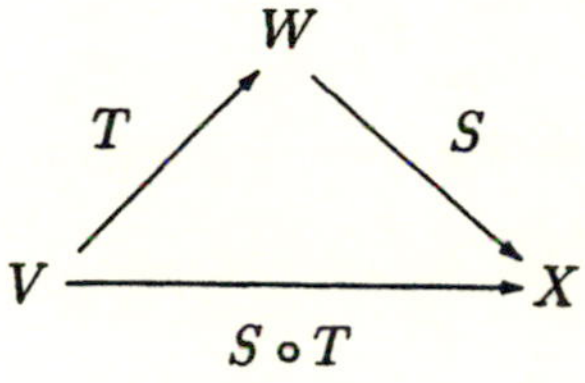

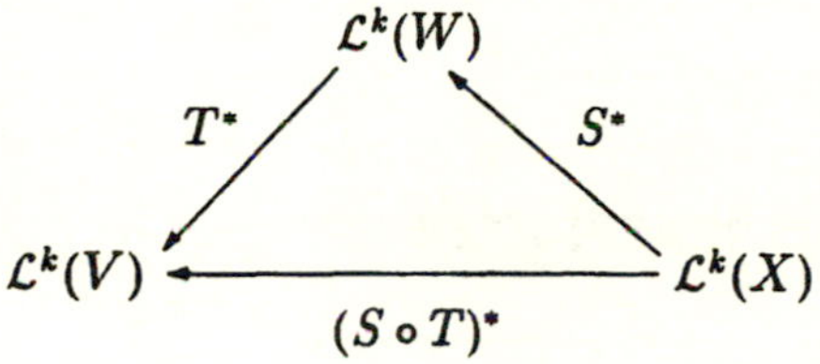

EXERCISES

1. (a) Show that if $f, g : V^k \to \mathbf{R}$ are multilinear, so is $af + bg$.
   (b) Check that $\mathcal{L}^k(V)$ satisfies the axioms of a vector space.
2. (a) Show that if f and g are multilinear, so is $f \otimes g$.
   (b) Check the basic properties of the tensor product (Theorem 26.4).
3. Verify (2) and (3) of Theorem 26.5.
4. Determine which of the following are tensors on $\mathbf{R}^4$, and express those that are in terms of the elementary tensors on $\mathbf{R}^4$:
$$f(\mathbf{x}, \mathbf{y}) = 3x_1y_2 + 5x_2x_3,$$
$$g(\mathbf{x}, \mathbf{y}) = x_1y_2 + x_2y_4 + 1,$$
$$h(\mathbf{x}, \mathbf{y}) = x_1y_1 - 7x_2y_3.$$
5. Repeat Exercise 4 for the functions
$$f(\mathbf{x}, \mathbf{y}, \mathbf{z}) = 3x_1x_2z_3 - x_3y_1z_4,$$
$$g(\mathbf{x}, \mathbf{y}, \mathbf{z}, \mathbf{u}, \mathbf{v}) = 5x_3y_2z_3u_4v_4,$$
$$h(\mathbf{x}, \mathbf{y}, \mathbf{z}) = x_1y_2z_4 + 2x_1z_3.$$
6. Let f and g be the following tensors on $\mathbf{R}^4$:
$$f(\mathbf{x}, \mathbf{y}, \mathbf{z}) = 2x_1y_2z_2 - x_2y_3z_1,$$
$$g = \phi_{2,1} - 5\phi_{3,1}.$$
   (a) Express $f \otimes g$ as a linear combination of elementary 5-tensors.
   (b) Express $(f \otimes g)(\mathbf{x}, \mathbf{y}, \mathbf{z}, \mathbf{u}, \mathbf{v})$ as a function.
7. Show that the four properties stated in Theorem 26.4 characterize the tensor product uniquely, for finite-dimensional spaces V.
8. Let f be a 1-tensor on $\mathbf{R}^n$; then $f(\mathbf{y}) = A \cdot \mathbf{y}$ for some matrix A of size 1 by n. If $T : \mathbf{R}^m \to \mathbf{R}^n$ is the linear transformation $T(\mathbf{x}) = B \cdot \mathbf{x}$, what is the matrix of the 1-tensor T^*f on $\mathbf{R}^m$?

§27. ALTERNATING TENSORS

In this section we introduce the particular kind of tensors with which we shall be concerned—the alternating tensors—and derive some of their properties. In order to do this, we need some basic facts about permutations.

Permutations

Definition. Let $k \geq 2$. A **permutation** of the set of integers $\{1, \ldots, k\}$ is a one-to-one function σ mapping this set onto itself. We denote the set of all such permutations by S_k. If σ and τ are elements of S_k, so are $\sigma \circ \tau$ and σ^{-1}. The set S_k thus forms a group, called the **symmetric group** (or the **permutation group**) on the set $\{1, \ldots, k\}$. There are $k!$ elements in this group.

Definition. Given $1 \leq i < k$, let e_i be the element of S_k defined by setting $e_i(j) = j$ for $j \neq i, i+1$; and

$$e_i(i) = i + 1 \quad \text{and} \quad e_i(i+1) = i.$$

We call e_i an **elementary permutation.** Note that $e_i \circ e_i$ equals the identity permutation, so that e_i is its own inverse.

Lemma 27.1. *If $\sigma \in S_k$, then σ equals a composite of elementary permutations.*

Proof. Given $0 \leq i \leq k$, we say that σ *fixes* the first i integers if $\sigma(j) = j$ for $1 \leq j \leq i$. If $i = 0$, then σ need not fix any integers at all. If $i = k$, then σ fixes *all* the integers $1, \ldots, k$, so that σ is the identity permutation. In this case the theorem holds, since the identity permutation equals $e_j \circ e_j$ for any j.

We show that if σ fixes the first $i-1$ integers (where $0 < i \leq k$), then σ can be written as the composite $\sigma = \pi \circ \sigma'$, where π is a composite of elementary permutations and σ' fixes the first i integers. The theorem then follows by induction.

The proof is easy. Since σ fixes the integers $1, \ldots, i-1$, and since σ is one-to-one, the value of σ on i must be a number different from $1, \ldots, i-1$. If $\sigma(i) = i$, then we set $\sigma' = \sigma$ and π equal to the identity permutation, and our result holds. If $\sigma(i) = \ell > i$, we set

$$\sigma' = e_i \circ \cdots \circ e_{\ell-1} \circ \sigma.$$

Then σ' fixes the integers $1, \ldots, i-1$ because σ fixes these integers and so do $e_i, \ldots, e_{\ell-1}$. And σ' also fixes i, since $\sigma(i) = \ell$ and

$$e_i(\cdots(e_{\ell-1}(\ell))\cdots) = i.$$

We can rewrite the equation defining σ' in the form

$$e_{\ell-1} \circ \cdots \circ e_i \circ \sigma' = \sigma;$$

thus our result holds. □

Definition. Let $\sigma \in S_k$. Consider the set of all pairs of integers i, j from the set $\{1, \ldots, k\}$ for which $i < j$ and $\sigma(i) > \sigma(j)$. Each such pair is called an **inversion in** σ. We define the **sign of** σ to be the number -1 if the number of inversions in σ is odd, and to be the number $+1$ if the number of inversions in σ is even. We call σ an **odd** or an **even permutation** according as the sign of σ equals -1 or $+1$, respectively. Denote the sign of σ by $\operatorname{sgn} \sigma$.

Lemma 27.2. *Let $\sigma, \tau \in S_k$.*

(a) *If σ equals a composite of m elementary permutations, then* $\operatorname{sgn} \sigma = (-1)^m$.

(b) $\operatorname{sgn}(\sigma \circ \tau) = (\operatorname{sgn} \sigma) \cdot (\operatorname{sgn} \tau)$.

(c) $\operatorname{sgn} \sigma^{-1} = \operatorname{sgn} \sigma$.

(d) *If $p \neq q$, and if τ is the permutation that exchanges p and q and leaves all other integers fixed, then* $\operatorname{sgn} \tau = -1$.

Proof. *Step 1.* We show that for any σ,

$$\operatorname{sgn}(\sigma \circ e_\ell) = -\operatorname{sgn} \sigma.$$

Given σ, let us write down the values of σ in order as follows:

$$(*) \qquad (\sigma(1), \sigma(2), \ldots, \sigma(\ell), \sigma(\ell+1), \ldots, \sigma(k)).$$

Let $\tau = \sigma \circ e_\ell$; then the corresponding sequence for τ is the k-tuple of numbers

$$(**) \qquad \begin{aligned} &(\tau(1), \tau(2), \ldots, \tau(\ell), \tau(\ell+1), \ldots, \tau(k)) \\ &= (\sigma(1), \sigma(2), \ldots, \sigma(\ell+1), \sigma(\ell), \ldots, \sigma(k)). \end{aligned}$$

The number of inversions in σ and τ, respectively, are the number of pairs of integers that appear in the sequences $(*)$ and $(**)$, respectively, in the reverse of their natural order. We compare inversions in these two sequences. Let $p \neq q$; we compare the positions of $\sigma(p)$ and $\sigma(q)$ in these two sequences. If neither p nor q equals ℓ or $\ell + 1$, then $\sigma(p)$ and $\sigma(q)$ appear in the same slots in both sequences, so they constitute an inversion in one sequence if and only if they constitute an inversion in the other. Now consider the case where one, say p, equals either ℓ or $\ell + 1$, and the other q is different from both ℓ and $\ell + 1$. Then $\sigma(q)$ appears in the same slot in both sequences, but $\sigma(p)$ appears in the two sequences in adjacent slots. Nevertheless, it is still true that $\sigma(p)$ and $\sigma(q)$ constitute an inversion in one sequence if and only if they constitute an inversion in the other.

So far the number of inversions in the two sequences are the same. But now we note that if $\sigma(\ell)$ and $\sigma(\ell + 1)$ form an inversion in the first sequence, they do not form an inversion in the second; and conversely. Hence sequence $(**)$ has either one more inversion, or one fewer inversion, than $(*)$.

Step 2. We prove the theorem. The identity permutation has sign +1; and composing it successively with m elementary permutation changes its sign m times, by Step 1. Thus (a) holds. To prove (b), we write σ as the composite of m elementary permutations, and τ as the composite of n elementary permutations. Then $\sigma \circ \tau$ is the composite of $m+n$ elementary permutations; and (b) follows from the equation $(-1)^{m+n} = (-1)^m(-1)^n$. To check (c), we note that since $\sigma^{-1} \circ \sigma$ equals the identity permutation, $(\text{sgn } \sigma^{-1})(\text{sgn } \sigma) = 1$.

To prove (d), one simply counts inversions. Suppose that $p < q$. We can write the values of τ in order as

$$(1, \ldots, p-1, \boxed{q}, p+1, \ldots, p+\ell-1, \boxed{p}, p+\ell+1, \ldots, k),$$

where $q = p+\ell$. Each of the pairs $\{q, p+1\}, \ldots, \{q, p+\ell-1\}$ constitutes an inversion in this sequence, and so does each of the pairs $\{p+1, p\}. \ldots, \{p+\ell-1, p\}$. Finally, $\{q, p\}$ is an inversion as well. Thus τ has $2\ell - 1$ inversions, so it is odd. $\square$

Alternating tensors

Definition. Let f be an arbitrary k-tensor on V. If σ is a permutation of $\{1, \ldots, k\}$, we define f^σ by the equation

$$f^\sigma(\mathbf{v}_1, \ldots, \mathbf{v}_k) = f(\mathbf{v}_{\sigma(1)}, \ldots, \mathbf{v}_{\sigma(k)}).$$

Because f is linear in each of its variables, so is f^σ; thus f^σ is a k-tensor on V. The tensor f is said to be **symmetric** if $f^e = f$ for each elementary permutation e, and it is said to be **alternating** if $f^e = -f$ for every elementary permutation e.

Said differently, f is symmetric if

$$f(\mathbf{v}_1, \ldots, \mathbf{v}_{i+1}, \mathbf{v}_i, \ldots, \mathbf{v}_k) = f(\mathbf{v}_1, \ldots, \mathbf{v}_i, \mathbf{v}_{i+1}, \ldots, \mathbf{v}_k)$$

for all i; and f is alternating if

$$f(\mathbf{v}_1, \ldots, \mathbf{v}_{i+1}, \mathbf{v}_i, \ldots, \mathbf{v}_k) = -f(\mathbf{v}_1, \ldots, \mathbf{v}_i, \mathbf{v}_{i+1}, \ldots, \mathbf{v}_k).$$

While symmetric tensors are important in mathematics, we shall not be concerned with them here. We shall be primarily interested in alternating tensors.

Definition. If V is a vector space, we denote the set of alternating k-tensors on V by $\mathcal{A}^k(V)$. It is easy to check that the sum of two alternating tensors is alternating, and that so is a scalar multiple of an alternating tensor. Then $\mathcal{A}^k(V)$ is a linear subspace of the space $\mathcal{L}^k(V)$ of all k-tensors on V. The condition that a 1-tensor be alternating is vacuous. Therefore we make the convention that $\mathcal{A}^1(V) = \mathcal{L}^1(V)$.

EXAMPLE 1. The elementary tensors of order $k > 1$ are not alternating, but certain linear combinations of them are alternating. For instance, the tensor

$$f = \phi_{i,j} - \phi_{j,i}$$

is alternating, as you can check. Indeed, if $V = \mathbf{R}^n$ and we use the usual basis for $\mathbf{R}^n$ and corresponding dual basis ϕ_i, the function f satisfies the equation

$$f(\mathbf{x}, \mathbf{y}) = x_i y_j - x_j y_i = \det \begin{bmatrix} x_i & y_i \\ x_j & y_j \end{bmatrix}.$$

Here it is obvious that $f(\mathbf{y}, \mathbf{x}) = -f(\mathbf{x}, \mathbf{y})$. Similarly, the function

$$g(\mathbf{x}, \mathbf{y}, \mathbf{z}) = \det \begin{bmatrix} x_i & y_i & z_i \\ x_j & y_j & z_j \\ x_k & y_k & z_k \end{bmatrix}$$

is an alternating 3-tensor on $\mathbf{R}^n$; one can also write g in the form

$$g = \phi_{i,j,k} + \phi_{j,k,i} + \phi_{k,i,j} - \phi_{j,i,k} - \phi_{i,k,j} - \phi_{k,j,i}.$$

This example suggests that alternating tensors and the determinant function are intimately related. This is in fact the case, as we shall see.

We now study the space $\mathcal{A}^k(V)$; in particular, we find a basis for it. Let us begin with a lemma:

Lemma 27.3. *Let f be a k-tensor on V; let $\sigma, \tau \in S_k$.*

(a) *The transformation $f \to f^\sigma$ is a linear transformation of $\mathcal{L}^k(V)$ to $\mathcal{L}^k(V)$. It has the property that for all σ, τ,*

$$(f^\sigma)^\tau = f^{\tau\circ\sigma}.$$

(b) *The tensor f is alternating if and only if $f^\sigma = (\operatorname{sgn} \sigma)f$ for all σ. If f is alternating and if $\mathbf{v}_p = \mathbf{v}_q$ with $p \neq q$, then $f(\mathbf{v}_1, \dots, \mathbf{v}_k) = 0$.*

Proof. (a) The linearity property is straightforward; it states simply that $(af + bg)^\sigma = af^\sigma + bg^\sigma$. To complete the proof of (a), we compute

$$\begin{aligned}(f^\sigma)^\tau(\mathbf{v}_1, \dots, \mathbf{v}_k) &= f^\sigma(\mathbf{v}_{\tau(1)}, \dots, \mathbf{v}_{\tau(k)}) \\ &= f^\sigma(\mathbf{w}_1, \dots, \mathbf{w}_k), \quad \text{where} \quad \mathbf{w}_i = \mathbf{v}_{\tau(i)}, \\ &= f(\mathbf{w}_{\sigma(1)}, \dots, \mathbf{w}_{\sigma(k)}) \\ &= f(\mathbf{v}_{\tau(\sigma(1))}, \dots, \mathbf{v}_{\tau(\sigma(k))}) \\ &= f^{\tau\circ\sigma}(\mathbf{v}_1, \dots, \mathbf{v}_k).\end{aligned}$$

(b) Given an arbitrary permutation σ, let us write it as the composite

$$\sigma = \sigma_1 \circ \sigma_2 \circ \cdots \circ \sigma_m,$$

where each σ_i is an elementary permutation. Then

$$\begin{aligned} f^\sigma &= f^{\sigma_1 \circ \cdots \circ \sigma_m} \\ &= ((\cdots(f^{\sigma_m})\cdots)^{\sigma_2})^{\sigma_1} \quad \text{by (a)}, \\ &= (-1)^m f \quad \text{because } f \text{ is alternating}, \\ &= (\operatorname{sgn} \sigma) f. \end{aligned}$$

Now suppose $\mathbf{v}_p = \mathbf{v}_q$ for $p \neq q$. Let τ be the permutation that exchanges p and q. Since $\mathbf{v}_p = \mathbf{v}_q$,

$$f^\tau(\mathbf{v}_1, \ldots, \mathbf{v}_k) = f(\mathbf{v}_1, \ldots, \mathbf{v}_k).$$

On the other hand,

$$f^\tau(\mathbf{v}_1, \ldots, \mathbf{v}_k) = -f(\mathbf{v}_1, \ldots, \mathbf{v}_k)$$

since $\operatorname{sgn} \tau = -1$. It follows that $f(\mathbf{v}_1, \ldots, \mathbf{v}_k) = 0$. □

We now obtain a basis for the space $\mathcal{A}^k(V)$. There is nothing to be done in the case $k = 1$, since $\mathcal{A}^1(V) = \mathcal{L}^1(V)$. And in the case where $k > n$, the space $\mathcal{A}^k(V)$ is trivial. For any k-tensor f is uniquely determined by its values on k-tuples of basis elements. If $k > n$, some basis element must appear in the k-tuple more than once, whence if f is alternating, the value of f on the k-tuple must be zero.

Finally, we consider the case $1 < k \leq n$. We show first that an alternating tensor f is entirely determined by its values on k-tuples of basis elements *whose indices are in ascending order.* Then we show that the value of f on such k-tuples may be specified arbitrarily.

Lemma 27.4. *Let* $\mathbf{a}_1, \ldots, \mathbf{a}_n$ *be a basis for* V. *If* f, g *are alternating* k*-tensors on* V, *and if*

$$f(\mathbf{a}_{i_1}, \ldots, \mathbf{a}_{i_k}) = g(\mathbf{a}_{i_1}, \ldots, \mathbf{a}_{i_k})$$

for every ascending k*-tuple of integers* $I = (i_1, \ldots, i_k)$ *from the set* $\{1, \ldots, n\}$, *then* $f = g$.

Proof. In view of Lemma 26.2, it suffices to prove that f and g have the same values on an arbitrary k-tuple $(\mathbf{a}_{j_1}, \ldots, \mathbf{a}_{j_k})$ of basis elements. Let $J = (j_1, \ldots, j_k)$.

If two of the indices, say, j_p and j_q, are the same, then the values of f and g on this tuple are zero, by the preceding lemma. If all the indices are distinct, let σ be the permutation of $\{1, \ldots, k\}$ such that the k-tuple $I = (j_{\sigma(1)}, \ldots, j_{\sigma(k)})$ is ascending. Then

$$\begin{aligned} f(\mathbf{a}_{i_1}, \ldots, \mathbf{a}_{i_k}) &= f^\sigma(\mathbf{a}_{j_1}, \ldots, \mathbf{a}_{j_k}) \quad \text{by definition of } f^\sigma, \\ &= (\operatorname{sgn} \sigma) f(\mathbf{a}_{j_1}, \ldots, \mathbf{a}_{j_k}) \quad \text{because } f \text{ is alternating.} \end{aligned}$$

A similar equation holds for g. Since f and g agree on the k-tuple $(\mathbf{a}_{i_1}, \ldots, \mathbf{a}_{i_k})$, they agree on the k-tuple $(\mathbf{a}_{j_1}, \ldots, \mathbf{a}_{j_k})$. □

Theorem 27.5. *Let V be a vector space with basis $\mathbf{a}_1, \ldots, \mathbf{a}_n$. Let $I = (i_1, \ldots, i_k)$ be an ascending k-tuple from the set $\{1, \ldots, n\}$. There is a unique alternating k-tensor ψ_I on V such that for every ascending k-tuple $J = (j_1, \ldots, j_k)$ from the set $\{1, \ldots, n\}$,*

$$\psi_I(\mathbf{a}_{j_1}, \ldots, \mathbf{a}_{j_k}) = \begin{cases} 0 & \text{if } I \neq J, \\ 1 & \text{if } I = J. \end{cases}$$

The tensors ψ_I form a basis for $\mathcal{A}^k(V)$. The tensor ψ_I in fact satisfies the formula

$$\psi_I = \sum_\sigma (\operatorname{sgn} \sigma)(\phi_I)^\sigma,$$

where the summation extends over all $\sigma \in S_k$.

The tensors ψ_I are called the **elementary alternating k-tensors** on V corresponding to the basis $\mathbf{a}_1, \ldots, \mathbf{a}_n$ for V.

Proof. Uniqueness follows from the preceding lemma. To prove existence, we define ψ_I by the formula given in the theorem, and show that ψ_I satisfies the requirements of the theorem.

First, we show ψ_I is alternating. If $\tau \in S_k$, we compute

$$\begin{aligned} (\psi_I)^\tau &= \sum_\sigma (\operatorname{sgn} \sigma) \left((\phi_I)^\sigma\right)^\tau \quad \text{by linearity,} \\ &= \sum_\sigma (\operatorname{sgn} \sigma) (\phi_I)^{\tau \circ \sigma} \\ &= (\operatorname{sgn} \tau) \sum_\sigma (\operatorname{sgn}(\tau \circ \sigma)) (\phi_I)^{\tau \circ \sigma} \\ &= (\operatorname{sgn} \tau) \psi_I; \end{aligned}$$

the last equation follows from the fact that $\tau \circ \sigma$ ranges over S_k as σ does.

We show ψ_I has the desired values. Given J, we have

$$\psi_I(\mathbf{a}_{j_1}, \ldots, \mathbf{a}_{j_k}) = \sum_{\sigma} (\operatorname{sgn} \sigma)\phi_I(\mathbf{a}_{j_{\sigma(1)}}, \ldots, \mathbf{a}_{j_{\sigma(k)}}).$$

Now at most one term of this summation can be non-zero, namely the term corresponding to the permutation σ for which $I = (j_{\sigma(1)}, \ldots, j_{\sigma(k)})$. Since both I and J are ascending, this occurs only if $I = J$ and σ is the identity permutation, in which case the value is 1. If $I \neq J$, then all terms vanish.

Now we show the ψ_I form a basis for $\mathcal{A}^k(V)$. Let f be an alternating k-tensor on V. We show that f can be written uniquely as a linear combination of the tensors ψ_I.

Given f, for each ascending k-tuple $I = (i_1, \ldots, i_k)$ from the set $\{1, \ldots, n\}$, let d_I be the scalar

$$d_I = f(\mathbf{a}_{i_1}, \ldots, \mathbf{a}_{i_k}).$$

Then consider the alternating k-tensor

$$g = \sum_{[J]} d_J \psi_J,$$

where the notation $[J]$ indicates that the summation extends over all ascending k-tuples from $\{1, \ldots, n\}$. If I is an ascending k-tuple, then the value of g on the k-tuple $(\mathbf{a}_{i_1}, \ldots, \mathbf{a}_{i_k})$ equals d_I; and the value of f on this k-tuple is the same. Hence $f = g$. Uniqueness of this representation of f follows from the preceding lemma. □

This theorem shows that once a basis $\mathbf{a}_1, \ldots, \mathbf{a}_n$ for V has been chosen, an arbitrary alternating k-tensor f can be written uniquely in the form

$$f = \sum_{[J]} d_J \psi_J.$$

The numbers d_J are called **components** of f relative to the basis $\{\psi_J\}$.

What is the dimension of the vector space $\mathcal{A}^k(V)$? If $k = 1$, then $\mathcal{A}^1(V)$ has dimension n, of course. In general, given $k > 1$ and given any subset of $\{1, \ldots, n\}$ having k elements, there is exactly one corresponding ascending k–tuple, and hence one corresponding elementary alternating k-tensor. Thus the number of basis elements for $\mathcal{A}^k(V)$ equals the number of combinations of n objects, taken k at a time. This number is the binomial coefficient

$$\binom{n}{k} = \frac{n!}{k!(n-k)!}.$$

The preceding theorem gives one formula for the elementary alternating tensor ψ_I. There is an alternative formula that expresses ψ_I directly in terms of the standard basis elements for the larger space $\mathcal{L}^k(V)$. It is given in Exercise 5.

Finally, we note that alternating tensors behave properly with respect to a linear transformation of their underlying vector spaces. The proof is left as an exercise.

Theorem 27.6. *Let $T : V \to W$ be a linear transformation. If f is an alternating tensor on W, then T^*f is an alternating tensor on V.* □

Determinants

We now (at long last!) construct the determinant function for matrices of size greater than 3 by 3.

Definition. Let $\mathbf{e}_1, \ldots, \mathbf{e}_n$ be the usual basis for $\mathbf{R}^n$; let $\phi_1, \ldots, \phi_n$ denote the dual basis for $\mathcal{L}^1(\mathbf{R}^n)$. The space $\mathcal{A}^n(\mathbf{R}^n)$ of alternating n-tensors on $\mathbf{R}^n$ has dimension 1; the unique elementary alternating n-tensor on $\mathbf{R}^n$ is the tensor $\psi_{1,\ldots,n}$. If $X = [\mathbf{x}_1 \cdots \mathbf{x}_n]$ is an n by n matrix, we define the **determinant** of X by the equation

$$\det X = \psi_{1,\ldots,n}(\mathbf{x}_1, \ldots, \mathbf{x}_n).$$

We show this function satisfies the axioms for the determinant function given in §2. For convenience, let us for the moment let g denote the function

$$g(X) = \psi_I(\mathbf{x}_1, \ldots, \mathbf{x}_n),$$

where $I = (1, \ldots, n)$. The function g is multilinear and alternating as a function of the columns of X, because ψ_I is an alternating tensor. Therefore the function f defined by the equation $f(A) = g(A^{\text{tr}})$ is multilinear and alternating as a function of the *rows* of the matrix A. Furthermore,

$$f(I_n) = g(I_n) = \psi_I(\mathbf{e}_1, \ldots, \mathbf{e}_n) = 1.$$

Hence the function f satisfies the axioms for the determinant function. In particular, it follows from Theorem 2.11 that $f(A) = f(A^{\text{tr}})$. Then $f(A) = f(A^{\text{tr}}) = g((A^{\text{tr}})^{\text{tr}}) = g(A)$, so that g also satisfies the axioms for the determinant function, as desired.

The formula for ψ_I given in Theorem 27.5 gives rise to a formula for the determinant function. If $I = (1, \ldots, n)$, we have

$$\det X = \sum_{\sigma} (\operatorname{sgn} \sigma)\phi_I(\mathbf{x}_{\sigma(1)}, \ldots, \mathbf{x}_{\sigma(n)})$$

$$= \sum_{\sigma} (\operatorname{sgn} \sigma) x_{1,\sigma(1)} \cdot x_{2,\sigma(2)} \cdots x_{n,\sigma(n)},$$

as you can check. This formula is sometimes used as the *definition* of the determinant function.

We can now obtain a formula for expressing ψ_I directly as a function of k-tuples of vectors of $\mathbf{R}^n$. It is the following:

Theorem 27.7. *Let ψ_I be an elementary alternating tensor on $\mathbf{R}^n$ corresponding to the usual basis for $\mathbf{R}^n$, where $I = (i_1, \ldots, i_k)$. Given vectors $\mathbf{x}_1, \ldots, \mathbf{x}_k$ of $\mathbf{R}^n$, let X be the matrix $X = [\mathbf{x}_1 \cdots \mathbf{x}_k]$. Then*

$$\psi_I(\mathbf{x}_1, \ldots, \mathbf{x}_k) = \det X_I,$$

where X_I denotes the matrix whose successive rows are rows $i_1, \ldots, i_k$ of X.

Proof. We compute

$$\psi_I(\mathbf{x}_1, \ldots, \mathbf{x}_k) = \sum_\sigma (\operatorname{sgn} \sigma)\phi_I(\mathbf{x}_{\sigma(1)}, \ldots, \mathbf{x}_{\sigma(k)})$$

$$= \sum_\sigma (\operatorname{sgn} \sigma) x_{i_1,\sigma(1)} \cdot x_{i_2,\sigma(2)} \cdots x_{i_k,\sigma(k)}.$$

This is just the formula for $\det X_I$. $\square$

EXAMPLE 2. Consider the space $\mathcal{A}^3(\mathbf{R}^4)$. The elementary alternating 3-tensors on $\mathbf{R}^4$, corresponding to the usual basis for $\mathbf{R}^4$, are the functions

$$\psi_{i,j,k}(\mathbf{x}, \mathbf{y}, \mathbf{z}) = \det \begin{bmatrix} x_i & y_i & z_i \\ x_j & y_j & z_j \\ x_k & y_k & z_k \end{bmatrix},$$

where (i, j, k) equals (1,2,3) or (1,2,4) or (1,3,4) or (2,3,4). The general element of $\mathcal{A}^3(\mathbf{R}^4)$ is a linear combination of these four functions.

A remark on notation. There is in the subject of multilinear algebra a standard construction called the exterior product operation. It assigns to any vector space W a certain quotient of the "k-fold tensor product" of W; this quotient is denoted $\Lambda^k(W)$ and is called the "k-fold exterior product" of W. (See [Gr], [N].) If V is a finite-dimensional vector space, then the exterior product operation, when applied to the dual space $V^* = \mathcal{L}^1(V)$, gives a space $\Lambda^k(V^*)$ that is isomorphic to the space of alternating k-tensors on V, in a natural way. For this reason, it is fairly common among mathematicians to abuse notation and denote the space of alternating k-tensors on V by $\Lambda^k(V^*)$. (See [B–G] and [G–P], for example.)

Unfortunately, others denote the space of alternating k-tensors on V by the symbol $\Lambda^k(V)$ rather than by $\Lambda^k(V^*)$. (See [A–M–R], [B], [D].) Other notations are also used. (See [F], [S].) Because of this notational confusion, we have settled on the neutral notation $\mathcal{A}^k(V)$ for use in this book.

EXERCISES

1. Which of the following are alternating tensors in $\mathbf{R}^4$?

$$f(\mathbf{x},\mathbf{y}) = x_1y_2 - x_2y_1 + x_1y_1.$$

$$g(\mathbf{x},\mathbf{y}) = x_1y_3 - x_3y_2.$$

$$h(\mathbf{x},\mathbf{y}) = (x_1)^3(y_2)^3 - (x_2)^3(y_1)^3.$$

2. Let $\sigma \in S_5$ be the permutation such that

$$\big(\sigma(1),\sigma(2),\sigma(3),\sigma(4),\sigma(5)\big) = (3,1,4,5,2).$$

Use the procedure given in the proof of Lemma 27.1 to write σ as a composite of elementary permutations.

3. Let ψ_I be an elementary alternating k-tensor on V corresponding to the basis
$\mathbf{a}_1, \ldots, \mathbf{a}_n$ for V. If $j_1, \ldots, j_k$ is an arbitrary k-tuple of integers from the set $\{1, \ldots, n\}$, what is the value of

$$\psi_I(\mathbf{a}_{j_1}, \ldots, \mathbf{a}_{j_k})?$$

4. Show that if $T : V \to W$ is a linear transformation and if $f \in \mathcal{A}^k(W)$, then $T^*f \in \mathcal{A}^k(V)$.

5. Show that

$$\psi_I = \sum_\sigma (\text{sgn }\sigma)\phi_{I_\sigma},$$

where if $I = (i_1, \ldots, i_k)$, we let $I_\sigma = (i_{\sigma(1)}, \ldots, i_{\sigma(k)})$. [*Hint:* Show first that $(\phi_{I_\sigma})^\sigma = \phi_I$.]

§28. THE WEDGE PRODUCT

Just as we did for general tensors, we seek to define a product operation in the set of alternating tensors. The product $f \otimes g$ is almost never alternating, even if f and g are alternating. So something else is needed. The actual definition of the product is not very important; what is important are the properties it satisfies. They are stated in the following theorem:

Theorem 28.1. *Let V be a vector space. There is an operation that assigns, to each $f \in \mathcal{A}^k(V)$ and each $g \in \mathcal{A}^\ell(V)$, an element $f \wedge g \in \mathcal{A}^{k+\ell}(V)$, such that the following properties hold:*

(1) *(Associativity).* $f \wedge (g \wedge h) = (f \wedge g) \wedge h$.

(2) *(Homogeneity).* $(cf) \wedge g = c(f \wedge g) = f \wedge (cg)$.

(3) *(Distributivity). If f and g have the same order,*

$$(f + g) \wedge h = f \wedge h + g \wedge h,$$

$$h \wedge (f + g) = h \wedge f + h \wedge g.$$

(4) *(Anticommutativity). If f and g have orders k and ℓ, respectively, then*

$$g \wedge f = (-1)^{k\ell} f \wedge g.$$

(5) *Given a basis $\mathbf{a}_1, \dots, \mathbf{a}_n$ for V, let ϕ_i denote the dual basis for V^*, and let ψ_I denote the corresponding elementary alternating tensors. If $I = (i_1, \dots, i_k)$ is an ascending k-tuple of integers from the set $\{1, \dots, n\}$, then*

$$\psi_I = \phi_{i_1} \wedge \phi_{i_2} \wedge \cdots \wedge \phi_{i_k}.$$

These five properties characterize the product $\wedge$ uniquely for finite-dimensional spaces V. Furthermore, it has the following additional property:

(6) *If $T : V \to W$ is a linear transformation, and if f and g are alternating tensors on W, then*

$$T^*(f \wedge g) = T^* f \wedge T^* g.$$

The tensor $f \wedge g$ is called the **wedge product** of f and g. Note that property (4) implies that for an alternating tensor f of odd order, $f \wedge f = 0$.

Proof. *Step 1.* Let F be a k-tensor on V (not necessarily alternating). For purposes of this proof, it is convenient to define a transformation $A : \mathcal{L}^k(V) \to \mathcal{L}^k(V)$ by the formula

$$AF = \sum_\sigma (\operatorname{sgn} \sigma) F^\sigma,$$

where the summation extends over all $\sigma \in S_k$. (Sometimes a factor of $1/k!$ is included in this formula, but that is not necessary for our purposes.) Note that in this notation, the definition of the elementary alternating tensors can be written as

$$\psi_I = A\phi_I.$$

The transformation A has the following properties:

(i) A is linear.

(ii) AF is an alternating tensor.

(iii) If F is already alternating, then $AF = (k!)F$.

Let us check these properties. The fact that A is linear comes from the fact that the map $F \to F^\sigma$ is linear. The fact that AF is alternating comes from the computation

$$\begin{aligned}
(AF)^\tau &= \sum_\sigma (\text{sgn } \sigma)(F^\sigma)^\tau \qquad \text{by linearity,} \\
&= \sum_\sigma (\text{sgn } \sigma)F^{\tau\circ\sigma} \\
&= (\text{sgn } \tau)\sum_\sigma (\text{sgn } \tau\circ\sigma)F^{\tau\circ\sigma} \\
&= (\text{sgn } \tau)AF.
\end{aligned}$$

(This is the same computation we made earlier in showing that ψ_I is alternating.) Finally, if F is already alternating, then $F^\sigma = (\text{sgn } \sigma)F$ for all σ. It follows that

$$AF = \sum_\sigma (\text{sgn } \sigma)^2 F = (k!)F.$$

Step 2. We now define the product $f \wedge g$. If f is an alternating k-tensor on V, and g is an alternating ℓ-tensor on V, we define

$$f \wedge g = \frac{1}{k!\ell!}A(f \otimes g).$$

Then $f \wedge g$ is an alternating tensor of order $k + \ell$.

It is not entirely clear why the coefficient $1/k!\ell!$ appears in this formula. Some such coefficient is in fact necessary if the wedge product is to be associative. One way of motivating the particular choice of the coefficient $1/k!\ell!$ is the following: Let us rewrite the definition of $f \wedge g$ in the form

$$(f \wedge g)(\mathbf{v}_1, \ldots, \mathbf{v}_{k+\ell}) =$$

$$\frac{1}{k!\ell!}\sum_\sigma (\text{sgn } \sigma) f(\mathbf{v}_{\sigma(1)}, \ldots, \mathbf{v}_{\sigma(k)}) \cdot g(\mathbf{v}_{\sigma(k+1)}, \ldots, \mathbf{v}_{\sigma(k+\ell)}).$$

Then let us consider a single term of the summation, say

$$(\operatorname{sgn}\sigma)f(\mathbf{v}_{\sigma(1)},\ldots,\mathbf{v}_{\sigma(k)})\cdot g(\mathbf{v}_{\sigma(k+1)},\ldots,\mathbf{v}_{\sigma(k+\ell)}).$$

A number of other terms of the summation can be obtained from this one by permuting the vectors $\mathbf{v}_{\sigma(1)},\ldots,\mathbf{v}_{\sigma(k)}$ among themselves, and permuting the vectors $\mathbf{v}_{\sigma(k+1)},\ldots,\mathbf{v}_{\sigma(k+\ell)}$ among themselves. Of course, the factor $(\operatorname{sgn}\sigma)$ changes as we carry out these permutations, but *because f and g are alternating,* the values of f and g change by being multiplied by the same sign. Hence all these terms have precisely the same value. There are $k!\ell!$ such terms, so it is reasonable to divide the sum by this number to eliminate the effect of this redundancy.

Step 3. Associativity is the most difficult of the properties to verify, so we postpone it for the moment. To check homogeneity, we compute

$$\begin{aligned}(cf)\wedge g &= A((cf)\otimes g)/k!\ell!\\ &= A(c(f\otimes g))/k!\ell! \quad \text{by homogeneity of } \otimes,\\ &= cA(f\otimes g)/k!\ell! \quad \text{by linearity of } A,\\ &= c(f\wedge g).\end{aligned}$$

A similar computation verifies the other part of homogeneity. Distributivity follows similarly from distributivity of $\otimes$ and linearity of A.

Step 4. We verify anticommutativity. In fact, we prove something slightly more general: Let F and G be tensors of orders k and ℓ, respectively (not necessarily alternating). We show that

$$A(F\otimes G)=(-1)^{k\ell}A(G\otimes F).$$

To begin, let π be the permutation of $(1,\ldots,k+\ell)$ such that

$$(\pi(1),\ldots,\pi(k+\ell))=(k+1,k+2,\ldots,k+\ell,1,2,\ldots,k).$$

Then $\operatorname{sgn}\pi=(-1)^{k\ell}$. (Count inversions!) It is easy to see that $(G\otimes F)^\pi = F\otimes G$, since

$$(G\otimes F)^\pi(\mathbf{v}_1,\ldots,\mathbf{v}_{k+\ell})=G(\mathbf{v}_{k+1},\ldots,\mathbf{v}_{k+\ell})\cdot F(\mathbf{v}_1,\ldots,\mathbf{v}_k),$$

$$(F\otimes G)(\mathbf{v}_1,\ldots,\mathbf{v}_{k+\ell})=F(\mathbf{v}_1,\ldots,\mathbf{v}_k)\cdot G(\mathbf{v}_{k+1},\ldots,\mathbf{v}_{k+\ell}).$$

We then compute

$$A(F \otimes G) = \sum_\sigma (\text{sgn } \sigma)(F \otimes G)^\sigma$$

$$= \sum_\sigma (\text{sgn } \sigma)((G \otimes F)^\pi)^\sigma$$

$$= (\text{sgn } \pi) \sum_\sigma (\text{sgn } \sigma \circ \pi)(G \otimes F)^{\sigma \circ \pi}$$

$$= (\text{sgn } \pi) A(G \otimes F),$$

since $\sigma \circ \pi$ runs over all elements of $S_{k+\ell}$ as σ does.

Step 5. Now we verify associativity. The proof requires several steps, of which the first is this:

Let F and G be tensors (not necessarily alternating) of orders k and ℓ, respectively, such that $AF = 0$. Then $A(F \otimes G) = 0$.

To prove that this result holds, let us consider one term of the expression for $A(F \otimes G)$, say the term

$$(\text{sgn } \sigma) F(\mathbf{v}_{\sigma(1)}, \ldots, \mathbf{v}_{\sigma(k)}) \cdot G(\mathbf{v}_{\sigma(k+1)}, \ldots, \mathbf{v}_{\sigma(k+\ell)}).$$

Let us group together all the terms in the expression for $A(F \otimes G)$ that involve the same *last* factor as this one. These terms can be written in the form

$$(\text{sgn } \sigma) [\sum_\tau (\text{sgn } \tau) F(\mathbf{v}_{\sigma(\tau(1))}, \ldots, \mathbf{v}_{\sigma(\tau(k))})] \cdot G(\mathbf{v}_{\sigma(k+1)}, \ldots, \mathbf{v}_{\sigma(k+\ell)}),$$

where τ ranges over all permutations of $\{1, \ldots, k\}$. Now the expression in brackets is just

$$AF(\mathbf{v}_{\sigma(1)}, \ldots, \mathbf{v}_{\sigma(k)}),$$

which vanishes by hypothesis. Thus the terms in this group cancel one another.

The same argument applies to each group of terms that involve the same last factor. We conclude that $A(F \otimes G) = 0$.

Step 6. Let F be an arbitrary tensor and let h be an alternating tensor of order m. We show that

$$(AF) \wedge h = \frac{1}{m!} A(F \otimes h).$$

Let F have order k. Our desired equation can be written as

$$\frac{1}{k!m!} A((AF) \otimes h) = \frac{1}{m!} A(F \otimes h).$$

Linearity of A and distributivity of $\otimes$ show this equation is equivalent to each of the equations

$$A\{(AF)\otimes h - (k!)F\otimes h\} = 0,$$

$$A\{\,[AF - (k!)F]\otimes h\} = 0.$$

In view of Step 5, this equation holds if we can show that

$$A[AF - (k!)F] = 0.$$

But this follows immediately from property (iii) of the transformation A, since AF is an alternating tensor of order k.

Step 7. Let f, g, h be alternating tensors of orders k, ℓ, m respectively. We show that

$$(f\wedge g)\wedge h = \frac{1}{k!\ell!m!}A\big((f\otimes g)\otimes h\big).$$

Let $F = f\otimes g$, for convenience. We have

$$f\wedge g = \frac{1}{k!\ell!}AF$$

by definition, so that

$$\begin{aligned}(f\wedge g)\wedge h &= \frac{1}{k!\ell!}(AF)\wedge h\\ &= \frac{1}{k!\ell!m!}A(F\otimes h) \quad \text{by Step 6,}\\ &= \frac{1}{k!\ell!m!}A\big((f\otimes g)\otimes h\big).\end{aligned}$$

Step 8. Finally, we verify associativity. Let f, g, h be as in Step 7. Then

$$\begin{aligned}(k!\ell!m!)(f\wedge g)\wedge h &= A\big((f\otimes g)\otimes h\big) &&\text{by Step 7,}\\ &= A\big(f\otimes(g\otimes h)\big) &&\text{by associativity of }\otimes,\\ &= (-1)^{k(\ell+m)}A\big((g\otimes h)\otimes f\big) &&\text{by Step 4,}\\ &= (-1)^{k(\ell+m)}(\ell!m!k!)(g\wedge h)\wedge f &&\text{by Step 7,}\\ &= (k!\ell!m!)f\wedge(g\wedge h) &&\text{by anticommutativity.}\end{aligned}$$

Step 9. We verify property (5). In fact, we prove something slightly more general. We show that for any collection $f_1, \dots, f_k$ of 1-tensors,

$$(*) \qquad A(f_1\otimes\cdots\otimes f_k) = f_1\wedge\cdots\wedge f_k.$$

Property (5) is an immediate consequence, since

$$\psi_I = A\phi_I = A(\phi_{i_1} \otimes \cdots \otimes \phi_{i_k}).$$

Formula (*) is trivial for $k = 1$. Supposing it true for $k - 1$, we prove it for k. Set $F = f_1 \otimes \cdots \otimes f_{k-1}$. Then

$$A(F \otimes f_k) = (1!)(AF) \wedge f_k \quad \text{by Step 6,}$$
$$= (f_1 \wedge \cdots \wedge f_{k-1}) \wedge f_k,$$

by the induction hypothesis.

Step 10. We verify uniqueness; indeed, we show how one can calculate wedge products, in the case of a finite-dimensional space V, using only properties (1)–(5). Let ϕ_i and ψ_I be as in property (5). Given alternating tensors f and g, we can write them uniquely in terms of the elementary alternating tensors as

$$f = \sum_{[I]} b_I \psi_I \quad \text{and} \quad g = \sum_{[J]} c_J \psi_J.$$

(Here I is an ascending k-tuple, and J is an ascending ℓ-tuple, from the set $\{1, \ldots, n\}$.) Distributivity and homogeneity imply that

$$f \wedge g = \sum_{[I]} \sum_{[J]} b_I c_J \psi_I \wedge \psi_J.$$

Therefore, to compute $f \wedge g$ we need only know how to compute wedge products of the form

$$\psi_I \wedge \psi_J = (\phi_{i_1} \wedge \cdots \wedge \phi_{i_k}) \wedge (\phi_{j_1} \wedge \cdots \wedge \phi_{j_\ell}).$$

For that, we use associativity and the simple rules

$$\phi_i \wedge \phi_j = -\phi_j \wedge \phi_i \quad \text{and} \quad \phi_i \wedge \phi_i = 0,$$

which follow from anticommutativity. It follows that the product $\psi_I \wedge \psi_J$ equals zero if two indices are the same. Otherwise it equals (sgn π) times the elementary alternating $k + \ell$ tensor ψ_K whose index is obtained by rearranging the indices in the $k + \ell$ tuple (I, J) in ascending order, where π is the permutation required to carry out this rearrangement.

Step 11. We complete the proof by verifying property (6). Let $T : V \to W$ be a linear transformation, and F be an arbitrary tensor on W (not necessarily alternating). It is easy to verify that $T^*(F^\sigma) = (T^*F)^\sigma$. Since T^* is linear, it then follows that $T^*(AF) = A(T^*F)$.

Now let f and g be alternating tensors on W of orders k and ℓ, respectively. We compute

$$\begin{aligned}
T^*(f \wedge g) &= \frac{1}{k!\ell!} T^*(A(f \otimes g)) \\
&= \frac{1}{k!\ell!} A(T^*(f \otimes g)) \\
&= \frac{1}{k!\ell!} A((T^* f) \otimes (T^* g)) \quad \text{by Theorem 26.5,} \\
&= (T^* f) \wedge (T^* g). \quad \square
\end{aligned}$$

With this theorem, we complete our study of multilinear algebra. There is, of course, much more to the subject (see [N] or [Gr], for example), but this is all we shall need. We shall in fact need only alternating tensors and their properties, as discussed in this section and the preceding one.

We remark that in some texts, such as [G–P], a slightly different definition of the wedge product is used; the coefficient $1/(k+\ell)!$ appears in the definition in place of the coefficient $1/k!\ell!$. This choice of coefficient also leads to an operation that is associative, as you can check. In fact, all the properties listed in Theorem 28.1 remain unchanged except for (5), which is altered by the insertion of a factor of $k!$ on the right side of the equation for ψ_I.

EXERCISES

1. Let $\mathbf{x}, \mathbf{y}, \mathbf{z} \in \mathbf{R}^5$. Let

$$F(\mathbf{x}, \mathbf{y}, \mathbf{z}) = 2x_2y_2z_1 + x_1y_5z_4,$$

$$G(\mathbf{x}, \mathbf{y}) = x_1y_3 + x_3y_1,$$

$$h(\mathbf{w}) = w_1 - 2w_3.$$

   (a) Write AF and AG in terms of elementary alternating tensors. [*Hint:* Write F and G in terms of elementary tensors and use Step 9 of the preceding proof to compute $A\phi_I$.]

   (b) Express $(AF) \wedge h$ in terms of elementary alternating tensors.

   (c) Express $(AF)(\mathbf{x}, \mathbf{y}, \mathbf{z})$ as a function.

2. If G is symmetric, show that $AG = 0$. Does the converse hold?

3. Show that if $f_1, \ldots, f_k$ are alternating tensors of orders $\ell_1, \ldots, \ell_k$, respectively, then

$$\frac{1}{\ell_1! \cdots \ell_k!} A(f_1 \otimes \cdots \otimes f_k) = f_1 \wedge \cdots \wedge f_k.$$

4. Let $\mathbf{x}_1, \ldots, \mathbf{x}_k$ be vectors in $\mathbf{R}^n$; let X be the matrix $X = [\mathbf{x}_1 \cdots \mathbf{x}_k]$. If $I = (i_1, \ldots, i_k)$ is an *arbitrary* k-tuple from the set $\{1, \ldots, n\}$, show that

$$\phi_{i_1} \wedge \cdots \wedge \phi_{i_k}(\mathbf{x}_1, \ldots, \mathbf{x}_k) = \det X_I.$$

5. Verify that $T^*(F^\sigma) = (T^*F)^\sigma$.

6. Let $T : \mathbf{R}^m \to \mathbf{R}^n$ be the linear transformation $T(\mathbf{x}) = B \cdot \mathbf{x}$.

   (a) If ψ_I is an elementary alternating k-tensor on $\mathbf{R}^n$, then $T^*\psi_I$ has the form

   $$T^*\psi_I = \sum_{[J]} c_J \psi_J,$$

   where the ψ_J are the elementary alternating k-tensors on $\mathbf{R}^m$. What are the coefficients c_J?

   (b) If $f = \sum_{[I]} d_I \psi_I$ is an alternating k-tensor on $\mathbf{R}^n$, express T^*f in terms of the elementary alternating k-tensors on $\mathbf{R}^m$.

§29. TANGENT VECTORS AND DIFFERENTIAL FORMS

In calculus, one studies vector algebra in $\mathbf{R}^3$—vector addition, dot products, cross products, and the like. Scalar and vector fields are introduced; and certain operators on scalar and vector fields are defined, namely, the operators

$$\text{grad } f = \vec{\nabla} f, \quad \text{curl } \vec{F} = \vec{\nabla} \times \vec{F}, \quad \text{and} \quad \text{div } \vec{G} = \vec{\nabla} \cdot \vec{G}.$$

These operators are crucial in the formulation of the basic theorems of the vector integral calculus.

Analogously, we have in this chapter studied tensor algebra in $\mathbf{R}^n$—tensor addition, alternating tensors, wedge products, and the like. Now we introduce the concept of a tensor field; more specifically, that of an alternating tensor field, which is called a "differential form." In the succeeding section, we shall introduce a certain operator on differential forms, called the "differential operator" d, which is the analogue of the operators grad, curl, and div. This operator is crucial in the formulation of the basic theorems concerning integrals of differential forms, which we shall study in the next chapter.

We begin by discussing vector fields in a somewhat more sophisticated manner than is done in calculus.

Tangent vectors and vector fields

Definition. Given $\mathbf{x} \in \mathbf{R}^n$, we define a **tangent vector** to $\mathbf{R}^n$ at $\mathbf{x}$ to be a pair $(\mathbf{x}; \mathbf{v})$, where $\mathbf{v} \in \mathbf{R}^n$. The set of all tangent vectors to $\mathbf{R}^n$ at $\mathbf{x}$ forms a vector space if we define

$$(\mathbf{x};\mathbf{v}) + (\mathbf{x};\mathbf{w}) = (\mathbf{x};\mathbf{v}+\mathbf{w}),$$

$$c(\mathbf{x};\mathbf{v}) = (\mathbf{x};c\mathbf{v}).$$

It is called the **tangent space** to $\mathbf{R}^n$ at $\mathbf{x}$, and is denoted $\mathcal{T}_{\mathbf{x}}(\mathbf{R}^n)$.

Although both $\mathbf{x}$ and $\mathbf{v}$ are elements of $\mathbf{R}^n$ in this definition, they play different roles. We think of $\mathbf{x}$ as a point of the metric space $\mathbf{R}^n$ and picture it as a "dot." We think of $\mathbf{v}$ as an element of the vector space $\mathbf{R}^n$ and picture it as an "arrow." We picture $(\mathbf{x}; \mathbf{v})$ as an arrow with its initial point at $\mathbf{x}$. The set $\mathcal{T}_{\mathbf{x}}(\mathbf{R}^n)$ is pictured as the set of all arrows with their initial points at $\mathbf{x}$; it is, of course, just the set $\mathbf{x} \times \mathbf{R}^n$.

We do not attempt to form the sum $(\mathbf{x};\mathbf{v}) + (\mathbf{y};\mathbf{w})$ if $\mathbf{x} \neq \mathbf{y}$.

Definition. Let (a,b) be an open interval in $\mathbf{R}$; let $\gamma : (a,b) \to \mathbf{R}^n$ be a map of class C^r. We define the **velocity vector** of γ, corresponding to the parameter value t, to be the vector $(\gamma(t);\ D\gamma(t))$.

This vector is pictured as an arrow in $\mathbf{R}^n$ with its initial point at the point $\mathbf{p} = \gamma(t)$. See Figure 29.1. This notion of a velocity vector is of course a reformulation of a familiar notion from calculus. If

$$\gamma(t) = x(t)\mathbf{e}_1 + y(t)\mathbf{e}_2 + z(t)\mathbf{e}_3$$

is a parametrized-curve in $\mathbf{R}^3$, then the velocity vector of γ is defined in calculus as the vector

$$D\gamma(t) = \frac{dx}{dt}\mathbf{e}_1 + \frac{dy}{dt}\mathbf{e}_2 + \frac{dz}{dt}\mathbf{e}_3.$$

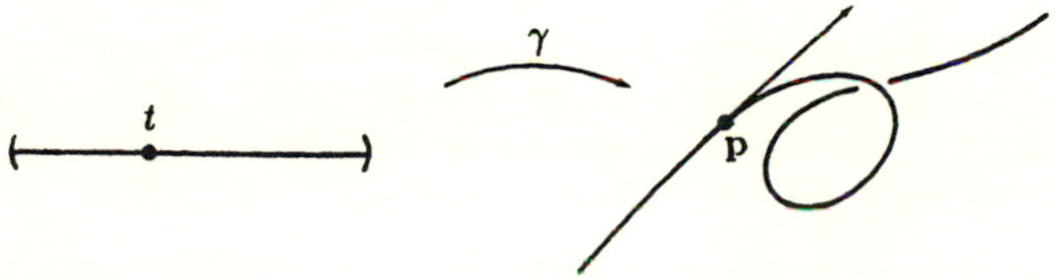

Figure 29.1

More generally, we make the following definition:

Definition. Let A be open in $\mathbf{R}^k$ or $\mathbf{H}^k$; let $\alpha : A \rightarrow \mathbf{R}^n$ be of class C^r. Let $\mathbf{x} \in A$, and let $\mathbf{p} = \alpha(\mathbf{x})$. We define a linear transformation

$$\alpha_* : \mathcal{T}_{\mathbf{x}}(\mathbf{R}^k) \longrightarrow \mathcal{T}_{\mathbf{p}}(\mathbf{R}^n)$$

by the equation

$$\alpha_*(\mathbf{x};\mathbf{v}) = (\mathbf{p}; D\alpha(\mathbf{x}) \cdot \mathbf{v}).$$

It is said to be the transformation **induced** by the differentiable map α.

Given $(\mathbf{x};\mathbf{v})$, the chain rule implies that the vector $\alpha_*(\mathbf{x};\mathbf{v})$ is in fact the velocity vector of the curve $\gamma(t) = \alpha(\mathbf{x}+t\mathbf{v})$, corresponding to the parameter value $t = 0$. See Figure 29.2.

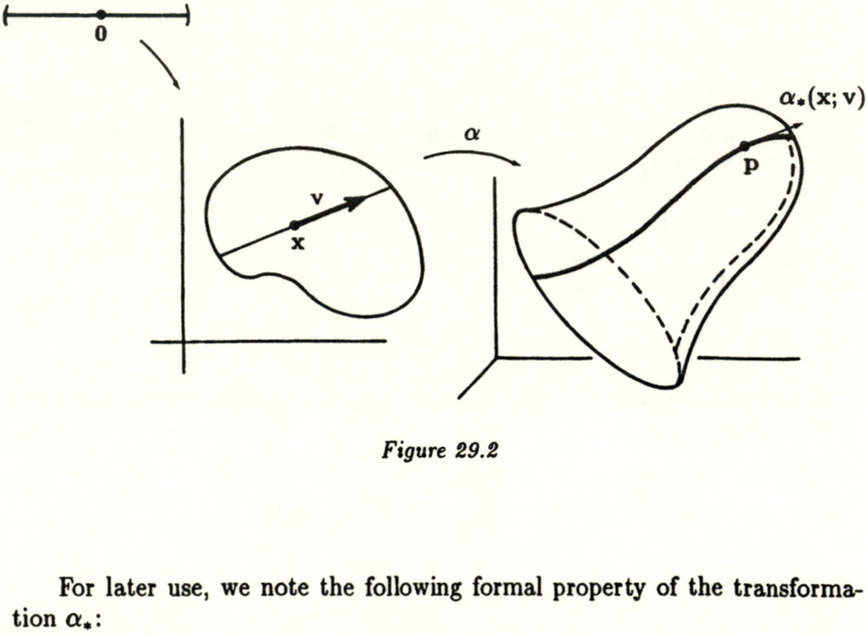

Figure 29.2

For later use, we note the following formal property of the transformation α_*:

Lemma 29.1. *Let A be open in $\mathbf{R}^k$ or $\mathbf{H}^k$; let $\alpha : A \rightarrow \mathbf{R}^m$ be of class C^r. Let B be an open set of $\mathbf{R}^m$ or $\mathbf{H}^m$ containing $\alpha(A)$; let $\beta : B \rightarrow \mathbf{R}^n$ be of class C^r. Then*

$$(\beta \circ \alpha)_* = \beta_* \circ \alpha_*.$$

Proof. This formula is just the chain rule. Let $\mathbf{y} = \alpha(\mathbf{x})$ and let $\mathbf{z} = \beta(\mathbf{y})$. We compute

$$\begin{aligned}(\beta \circ \alpha)_*(\mathbf{x};\mathbf{v}) &= (\beta(\alpha(\mathbf{x})); D(\beta \circ \alpha)(\mathbf{x}) \cdot \mathbf{v})\\ &= (\beta(\mathbf{y}); D\beta(\mathbf{y}) \cdot D\alpha(\mathbf{x}) \cdot \mathbf{v})\\ &= \beta_*(\mathbf{y}; D\alpha(\mathbf{x}) \cdot \mathbf{v})\\ &= \beta_*(\alpha_*(\mathbf{x};\mathbf{v})). \quad \square\end{aligned}$$

These maps and their induced transformations are indicated in the following diagrams:

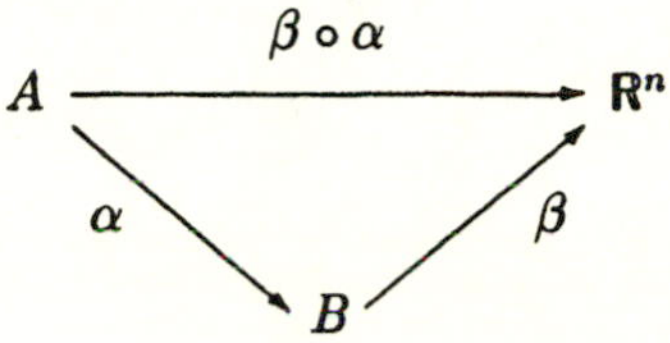

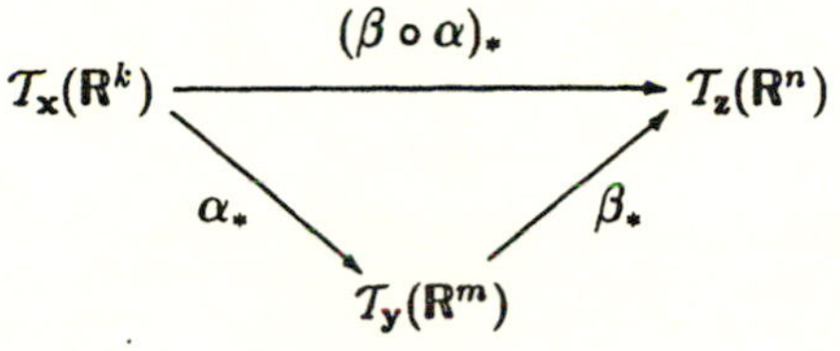

Definition. If A is an open set in $\mathbf{R}^n$, a **tangent vector field** in A is a continuous function $F : A \to \mathbf{R}^n \times \mathbf{R}^n$ such that $F(\mathbf{x}) \in \mathcal{T}_{\mathbf{x}}(\mathbf{R}^n)$, for each $\mathbf{x} \in A$. Then F has the form $F(\mathbf{x}) = (\mathbf{x}; f(\mathbf{x}))$, where $f : A \to \mathbf{R}^n$. If F is of class C^r, we say that it is a tangent vector field of class C^r.

Now we define tangent vectors to manifolds. We shall use these notions in Chapter 7.

Definition. Let M be a k-manifold of class C^r in $\mathbf{R}^n$. If $\mathbf{p} \in M$, choose a coordinate patch $\alpha : U \to V$ about $\mathbf{p}$, where U is open in $\mathbf{R}^k$ or $\mathbf{H}^k$. Let $\mathbf{x}$ be the point of U such that $\alpha(\mathbf{x}) = \mathbf{p}$. The set of all vectors of the form $\alpha_*(\mathbf{x};\mathbf{v})$, where $\mathbf{v}$ is a vector in $\mathbf{R}^k$, is called the **tangent space to** M **at** $\mathbf{p}$, and is denoted $\mathcal{T}_{\mathbf{p}}(M)$. Said differently,

$$\mathcal{T}_{\mathbf{p}}(M) = \alpha_*(\mathcal{T}_{\mathbf{x}}(\mathbf{R}^k)).$$

It is not hard to show that $\mathcal{T}_{\mathbf{p}}(M)$ is a linear subspace of $\mathcal{T}_{\mathbf{p}}(\mathbf{R}^n)$ that is well-defined, independent of the choice of α. Because $\mathbf{R}^k$ is spanned by the vectors $\mathbf{e}_1, \ldots, \mathbf{e}_k$, the space $\mathcal{T}_{\mathbf{p}}(M)$ is spanned by the vectors

$$(\mathbf{p}; D\alpha(\mathbf{x}) \cdot \mathbf{e}_j) = (\mathbf{p}; \partial\alpha/\partial x_j),$$

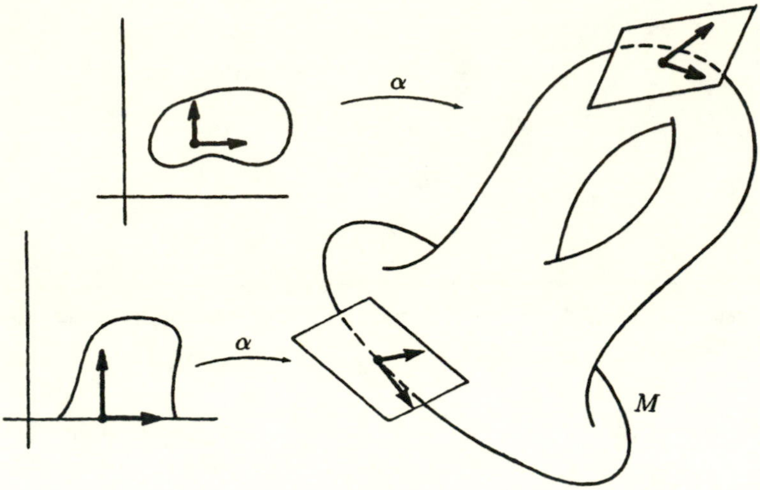

Figure 29.3

for $j = 1, \ldots, k$. Since $D\alpha$ has rank k, these vectors are independent; hence they form a basis for $\mathcal{T}_{\mathbf{p}}(M)$. Typical cases are pictured in Figure 29.3.

We denote the union of the tangent spaces $\mathcal{T}_{\mathbf{p}}(M)$, for $\mathbf{p} \in M$, by $\mathcal{T}(M)$; and we call it the **tangent bundle of** M. A **tangent vector field to** M is a continuous function $F : M \to \mathcal{T}(M)$ such that $F(\mathbf{p}) \in \mathcal{T}_{\mathbf{p}}(M)$ for each $\mathbf{p} \in M$.

Tensor fields

Definition. Let A be an open set in $\mathbb{R}^n$. A **k-tensor field in A** is a function ω assigning, to each $\mathbf{x} \in A$, a k-tensor defined on the vector space $\mathcal{T}_{\mathbf{x}}(\mathbb{R}^n)$. That is,

$$\omega(\mathbf{x}) \in \mathcal{L}^k(\mathcal{T}_{\mathbf{x}}(\mathbb{R}^n))$$

for each $\mathbf{x}$. Thus $\omega(\mathbf{x})$ is a function mapping k-tuples of tangent vectors to $\mathbb{R}^n$ at $\mathbf{x}$ into $\mathbb{R}$; as such, its value on a given k-tuple can be written in the form

$$\omega(\mathbf{x})\big((\mathbf{x};\mathbf{v}_1), \ldots, (\mathbf{x};\mathbf{v}_k)\big).$$

We require this function to be continuous as a function of $(\mathbf{x}, \mathbf{v}_1, \ldots, \mathbf{v}_k)$; if it is of class C^r, we say that ω is a tensor field of class C^r. If it happens that $\omega(\mathbf{x})$ is an alternating k-tensor for each $\mathbf{x}$, then ω is called a **differential form** (or simply, a **form**) **of order** k, on A.

More generally, if M is an m-manifold in $\mathbf{R}^n$, then we define a k-tensor field on M to be a function ω assigning to each $\mathbf{p} \in M$ an element of $\mathcal{L}^k(\mathcal{T}_{\mathbf{p}}(M))$. If in fact $\omega(\mathbf{p})$ is alternating for each $\mathbf{p}$, then ω is called a differential form on M.

If ω is a tensor field defined on an open set of $\mathbf{R}^n$ containing M, then ω of course restricts to a tensor field defined on M, since every tangent vector to M is also a tangent vector to $\mathbf{R}^n$. Conversely, any tensor field on M can be extended to a tensor field defined on an open set of $\mathbf{R}^n$ containing M; the proof, however, is decidedly non-trivial. For simplicity, we shall restrict ourselves in this book to tensor fields that are defined on open sets of $\mathbf{R}^n$.

Definition. Let $\mathbf{e}_1, \dots, \mathbf{e}_n$ be the usual basis for $\mathbf{R}^n$. Then $(\mathbf{x};\mathbf{e}_1), \dots, (\mathbf{x};\mathbf{e}_n)$ is called the **usual basis** for $\mathcal{T}_{\mathbf{x}}(\mathbf{R}^n)$. We define a 1-form $\tilde{\phi}_i$ on $\mathbf{R}^n$ by the equation

$$\tilde{\phi}_i(\mathbf{x})(\mathbf{x};\mathbf{e}_j) = \begin{cases} 0 & \text{if } i \neq j, \\ 1 & \text{if } i = j. \end{cases}$$

The forms $\tilde{\phi}_1, \dots, \tilde{\phi}_n$ are called the **elementary 1-forms** on $\mathbf{R}^n$. Similarly, given an ascending k-tuple $I = (i_1, \dots, i_k)$ from the set $\{1, \dots, n\}$, we define a k-form $\tilde{\psi}_I$ on $\mathbf{R}^n$ by the equation

$$\tilde{\psi}_I(\mathbf{x}) = \tilde{\phi}_{i_1}(\mathbf{x}) \wedge \cdots \wedge \tilde{\phi}_{i_k}(\mathbf{x}).$$

The forms $\tilde{\psi}_I$ are called the **elementary k-forms** on $\mathbf{R}^n$.

Note that for each $\mathbf{x}$, the 1-tensors $\tilde{\phi}_1(\mathbf{x}), \dots, \tilde{\phi}_n(\mathbf{x})$ constitute the basis for $\mathcal{L}^1(\mathcal{T}_{\mathbf{x}}(\mathbf{R}^n))$ dual to the usual basis for $\mathcal{T}_{\mathbf{x}}(\mathbf{R}^n)$, and the k-tensor $\tilde{\psi}_I(\mathbf{x})$ is the corresponding elementary alternating tensor on $\mathcal{T}_{\mathbf{x}}(\mathbf{R}^n)$.

The fact that $\tilde{\phi}_i$ and $\tilde{\psi}_I$ are of class C^∞ follows at once from the equations

$$\tilde{\phi}_i(\mathbf{x})(\mathbf{x};\mathbf{v}) = v_i,$$

$$\tilde{\psi}_I(\mathbf{x})((\mathbf{x};\mathbf{v}_1), \dots, (\mathbf{x};\mathbf{v}_k)) = \det X_I,$$

where X is the matrix $X = [\mathbf{v}_1 \cdots \mathbf{v}_k]$.

If ω is a k-form defined on an open set A of $\mathbf{R}^n$, then the k-tensor $\omega(\mathbf{x})$ can be written uniquely in the form

$$\omega(\mathbf{x}) = \sum_{[I]} b_I(\mathbf{x})\tilde{\psi}_I(\mathbf{x}),$$

for some scalar functions $b_I(\mathbf{x})$. These functions are called the **components** of ω relative to the standard elementary forms in $\mathbf{R}^n$.

Lemma 29.2. *Let ω be a k-form on the open set A of $\mathbf{R}^n$. Then ω is of class C^r if and only if its component functions b_I are of class C^r on A.*

Proof. Given ω, let us express it in terms of elementary forms by the equation

$$\omega = \sum_{[I]} b_I \tilde{\psi}_I.$$

The functions $\tilde{\psi}_I$ are of class C^∞. Therefore, if the functions b_I are of class C^r, so is the function ω. Conversely, if ω is of class C^r as a function of $(\mathbf{x}, \mathbf{v}_1, \ldots, \mathbf{v}_k)$, then in particular, given an ascending k-tuple $J = (j_1, \ldots, j_k)$ from the set $\{1, \ldots, n\}$, the function

$$\omega(\mathbf{x})\big((\mathbf{x}; \mathbf{e}_{j_1}), \ldots, (\mathbf{x}; \mathbf{e}_{j_k})\big)$$

is of class C^r as a function of $\mathbf{x}$. But this function equals $b_J(\mathbf{x})$. □

Lemma 29.3. *Let ω and η be k-forms, and let θ be an ℓ-form, on the open set A of $\mathbf{R}^n$. If ω and η and θ are of class C^r, so are $a\omega + b\eta$ and $\omega \wedge \theta$.*

Proof. It is immediate that $a\omega + b\eta$ is of class C^r, since it is a linear combination of C^r functions. To show that $\omega \wedge \theta$ is of class C^r, one could use the formula for the wedge product given in the proof of Theorem 28.1. Alternatively, one can use the preceding theorem: Let us write

$$\omega = \sum_{[I]} b_I \tilde{\psi}_I \quad \text{and} \quad \theta = \sum_{[J]} c_J \tilde{\psi}_J,$$

where I and J are ascending k- and ℓ-tuples, respectively, from the set $\{1, \ldots, n\}$. Then

$$\omega \wedge \theta = \sum_{[I]} \sum_{[J]} b_I c_J \tilde{\psi}_I \wedge \tilde{\psi}_J.$$

To write $(\omega \wedge \theta)(\mathbf{x})$ in terms of elementary alternating tensors, we drop all terms with repeated indices, rewrite the remaining terms with indices in ascending order, and collect like terms. We see thus that each component of $\omega \wedge \theta$ is the sum (with signs ± 1) of functions of the form $b_I c_J$. Thus the component functions of $\omega \wedge \theta$ are of class C^r. □

Differential forms of order zero

In what follows, we shall need to deal not only with tensor fields in $\mathbf{R}^n$, but with scalar fields as well. It is convenient to treat scalar fields as differential forms of order 0.

Definition. If A is open in $\mathbf{R}^n$, and if $f : A \to \mathbf{R}$ is a map of class C^r, then f is called a **scalar field** in A. We also call f a **differential form of order** 0.

The sum of two such functions is another such, and so is the product by a scalar. We define the wedge product of two 0-forms f and g by the rule $f \wedge g = f \cdot g$, which is just the usual product of real-valued functions. More generally, we define the wedge product of the 0-form f and the k-form ω by the rule

$$(\omega \wedge f)(\mathbf{x}) = (f \wedge \omega)(\mathbf{x}) = f(\mathbf{x}) \cdot \omega(\mathbf{x});$$

this is just the usual product of the tensor $\omega(\mathbf{x})$ and the scalar $f(\mathbf{x})$.

Note that all the formal algebraic properties of the wedge product hold. Associativity, homogeneity, and distributivity are immediate; and anticommutativity holds because scalar fields are forms of order 0:

$$f \wedge g = (-1)^0 g \wedge f \quad \text{and} \quad f \wedge \omega = (-1)^0 \omega \wedge f.$$

Convention. *Henceforth, we shall use Roman letters such as* f, g, h *to denote* 0*-forms, and Greek letters such as* ω, η, θ *to denote* k*-forms for* $k > 0$.

EXERCISES

1. Let $\gamma : \mathbf{R} \to \mathbf{R}^n$ be of class C^r. Show that the velocity vector of γ corresponding to the parameter value t is the vector $\gamma_*(t; \mathbf{e}_1)$.
2. If A is open in $\mathbf{R}^k$ and $\alpha : A \to \mathbf{R}^n$ is of class C^r, show that $\alpha_*(\mathbf{x}; \mathbf{v})$ is the velocity vector of the curve $\gamma(t) = \alpha(\mathbf{x} + t\mathbf{v})$ corresponding to parameter value $t = 0$.
3. Let M be a k-manifold of class C^r in $\mathbf{R}^n$. Let $\mathbf{p} \in M$. Show that the tangent space to M at $\mathbf{p}$ is well-defined, independent of the choice of the coordinate patch.
4. Let M be a k-manifold in $\mathbf{R}^n$ of class C^r. Let $\mathbf{p} \in M - \partial M$.
   (a) Show that if $(\mathbf{p}; \mathbf{v})$ is a tangent vector to M, then there is a parametrized-curve $\gamma : (-\epsilon, \epsilon) \to \mathbf{R}^n$ whose image set lies in M, such that $(\mathbf{p}; \mathbf{v})$ equals the velocity vector of γ corresponding to parameter value $t = 0$. See Figure 29.4.
   (b) Prove the converse. [*Hint:* Recall that for any coordinate patch α, the map α^{-1} is of class C^r. See Theorem 24.1.]
5. Let M be a k-manifold in $\mathbf{R}^n$ of class C^r. Let $\mathbf{q} \in \partial M$.
   (a) Show that if $(\mathbf{q}; \mathbf{v})$ is a tangent vector to M at $\mathbf{q}$, then there is a parametrized-curve $\gamma : (-\epsilon, \epsilon) \to \mathbf{R}^n$, where γ carries either $(-\epsilon, 0]$ or $[0, \epsilon)$ into M, such that $(\mathbf{q}; \mathbf{v})$ equals the velocity vector of γ corresponding to parameter value $t = 0$.

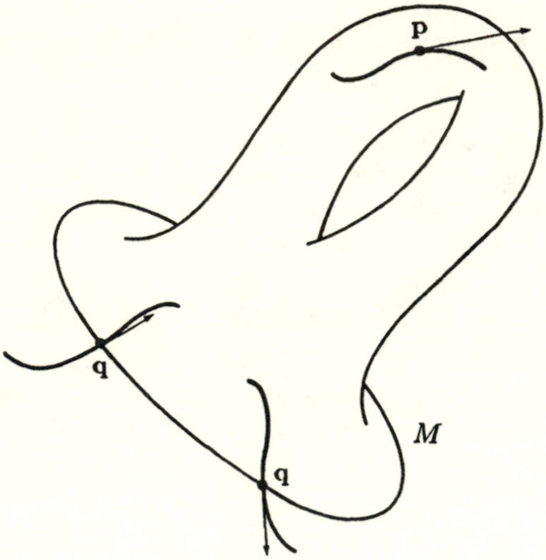

Figure 29.4

(b) Prove the converse.

§30. THE DIFFERENTIAL OPERATOR

We now introduce a certain operator d on differential forms. In general, the operator d, when applied to a k-form, gives a $k+1$ form. We begin by defining d for 0-forms.

The differential of a 0-form

A 0-form on an open set A of $\mathbf{R}^n$ is a function $f : A \to \mathbf{R}$. The differential df of f is to be a 1-form on A, that is, a linear transformation of $\mathcal{T}_{\mathbf{x}}(\mathbf{R}^n)$ into $\mathbf{R}$, for each $\mathbf{x} \in A$. We studied such a linear transformation in Chapter 2. We called it the "derivative of f at $\mathbf{x}$ with respect to the vector $\mathbf{v}$." We now look at this notion as defining a 1-form on A.

Definition. Let A be open in $\mathbf{R}^n$; let $f : A \to \mathbf{R}$ be a function of class C^r. We define a 1-form df on A by the formula

$$df(\mathbf{x})(\mathbf{x};\mathbf{v}) = f'(\mathbf{x};\mathbf{v}) = Df(\mathbf{x}) \cdot \mathbf{v}.$$

The 1-form df is called the **differential** of f. It is of class C^{r-1} as a function of $\mathbf{x}$ and $\mathbf{v}$.

Theorem 30.1. *The operator d is linear on 0-forms.*

Proof. Let $f, g : A \to \mathbb{R}$ be of class C^r. Let $h = af + bg$. Then

$$Dh(\mathbf{x}) = a\ Df(\mathbf{x}) + b\ Dg(\mathbf{x}),$$

so that

$$dh(\mathbf{x})(\mathbf{x};\mathbf{v}) = a\ df(\mathbf{x})(\mathbf{x};\mathbf{v}) + b\ dg(\mathbf{x})(\mathbf{x};\mathbf{v}).$$

Thus $dh = a(df) + b(dg)$, as desired. □

Using the operator d, we can obtain a new way of expressing the elementary 1-forms $\tilde{\phi}_i$ in $\mathbb{R}^n$:

Lemma 30.2. *Let $\tilde{\phi}_1, \ldots, \tilde{\phi}_n$ be the elementary 1-forms in $\mathbb{R}^n$. Let $\pi_i : \mathbb{R}^n \to \mathbb{R}$ be the ith projection function, defined by the equation*

$$\pi_i(x_1, \ldots, x_n) = x_i.$$

Then $d\pi_i = \tilde{\phi}_i$.

Proof. Since π_i is a C^∞ function, $d\pi_i$ is a 1-form of class C^∞. We compute

$$\begin{aligned} d\pi_i(\mathbf{x})(\mathbf{x};\mathbf{v}) &= D\pi_i(\mathbf{x}) \cdot \mathbf{v} \\ &= [0 \cdots 0\ 1\ 0 \cdots 0] \begin{bmatrix} v_1 \\ \vdots \\ v_n \end{bmatrix} = v_i. \end{aligned}$$

Thus $d\pi_i = \tilde{\phi}_i$. □

Now it is common in this subject to abuse notation slightly, denoting the ith projection function not by π_i but by x_i. Then in this notation, $\tilde{\phi}_i$ is equal to dx_i. We shall use this notation henceforth:

Convention. *If $\mathbf{x}$ denotes the general point of $\mathbb{R}^n$, we denote the ith projection function mapping $\mathbb{R}^n$ to $\mathbb{R}$ by the symbol x_i. Then dx_i equals* *the elementary 1-form* $\widetilde{\phi}_i$ *in* $\mathbf{R}^n$. *If* $I = (i_1, \ldots, i_k)$ *is an ascending k-tuple from the set* $\{1, \ldots, n\}$, *then we introduce the notation*

$$dx_I = dx_{i_1} \wedge \cdots \wedge dx_{i_k}$$

for the elementary k-form $\widetilde{\psi}_I$ *in* $\mathbf{R}^n$. *The general k-form can then be written uniquely in the form*

$$\omega = \sum_{[I]} b_I \, dx_I,$$

for some scalar functions b_I.

The forms dx_i and dx_I are of course characterized by the equations

$$dx_i(\mathbf{x})(\mathbf{x}; \mathbf{v}) = v_i,$$

$$dx_I(\mathbf{x})\big((\mathbf{x}; \mathbf{v}_1), \ldots, (\mathbf{x}; \mathbf{v}_k)\big) = \det X_I,$$

where X is the matrix $X = [\mathbf{v}_1 \cdots \mathbf{v}_k]$.

For convenience, we extend this notation to an *arbitrary* k-tuple $J = (j_1, \ldots, j_k)$ from the set $\{1, \ldots, n\}$, setting

$$dx_J = dx_{j_1} \wedge \cdots \wedge dx_{j_k}.$$

Note that whereas dx_j is the differential of a 0-form, dx_J does not denote the differential of a form, but rather a wedge product of elementary 1-forms.

REMARK. Why do we call the use of x_i for π_i an abuse of notation? The reason is this: Normally, we use a single letter such as f to denote a *function*, and we use the symbol $f(x)$ to denote the *value* of the function at the point x. That is, f stands for the rule defining the function, and $f(x)$ denotes an element of the range of f. It is an abuse of notation to confuse the *function* with the *value* of the function.

However, this abuse is fairly common. We often speak of "the function $x^3 + 2x + 1$" when we should instead speak of "the function f defined by the equation $f(x) = x^3 + 2x + 1$," and we speak of "the function e^x" when we should speak of "the exponential function."

We are doing the same thing here. The *value* of the ith projection function at the point $\mathbf{x}$ is the number x_i; we abuse notation when we use x_i to denote the function itself. This usage is standard, however, and we shall conform to it.

If f is a 0-form, then df is a 1-form, so it can be expressed as a linear combination of elementary 1-forms. The expression is a familiar one:

Theorem 30.3. *Let A be open in $\mathbf{R}^n$; let $f : A \to \mathbf{R}$ be of class C^r. Then*

$$df = (D_1 f)dx_1 + \cdots + (D_n f)dx_n.$$

In particular, $df = 0$ if f is a constant function.

In Leibnitz's notation, this equation takes the form

$$df = \frac{\partial f}{\partial x_1} dx_1 + \cdots + \frac{\partial f}{\partial x_n} dx_n.$$

This formula sometimes appears in calculus books, but its meaning is not explained there.

Proof. We evaluate both sides of the equation on the tangent vector $(\mathbf{x}; \mathbf{v})$. We have

$$df(\mathbf{x})(\mathbf{x}; \mathbf{v}) = Df(\mathbf{x}) \cdot \mathbf{v}$$

by definition, whereas

$$\sum_{i=1}^{n} D_i f(\mathbf{x})\, dx_i(\mathbf{x})(\mathbf{x}; \mathbf{v}) = \sum_{i=1}^{n} D_i f(\mathbf{x}) v_i.$$

The theorem follows. □

The fact that df is only of class C^{r-1} if f is of class C^r is very inconvenient. It means that we must keep track of how many degrees of differentiability are needed in any given argument. In order to avoid these difficulties, we make the following convention:

Convention. *Henceforth, we restrict ourselves to manifolds, maps, vector fields, and forms that are of class C^∞.*

The differential of a k-form

We now define the differential operator d in general. It is in some sense a generalized directional derivative. A formula that makes this fact explicit appears in the exercises. Rather than using this formula to define d, we shall instead characterize d by its formal properties, as given in the theorem that follows.

Definition. If A is an open set in $\mathbf{R}^n$, let $\Omega^k(A)$ denote the set of all k-forms on A (of class C^∞). The sum of two such k-forms is another k-form, and so is the product of a k-form by a scalar. It is easy to see that $\Omega^k(A)$ satisfies the axioms for a vector space; we call it the **linear space of k-forms** on A.

Theorem 30.4. *Let A be an open set in $\mathbf{R}^n$. There exists a unique linear transformation*

$$d : \Omega^k(A) \to \Omega^{k+1}(A),$$

defined for $k \geq 0$, such that:

(1) *If f is a 0-form, then df is the 1-form*

$$df(\mathbf{x})(\mathbf{x}; \mathbf{v}) = Df(\mathbf{x}) \cdot \mathbf{v}.$$

(2) *If ω and η are forms of orders k and ℓ, respectively, then*

$$d(\omega \wedge \eta) = d\omega \wedge \eta + (-1)^k \omega \wedge d\eta.$$

(3) *For every form ω,*

$$d(d\omega) = 0.$$

We call d the **differential operator**, and we call $d\omega$ the **differential** of ω.

Proof. Step 1. We verify uniqueness. First, we show that conditions (2) and (3) imply that for any forms $\omega_1, \ldots, \omega_k$, we have

$$d(d\omega_1 \wedge \cdots \wedge d\omega_k) = 0.$$

If $k = 1$, this equation is a consequence of (3). Supposing it true for $k - 1$, we set $\eta = (d\omega_2 \wedge \cdots \wedge d\omega_k)$ and use (2) to compute

$$d(d\omega_1 \wedge \eta) = d(d\omega_1) \wedge \eta \pm d\omega_1 \wedge d\eta.$$

The first term vanishes by (3) and the second vanishes by the induction hypothesis.

Now we show that for any k-form ω, the form $d\omega$ is entirely determined by the value of d on 0-forms, which is specified by (1). Since d is linear, it suffices to consider the case $\omega = f\, dx_I$. We compute

$$\begin{aligned} d\omega = d(f\, dx_I) &= d(f \wedge dx_I) \\ &= df \wedge dx_I + f \wedge d(dx_I) \qquad \text{by (2),} \\ &= df \wedge dx_I, \end{aligned}$$

by the result just proved. Thus $d\omega$ is determined by the value of d on the 0-form f.

Step 2. We now define d. Its value for 0-forms is specified by (1). The computation just made tells us how to define it for forms of positive order: If A is an open set in $\mathbf{R}^n$ and if ω is a k-form on A, we write ω uniquely in the form

$$\omega = \sum_{[I]} f_I \, dx_I,$$

and define

$$d\omega = \sum_{[I]} df_I \wedge dx_I.$$

We check that $d\omega$ is of class C^∞. For this purpose, we first compute

$$d\omega = \sum_{[I]} [\sum_{j=1}^{n} (D_j f_I) dx_j] \wedge dx_I.$$

To express $d\omega$ as a linear combination of elementary $k+1$ forms, one proceeds as follows: First, delete all terms for which j is the same as one of the indices in the k-tuple I. Second, take the remaining terms and rearrange the dx_i so the indices are in ascending order. Third, collect like terms. One sees in this way that each component of $d\omega$ is a linear combination of the functions $D_j f_I$, so that it is of class C^∞. Thus $d\omega$ is of class C^∞. (Note that if ω were only of class C^r, then $d\omega$ would be of class C^{r-1}.)

We show d is linear on k-forms with $k > 0$. Let

$$\omega = \sum_{[I]} f_I \, dx_I \quad \text{and} \quad \eta = \sum_{[I]} g_I \, dx_I$$

be k-forms. Then

$$\begin{aligned} d(a\omega + b\eta) &= d \sum_{[I]} (a f_I + b g_I) dx_I \\ &= \sum_{[I]} d(a f_I + b g_I) \wedge dx_I && \text{by definition,} \\ &= \sum_{[I]} (a \, df_I + b \, dg_I) \wedge dx_I && \text{since } d \text{ is linear on 0-forms,} \\ &= a \, d\omega + b \, d\eta. \end{aligned}$$

Step 3. We now show that if J is an *arbitrary* k-tuple of integers from the set $\{1, \ldots, n\}$, then

$$d(f \wedge dx_J) = df \wedge dx_J.$$

Certainly this formula holds if two of the indices in J are the same, since $dx_J = 0$ in this case. So suppose the indices in J are distinct. Let I be the k-tuple obtained by rearranging the indices in J in ascending order; let π be the permutation involved. Anticommutativity of the wedge product implies that $dx_I = (\text{sgn } \pi) dx_J$. Because d is linear and the wedge product is homogeneous, the formula $d(f \wedge dx_I) = df \wedge dx_I$, which holds by definition, implies that

$$(\text{sgn } \pi)\, d(f \wedge dx_J) = (\text{sgn } \pi)\, df \wedge dx_J.$$

Our desired result follows.

Step 4. We verify property (2), in the case $k = 0$ and $\ell = 0$. We compute

$$\begin{aligned} d(f \wedge g) &= \sum_{j=1}^{n} D_j(f \cdot g) dx_j \\ &= \sum_{j=1}^{n} (D_j f) \cdot g\, dx_j + \sum_{j=1}^{n} f \cdot (D_j g) dx_j \\ &= (df) \wedge g + f \wedge (dg). \end{aligned}$$

Step 5. We verify property (2) in general. First, we consider the case where both forms have positive order. Since both sides of our desired equation are linear in ω and in η, it suffices to consider the case

$$\omega = f\, dx_I \quad \text{and} \quad \eta = g\, dx_J.$$

We compute

$$\begin{aligned} d(\omega \wedge \eta) &= d(fg\ dx_I \wedge dx_J) \\ &= d(fg) \wedge dx_I \wedge dx_J \qquad \text{by Step 3,} \\ &= (df \wedge g + f \wedge dg) \wedge dx_I \wedge dx_J \quad \text{by Step 4,} \\ &= (df \wedge dx_I) \wedge (g \wedge dx_J) + (-1)^k (f \wedge dx_I) \wedge (dg \wedge dx_J) \\ &= d\omega \wedge \eta + (-1)^k \omega \wedge d\eta. \end{aligned}$$

The sign $(-1)^k$ comes from the fact that dx_I is a k-form and dg is a 1-form.

Finally, the proof in the case where one of k or ℓ is zero proceeds as in the argument just given. If $k = 0$, the term dx_I is missing from the equations, while if $\ell = 0$, the term dx_J is missing. We leave the details to you.

Step 6. We show that if f is a 0-form, then $d(df) = 0$. We have

$$d(df) = d\sum_{j=1}^{n} D_j f\, dx_j,$$

$$= \sum_{j=1}^{n} d(D_j f) \wedge dx_j \quad \text{by definition,}$$

$$= \sum_{j=1}^{n} \sum_{i=1}^{n} D_i D_j f\, dx_i \wedge dx_j.$$

To write this expression in standard form, we delete all terms for which $i = j$, and collect the remaining terms as follows:

$$d(df) = \sum_{i<j} (D_i D_j f - D_j D_i f) dx_i \wedge dx_j.$$

The equality of the mixed partial derivatives implies that $d(df) = 0$.

Step 7. We show that if ω is a k-form with $k > 0$, then $d(d\omega) = 0$. Since d is linear, it suffices to consider the case $\omega = f\, dx_I$. Then

$$d(d\omega) = d(df \wedge dx_I)$$

$$= d(df) \wedge dx_I - df \wedge d(dx_I),$$

by property (2). Now $d(df) = 0$ by Step 6, and

$$d(dx_I) = d(1) \wedge dx_I = 0$$

by definition. Hence $d(d\omega) = 0$. □

Definition. Let A be an open set in $\mathbf{R}^n$. A 0-form f on A is said to be **exact** on A if it is constant on A; a k-form ω on A with $k > 0$ is said to be **exact** on A if there is a $k-1$ form θ on A such that $\omega = d\theta$. A k-form ω on A with $k \geq 0$ is said to be **closed** if $d\omega = 0$.

Every exact form is closed; for if f is constant, then $df = 0$, while if $\omega = d\theta$, then $d\omega = d(d\theta) = 0$. Conversely, every closed form on A is exact on A if A equals all of $\mathbf{R}^n$, or more generally, if A is a "star-convex" subset of $\mathbf{R}^n$. (See Chapter 8.) But the converse does not hold in general, as we shall see. If every closed k-form on A is exact on A, then we say that A is **homologically trivial** in dimension k. We shall explore this notion further in Chapter 8.

EXAMPLE 1. Let A be the open set in $\mathbf{R}^2$ consisting of all points (x, y) for which $x \neq 0$. Set $f(x,y) = x/|x|$ for $(x,y) \in A$. Then f is of class C^∞ on A, and $df = 0$ on A. But f is not exact on A because f is not constant on A.

EXAMPLE 2. Exactness is a notion you have seen before. In differential equations, for example, the equation

$$P(x,y)\,dx + Q(x,y)\,dy = 0$$

is said to be exact if there is a function f such that $P = \partial f/\partial x$ and $Q = \partial f/\partial y$. In our terminology, this means simply that the 1-form $P\,dx + Q\,dy$ is the differential of the 0-form f, so that it is exact.

Exactness is also related to the notion of conservative vector fields. In $\mathbf{R}^3$, for example, the vector field

$$\vec{F} = P\vec{i} + Q\vec{j} + R\vec{k}$$

is said to be conservative if it is the gradient of a scalar field f, that is, if

$$P = \partial f/\partial x \quad \text{and} \quad Q = \partial f/\partial y \quad \text{and} \quad R = \partial f/\partial z.$$

This is precisely the same as saying that the form $P\,dx + Q\,dy + R\,dz$ is the differential of the 0-form f.

We shall explore further the connection between forms and vector fields in the next section.

EXERCISES

1. Let A be open in $\mathbf{R}^n$.
   (a) Show that $\Omega^k(A)$ is a vector space.
   (b) Show that the set of all C^∞ vector fields on A is a vector space.
2. Consider the forms

$$\omega = xy\ dx + 3\ dy - yz\ dz,$$

$$\eta = x\ dx - yz^2\ dy + 2x\ dz,$$

   in $\mathbf{R}^3$. Verify by direct computation that

$$d(d\omega) = 0 \quad \text{and} \quad d(\omega \wedge \eta) = (d\omega) \wedge \eta - \omega \wedge d\eta.$$

3. Let ω be a k-form defined in an open set A of $\mathbf{R}^n$. We say that ω vanishes at x if $\omega(\mathbf{x})$ is the zero tensor.
   (a) Show that if ω vanishes at each $\mathbf{x}$ in a neighborhood of $\mathbf{x}_0$, then $d\omega$ vanishes at $\mathbf{x}_0$.

(b) Give an example to show that if ω vanishes at $\mathbf{x}_0$, then $d\omega$ need not vanish at $\mathbf{x}_0$.

4. Let $A = \mathbf{R}^2 - 0$; consider the 1-form in A defined the equation

$$\omega = (x\ dx + y\ dy)/(x^2 + y^2).$$

(a) Show ω is closed.

(b) Show that ω is exact on A.

*5. Prove the following:

Theorem. *Let* $A = \mathbf{R}^2 - 0$; *let*

$$\omega = (-y\ dx + x\ dy)/(x^2 + y^2)$$

in A. *Then* ω *is closed, but not exact, in* A.

Proof. (a) Show ω is closed.

(b) Let B consist of $\mathbf{R}^2$ with the non-negative x-axis deleted. Show that for each $(x, y) \in B$, there is a unique t with $0 < t < 2\pi$ such that

$$x = (x^2 + y^2)^{1/2} \cos t \quad \text{and} \quad y = (x^2 + y^2)^{1/2} \sin t;$$

denote this value of t by $\phi(x, y)$.

(c) Show that ϕ is of class C^∞. [*Hint:* The inverse sine and inverse cosine functions are of class C^∞ on the interval (-1, 1).]

(d) Show that $\omega = d\phi$ in B. [*Hint:* We have $\tan \phi = y/x$ if $x \neq 0$ and $\cot \phi = x/y$ if $y \neq 0$.]

(e) Show that if g is a closed 0-form in B, then g is constant in B. [*Hint:* Use the mean-value theorem to show that if $\mathbf{a}$ is the point $(-1, 0)$ of $\mathbf{R}^2$, then $g(\mathbf{x}) = g(\mathbf{a})$ for all $\mathbf{x} \in B$.]

(f) Show that ω is not exact in A. [*Hint:* If $\omega = df$ in A, then $f - \phi$ is constant in B. Evaluate the limit of $f(1, y)$ as y approaches 0 through positive and negative values.]

6. Let $A = \mathbf{R}^n - 0$. Let m be a fixed positive integer. Consider the following $n - 1$ form in A:

$$\eta = \sum_{i=1}^{n} (-1)^{i-1} f_i\ dx_1 \wedge \cdots \wedge \widehat{dx_i} \wedge \cdots \wedge dx_n,$$

where $f_i(\mathbf{x}) = x_i / \|\mathbf{x}\|^m$, and where $\widehat{dx_i}$ means that the factor dx_i is to be omitted.

(a) Calculate $d\eta$.

(b) For what values of m is it true that $d\eta = 0$? (We show later that η is not exact.)

*7. Prove the following, which expresses d as a generalized "directional derivative":

Theorem. *Let A be open in $\mathbf{R}^n$; let ω be a k-1 form in A. Given $\mathbf{v}_1, \ldots, \mathbf{v}_k \in \mathbf{R}^n$, define*

$$h(\mathbf{x}) = d\omega(\mathbf{x})\big((\mathbf{x};\mathbf{v}_1), \ldots, (\mathbf{x};\mathbf{v}_k)\big),$$

$$g_j(\mathbf{x}) = \omega(\mathbf{x})\big((\mathbf{x};\mathbf{v}_1), \ldots, \widehat{(\mathbf{x};\mathbf{v}_j)}, \ldots, (\mathbf{x};\mathbf{v}_k)\big),$$

where $\widehat{a}$ means that the component a is to be omitted. Then

$$h(\mathbf{x}) = \sum_{j=1}^{k} (-1)^{j-1} Dg_j(\mathbf{x}) \cdot \mathbf{v}_j.$$

Proof. (a) Let $X = [\mathbf{v}_1 \cdots \mathbf{v}_k]$. For each j, let $Y_j = [\mathbf{v}_1 \cdots \widehat{\mathbf{v}}_j \cdots \mathbf{v}_k]$. Given $(i, i_1, \ldots, i_{k-1})$, show that

$$\det X(i, i_1, \ldots, i_{k-1}) = \sum_{j=1}^{k} (-1)^{j-1} v_{ij} \det Y_j(i_1, \ldots, i_{k-1}).$$

(b) Verify the theorem in the case $\omega = f\, dx_I$.

(c) Complete the proof.

*§31. APPLICATION TO VECTOR AND SCALAR FIELDS

Finally, it is time to show that what we have been doing with tensor fields and forms and the differential operator is a true generalization to $\mathbf{R}^n$ of familiar facts about vector analysis in $\mathbf{R}^3$. We will use these results in §38, when we prove the classical versions of Stokes' theorem and the divergence theorem.

We know that if A is an open set in $\mathbf{R}^n$, then the set $\Omega^k(A)$ of k-forms on A is a linear space. It is also easy to check that the set of all C^∞ vector fields on A is a linear space. We define here a sequence of linear transformations from *scalar fields* and *vector fields* to forms. These transformations act as operators that "translate" theorems written in the language of scalar and vector fields to theorems written in the language of forms, and conversely.

We begin by defining the gradient and the divergence operators in $\mathbf{R}^n$.

Definition. Let A be open in $\mathbf{R}^n$. Let $f : A \to \mathbf{R}$ be a scalar field in A. We define a corresponding vector field in A, called the **gradient** of f, by the equation

$$(\text{grad } f)(\mathbf{x}) = (\mathbf{x};\ D_1 f(\mathbf{x})\mathbf{e}_1 + \cdots + D_n f(\mathbf{x})\mathbf{e}_n).$$

If $G(\mathbf{x}) = (\mathbf{x};\ g(\mathbf{x}))$ is a vector field in A, where $g : A \to \mathbf{R}^n$ is given by the equation

$$g(\mathbf{x}) = g_1(\mathbf{x})\mathbf{e}_1 + \cdots + g_n(\mathbf{x})\mathbf{e}_n,$$

then we define a corresponding scalar field in A, called the **divergence** of G, by the equation

$$(\text{div } G)(\mathbf{x}) = D_1 g_1(\mathbf{x}) + \cdots + D_n g_n(\mathbf{x}).$$

These operators are of course familiar from calculus in the case $n = 3$. The following theorem shows how these operators correspond to the operator d:

Theorem 31.1. *Let A be an open set in $\mathbf{R}^n$. There exist vector space isomorphisms α_i and β_j as in the following diagram:*

$$\begin{array}{ccc} \text{Scalar fields in } A & \xrightarrow{\alpha_0} & \Omega^0(A) \\ \downarrow \text{grad} & & \downarrow d \\ \text{Vector fields in } A & \xrightarrow{\alpha_1} & \Omega^1(A) \end{array}$$

$$\begin{array}{ccc} \text{Vector fields in } A & \xrightarrow{\beta_{n-1}} & \Omega^{n-1}(A) \\ \downarrow \text{div} & & \downarrow d \\ \text{Scalar fields in } A & \xrightarrow{\beta_n} & \Omega^n(A) \end{array}$$

such that

$$d \circ \alpha_0 = \alpha_1 \circ \text{grad} \quad \textit{and} \quad d \circ \beta_{n-1} = \beta_n \circ \text{div}.$$

Proof. Let f and h be scalar fields in A; let

$$F(\mathbf{x}) = (\mathbf{x};\ \textstyle\sum f_i(\mathbf{x})\mathbf{e}_i) \quad \text{and} \quad G(\mathbf{x}) = (\mathbf{x};\ \textstyle\sum g_i(\mathbf{x})\mathbf{e}_i)$$

be vector fields in A. We define the transformations α_i and β_j by the equations

$$\alpha_0 f = f,$$

$$\alpha_1 F = \sum_{i=1}^{n} f_i dx_i,$$

$$\beta_{n-1} G = \sum_{i=1}^{n} (-1)^{i-1} g_i \, dx_1 \wedge \cdots \wedge \widehat{dx_i} \wedge \cdots \wedge dx_n,$$

$$\beta_n h = h \, dx_1 \wedge \cdots \wedge dx_n.$$

(As usual, the notation $\widehat{a}$ means that the factor a is to be omitted.) The fact that each α_i and β_j is a linear isomorphism, and that the two equations hold, is left as an exercise. □

This theorem is all that can be said about applications to vector fields in general. However, in the case of $\mathbf{R}^3$, we have a "curl" operator, and something more can be said.

Definition. Let A be open in $\mathbf{R}^3$; let

$$F(\mathbf{x}) = (\mathbf{x}; \sum f_i(\mathbf{x})\mathbf{e}_i)$$

be a vector field in A. We define another vector field in A, called the **curl** of F, by the equation

$$(\text{curl } F)(\mathbf{x}) = (\mathbf{x}; (D_2 f_3 - D_3 f_2)\mathbf{e}_1 + (D_3 f_1 - D_1 f_3)\mathbf{e}_2 + (D_1 f_2 - D_2 f_1)\mathbf{e}_3).$$

A convenient trick for remembering the definition of the curl operator is to think of it as obtained by evaluation of the symbolic determinant

$$\det \begin{bmatrix} \mathbf{e}_1 & \mathbf{e}_2 & \mathbf{e}_3 \\ D_1 & D_2 & D_3 \\ f_1 & f_2 & f_3 \end{bmatrix}.$$

For $\mathbf{R}^3$, we have the following strengthened version of the preceding theorem:

Theorem 31.2. *Let A be an open set in $\mathbf{R}^3$. There exist vector space isomorphisms α_i and β_j as in the following diagram:*

$$\begin{array}{ccc}
\text{Scalar fields in } A & \xrightarrow{\alpha_0} & \Omega^0(A) \\
\big\downarrow \text{grad} & & \big\downarrow d \\
\text{Vector fields in } A & \xrightarrow{\alpha_1} & \Omega^1(A) \\
\big\downarrow \text{curl} & & \big\downarrow d \\
\text{Vector fields in } A & \xrightarrow{\beta_2} & \Omega^2(A) \\
\big\downarrow \text{div} & & \big\downarrow d \\
\text{Scalar fields in } A & \xrightarrow{\beta_3} & \Omega^3(A)
\end{array}$$

such that

$$d \circ \alpha_0 = \alpha_1 \circ \text{grad} \quad \textit{and} \quad d \circ \alpha_1 = \beta_2 \circ \text{curl} \quad \textit{and} \quad d \circ \beta_2 = \beta_3 \circ \text{div}.$$

Proof. The maps α_i and β_j are those defined in the proof of the preceding theorem. Only the second equation needs checking; we leave it to you. □

EXERCISES

1. Prove Theorems 31.1 and 31.2.
2. Note that in the case $n = 2$, Theorem 31.1 gives us two maps α_1 and β_1 from vector fields to 1-forms. Compare them.
3. Let A be an open set in $\mathbf{R}^3$.
   (a) Translate the equation $d(d\omega) = 0$ into two theorems about vector and scalar fields in $\mathbf{R}^3$.
   (b) Translate the condition that A is homologically trivial in dimension k into a statement about vector and scalar fields in A. Consider the cases $k = 0, 1, 2$.
4. For $\mathbf{R}^4$, there is a way of translating theorems about forms into more familiar language, if one allows oneself to use matrix fields as well as vector fields and scalar fields. We outline it here. The complications involved may help you understand why the language of forms was invented to deal with $\mathbf{R}^n$ in general.

   A square matrix B is said to be skew-symmetric if $B^{\text{tr}} = -B$. Let A be an open set in $\mathbf{R}^4$. Let $S(A)$ be the set of all C^∞ functions H mapping A into the set of 4 by 4 skew-symmetric matrices. If $h_{ij}(\mathbf{x})$ denotes the entry of $H(\mathbf{x})$ in row i and column j, define $\gamma_2 : S(A) \to \Omega^2(A)$ by the equation

$$\gamma_2(H) = \sum_{i<j} h_{ij}(\mathbf{x})\, dx_i \wedge dx_j.$$

(a) Show γ_2 is a linear isomorphism.

(b) Let $\alpha_0, \alpha_1, \beta_3, \beta_4$ be defined as in Theorem 31.1. Define operators "twist" and "spin" as in the following diagram:

$$\begin{array}{ccc} \text{Vector fields in } A & \xrightarrow{\alpha_1} & \Omega^1(A) \\ \downarrow{\scriptstyle \text{twist}} & & \downarrow{\scriptstyle d} \\ S(A) & \xrightarrow{\gamma_2} & \Omega^2(A) \\ \downarrow{\scriptstyle \text{spin}} & & \downarrow{\scriptstyle d} \\ \text{Vector fields in } A & \xrightarrow{\beta_3} & \Omega^3(A) \end{array}$$

such that

$$d \circ \alpha_1 = \gamma_2 \circ \text{twist} \quad \text{and} \quad d \circ \gamma_2 = \beta_3 \circ \text{spin}.$$

(These operators are facetious analogues in $\mathbf{R}^4$ of the operator "curl" in $\mathbf{R}^3$.)

5. The operators grad, curl, and div, and the translation operators α_i and β_j, seem to depend on the choice of a basis in $\mathbf{R}^n$, since the formulas defining them involve the components of the vectors involved relative to the basis $\mathbf{e}_1, \cdots, \mathbf{e}_n$ in $\mathbf{R}^n$. However, they in fact depend only on the inner product in $\mathbf{R}^n$ and the notion of right-handedness, as the following exercise shows.

Recall that the k-volume function $V(\mathbf{x}_1, \ldots, \mathbf{x}_k)$ depends only on the inner product in $\mathbf{R}^n$. (See the exercises of §21.)

(a) Let $F(\mathbf{x}) = (\mathbf{x}; f(\mathbf{x}))$ be a vector field defined in an open set of $\mathbf{R}^n$. Show that $\alpha_1 F$ is the unique 1-form such that

$$\alpha_1 F(\mathbf{x})(\mathbf{x}; \mathbf{v}) = \langle f(\mathbf{x}), \mathbf{v} \rangle.$$

(b) Let $G(\mathbf{x}) = (\mathbf{x}; g(\mathbf{x}))$ be a vector field defined in an open set of $\mathbf{R}^n$. Show that $\beta_{n-1} G$ is the unique $n-1$ form such that

$$\beta_{n-1} G(\mathbf{x})\big((\mathbf{x}; \mathbf{v}_1), \ldots, (\mathbf{x}; \mathbf{v}_{n-1})\big) = \epsilon \cdot V(g(\mathbf{x}), \mathbf{v}_1, \ldots, \mathbf{v}_{n-1}),$$

where $\epsilon = +1$ if the frame $(g(\mathbf{x}), \mathbf{v}_1, \ldots, \mathbf{v}_{n-1})$ is right-handed, and $\epsilon = -1$ otherwise.

(c) Let h be a scalar field defined in an open set of $\mathbf{R}^n$. Show that $\beta_n h$ is the unique n-form such that

$$\beta_n h(\mathbf{x})\big((\mathbf{x}; \mathbf{v}_1), \ldots, (\mathbf{x}; \mathbf{v}_n)\big) = \epsilon \cdot h(\mathbf{x}) \cdot V(\mathbf{v}_1, \ldots, \mathbf{v}_n),$$

where $\epsilon = +1$ if $(\mathbf{v}_1, \ldots, \mathbf{v}_n)$ is right-handed, and $\epsilon = -1$ otherwise.

(d) Conclude that the operators grad and div (and curl if $n = 3$) depend only on the inner product in $\mathbf{R}^n$ and the notion of right-handedness in $\mathbf{R}^n$. [*Hint:* The operator d depends only on the vector space structure of $\mathbf{R}^n$.]

§32. THE ACTION OF A DIFFERENTIABLE MAP

If $\alpha : A \to \mathbf{R}^n$ is a C^∞ map, where A is open in $\mathbf{R}^k$, then α gives rise to a linear transformation α_* mapping the tangent space to $\mathbf{R}^k$ at $\mathbf{x}$ into the tangent space to $\mathbf{R}^n$ at $\alpha(\mathbf{x})$. Furthermore, we know that any linear transformation $T : V \to W$ of vector spaces gives rise to a dual transformation $T^* : \mathcal{A}^\ell(W) \to \mathcal{A}^\ell(V)$ of alternating tensors. We combine these two facts to show how a C^∞ map α gives rise to a dual transformation of forms, which we denote by α^*. The transformation α^* preserves all the structure we have imposed on the space of forms—the vector space structure, the wedge product, and the differential operator.

Definition. Let A be open in $\mathbf{R}^k$; let $\alpha : A \to \mathbf{R}^n$ be of class C^∞; let B be an open set of $\mathbf{R}^n$ containing $\alpha(A)$. We define a **dual transformation of forms**

$$\alpha^* : \Omega^\ell(B) \to \Omega^\ell(A)$$

as follows: Given a 0-form $f : B \to \mathbf{R}$ on B, we define a 0-form $\alpha^* f$ on A by setting $(\alpha^* f)(\mathbf{x}) = f(\alpha(\mathbf{x}))$ for each $\mathbf{x} \in A$. Then, given an ℓ-form ω on B with $\ell > 0$, we define an ℓ-form $\alpha^*\omega$ on A by the equation

$$(\alpha^*\omega)(\mathbf{x})((\mathbf{x};\mathbf{v}_1), \ldots, (\mathbf{x};\mathbf{v}_\ell)) = \omega(\alpha(\mathbf{x}))\,(\alpha_*(\mathbf{x};\mathbf{v}_1), \ldots, \alpha_*(\mathbf{x};\mathbf{v}_\ell)).$$

Since f and ω and α and $D\alpha$ are all of class C^∞, so are the forms $\alpha^* f$ and $\alpha^*\omega$. Note that if f and ω and α were of class C^r, then $\alpha^* f$ would be of class C^r but $\alpha^*\omega$ would only be of class C^{r-1}. Here again it is convenient to have restricted ourselves to C^∞ maps.

Note that if α is a constant map, then $\alpha^* f$ is also constant, and $\alpha^*\omega$ is the 0-tensor.

The relation between α^* and the dual of the linear transformation α_* is the following: Given $\alpha : A \to \mathbf{R}^n$ of class C^∞, with $\alpha(\mathbf{x}) = \mathbf{y}$, it induces the linear transformation

$$T = \alpha_* : \mathcal{T}_{\mathbf{x}}(\mathbf{R}^k) \to \mathcal{T}_{\mathbf{y}}(\mathbf{R}^n);$$

this transformation in turn gives rise to a dual transformation of alternating tensors,

$$T^* : \mathcal{A}^\ell(\mathcal{T}_{\mathbf{y}}(\mathbf{R}^n)) \to \mathcal{A}^\ell(\mathcal{T}_{\mathbf{x}}(\mathbf{R}^k)).$$

If ω is an ℓ-form on B, then $\omega(\mathbf{y})$ is an alternating tensor on $\mathcal{T}_{\mathbf{y}}(\mathbf{R}^n)$, so that $T^*(\omega(\mathbf{y}))$ is an alternating tensor on $\mathcal{T}_{\mathbf{x}}(\mathbf{R}^k)$. It satisfies the equation

$$T^*(\omega(\mathbf{y})) = (\alpha^*\omega)(\mathbf{x});$$

for

$$T^*(\omega(\mathbf{y}))\,((\mathbf{x};\mathbf{v}_1),\ldots,(\mathbf{x};\mathbf{v}_\ell)) = \omega(\alpha(\mathbf{x}))\,(\alpha_*(\mathbf{x};\mathbf{v}_1),\ldots,\alpha_*(\mathbf{x};\mathbf{v}_\ell))$$
$$= (\alpha^*\omega)(\mathbf{x})\,((\mathbf{x};\mathbf{v}_1),\ldots,(\mathbf{x};\mathbf{v}_\ell)).$$

This fact enables us to rewrite earlier results concerning the dual transformation T^* as results about forms:

Theorem 32.1. *Let A be open in $\mathbf{R}^k$; let $\alpha : A \to \mathbf{R}^m$ be a C^∞ map. Let B be open in $\mathbf{R}^m$ and contain $\alpha(A)$; let $\beta : B \to \mathbf{R}^n$ be a C^∞ map. Let ω, η, θ be forms defined in an open set C of $\mathbf{R}^n$ containing $\beta(B)$; assume ω and η have the same order. The transformations α^* and β^* have the following properties:*

(1) $\beta^*(a\omega + b\eta) = a(\beta^*\omega) + b(\beta^*\eta)$.

(2) $\beta^*(\omega \wedge \theta) = \beta^*\omega \wedge \beta^*\theta$.

(3) $(\beta \circ \alpha)^*\omega = \alpha^*(\beta^*\omega)$.

Proof. See Figure 32.1. In the case of forms of positive order, properties (1) and (3) are merely restatements, in the language of forms, of Theorem 26.5, and (2) is a restatement of (6) of Theorem 28.1.

Checking the properties when some or all of the forms have order zero is a computation we leave to you. □

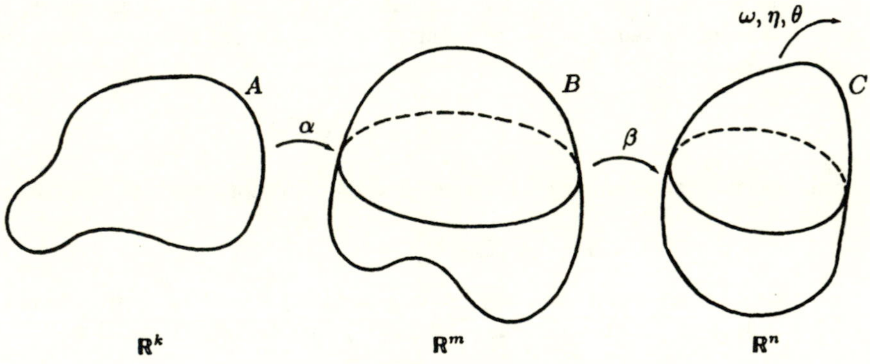

Figure 32.1

This theorem shows that α^* preserves the vector space structure and the wedge product. We now show it preserves the operator d. For this purpose (and later purposes as well), we obtain a formula for computing $\alpha^*\omega$. If A is open in $\mathbf{R}^k$ and $\alpha : A \to \mathbf{R}^n$, we derive this formula in two cases—when ω is a 1-form and when ω is a k-form. This is all we shall need. The general case is treated in the exercises.

Since α^* is linear and preserves wedge products, and since $\alpha^* f$ equals $f \circ \alpha$, it remains only to compute α^* for elementary 1-forms and elementary k-forms. Here is the required formula:

Theorem 32.2. *Let A be open in $\mathbf{R}^k$; let $\alpha : A \to \mathbf{R}^n$ be a C^∞ map. Let $\mathbf{x}$ denote the general point of $\mathbf{R}^k$; let $\mathbf{y}$ denote the general point of $\mathbf{R}^n$. Then dx_i and dy_i denote the elementary 1-forms in $\mathbf{R}^k$ and $\mathbf{R}^n$, respectively.*

(a) $\alpha^*(dy_i) = d\alpha_i$.

(b) *If $I = (i_1, \ldots, i_k)$ is an ascending k-tuple from the set $\{1, \ldots, n\}$, then*

$$\alpha^*(dy_I) = (\det \frac{\partial \alpha_I}{\partial \mathbf{x}})\, dx_1 \wedge \cdots \wedge dx_k,$$

where

$$\frac{\partial \alpha_I}{\partial \mathbf{x}} = \frac{\partial(\alpha_{i_1}, \ldots, \alpha_{i_k})}{\partial(x_1, \ldots, x_k)}.$$

Proof. (a) Set $\mathbf{y} = \alpha(\mathbf{x})$. We compute the value of $\alpha^*(dy_i)$ on a typical tangent vector as follows:

$$\begin{aligned}
\big(\alpha^*(dy_i)\big)(\mathbf{x})(\mathbf{x};\mathbf{v}) &= dy_i(\mathbf{y})\big(\alpha_*(\mathbf{x};\mathbf{v})\big) \\
&= i^{\text{th}} \text{ component of } \quad (D\alpha(\mathbf{x}) \cdot \mathbf{v}) \\
&= \sum_{j=1}^{k} D_j\alpha_i(\mathbf{x}) \cdot v_j \\
&= \sum_{j=1}^{k} \frac{\partial \alpha_i}{\partial x_j}(\mathbf{x})\, dx_j(\mathbf{x})(\mathbf{x};\mathbf{v}).
\end{aligned}$$

It follows that

$$\alpha^*(dy_i) = \sum_{j=1}^{k} \frac{\partial \alpha_i}{\partial x_j}\, dx_j.$$

By Theorem 30.3, the latter expression equals $d\alpha_i$.

(b) The form $\alpha^*(dy_I)$ is a k-form defined in an open set of $\mathbf{R}^k$, so it has the form

$$\alpha^*(dy_I) = h\, dx_1 \wedge \cdots \wedge dx_k$$

for some scalar function h. If we evaluate the right side of this equation on the k-tuple $(\mathbf{x};\mathbf{e}_1), \ldots, (\mathbf{x};\mathbf{e}_k)$, we obtain the function $h(\mathbf{x})$. The theorem then follows from the following computation:

$$\begin{aligned} h(\mathbf{x}) &= \big(\alpha^*(dy_I)\big)(\mathbf{x})\big((\mathbf{x};\mathbf{e}_1), \ldots, (\mathbf{x};\mathbf{e}_k)\big) \\ &= dy_I(\mathbf{y})\big(\alpha_*(\mathbf{x};\mathbf{e}_1), \ldots, \alpha_*(\mathbf{x};\mathbf{e}_k)\big) \\ &= dy_I(\mathbf{y})\big((\mathbf{y};\partial\alpha/\partial x_1), \ldots, (\mathbf{y};\partial\alpha/\partial x_k)\big) \\ &= \det[D\alpha(\mathbf{x})]_I \\ &= \det \frac{\partial \alpha_I}{\partial \mathbf{x}}. \quad \square \end{aligned}$$

It is easy to remember the formula (a); to compute $\alpha^*(dy_i)$, one simply takes the form dy_i and makes the substitution $y_i = \alpha_i(\mathbf{x})$!

Note that one could compute $\alpha^*(dy_I)$ by the formula

$$\begin{aligned} \alpha^*(dy_I) &= \alpha^*(dy_{i_1}) \wedge \cdots \wedge \alpha^*(dy_{i_k}) \\ &= d\alpha_{i_1} \wedge \cdots \wedge d\alpha_{i_k}, \end{aligned}$$

but the computation of this wedge product is laborious if $k > 2$.

Theorem 32.3. *Let A be open in $\mathbf{R}^k$; let $\alpha : A \to \mathbf{R}^n$ be of class C^∞. If ω is an ℓ-form defined in an open set of $\mathbf{R}^n$ containing $\alpha(A)$, then*

$$\alpha^*(d\omega) = d(\alpha^*\omega).$$

Proof. Let $\mathbf{x}$ denote the general point of $\mathbf{R}^k$; let $\mathbf{y}$ denote the general point of $\mathbf{R}^n$.

Step 1. We verify the theorem first for a 0-form f. We compute the left side of the equation as follows:

$$(*) \qquad \begin{aligned} \alpha^*(df) &= \alpha^*\Big(\sum_{i=1}^{n} (D_i f)\, dy_i\Big) \\ &= \sum_{i=1}^{n} ((D_i f) \circ \alpha)\, d\alpha_i. \end{aligned}$$

Then we compute the right side of the equation. We have

$$(**)\qquad d(\alpha^* f) = d(f \circ \alpha) = \sum_{j=1}^{k} D_j(f \circ \alpha)\, dx_j.$$

We now apply the chain rule. Setting $\mathbf{y} = \alpha(\mathbf{x})$, we have

$$D(f \circ \alpha)(\mathbf{x}) = Df(\mathbf{y}) \cdot D\alpha(\mathbf{x});$$

since $D(f \circ \alpha)$ and Df are row matrices, it follows that

$$D_j(f \circ \alpha)(\mathbf{x}) = Df(\mathbf{y}) \cdot (j^{\text{th}} \text{ column of } D\alpha(\mathbf{x})) = \sum_{i=1}^{n} D_i f(\mathbf{y}) \cdot D_j \alpha_i(\mathbf{x}).$$

Thus

$$D_j(f \circ \alpha) = \sum_{i=1}^{n} ((D_i f) \circ \alpha) \cdot D_j \alpha_i.$$

Substituting this result in the equation $(**)$, we have

$$(***)\qquad d(\alpha^* f) = \sum_{j=1}^{k} \sum_{i=1}^{n} ((D_i f) \circ \alpha) \cdot D_j \alpha_i\, dx_j = \sum_{i=1}^{n} ((D_i f) \circ \alpha)\, d\alpha_i.$$

Comparing $(*)$ and $(***)$, we see that $\alpha^*(df) = d(\alpha^* f)$.

Step 2. We prove the theorem for forms of positive order. Since α^* and d are linear, it suffices to treat the case $\omega = f\, dy_I$, where $I = (i_1, \ldots, i_\ell)$ is an ascending ℓ-tuple from the set $\{1, \ldots, n\}$. We first compute

$$(\dagger)\qquad \alpha^*(d\omega) = \alpha^*(df \wedge dy_I) = \alpha^*(df) \wedge \alpha^*(dy_I).$$

On the other hand,

$$(\dagger\dagger)\qquad d(\alpha^*\omega) = d[\alpha^*(f \wedge dy_I)] = d[(\alpha^* f) \wedge \alpha^*(dy_I)] = d(\alpha^* f) \wedge \alpha^*(dy_I) + (\alpha^* f) \wedge 0,$$

since

$$d(\alpha^*(dy_I)) = d(d\alpha_{i_1} \wedge \cdots \wedge d\alpha_{i_l}) = 0.$$

The theorem follows by comparing (†) and (††) and using the result of Step 1. □

We now have the algebra of differential forms at our disposal, along with differential operator d. The basic properties of this algebra and the operator d, as summarized in this section and §30, are all we shall need in the sequel.

It is at this point, where one is dealing with the action of a differentiable map, that one begins to see that forms are in some sense more natural objects to deal with than are vector fields. A C^∞ map $\alpha : A \to \mathbf{R}^n$, where A is open in $\mathbf{R}^k$, gives rise to a linear transformation α_* on tangent vectors. But there is no way to obtain from α a transformation that carries a vector *field* on A to a vector *field* on $\alpha(A)$. Suppose for instance that $F(\mathbf{x}) = (\mathbf{x}; f(\mathbf{x}))$ is a vector field in A. If $\mathbf{y}$ is a point of the set $B = \alpha(A)$ such that $\mathbf{y} = \alpha(\mathbf{x}_1) = \alpha(\mathbf{x}_2)$ for two distinct points $\mathbf{x}_1, \mathbf{x}_2$ of A, then α_* gives rise to two (possibly different) tangent vectors $\alpha_*(\mathbf{x}_1; f(\mathbf{x}_1))$ and $\alpha_*(\mathbf{x}_2; f(\mathbf{x}_2))$ at $\mathbf{y}$! See Figure 32.2.

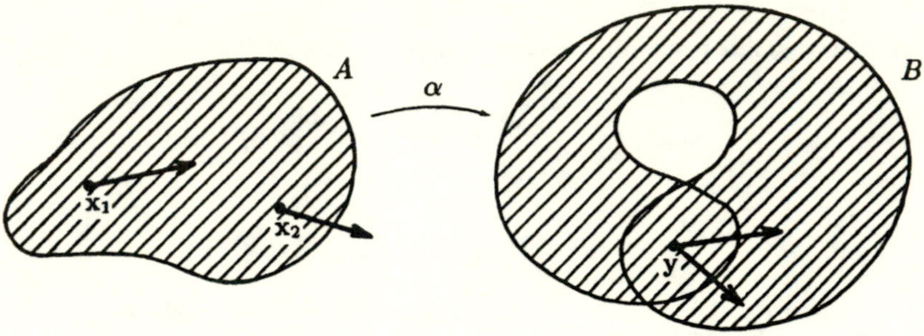

Figure 32.2

This problem does not occur if $\alpha : A \to B$ is a diffeomorphism. In this case, one can obtain an induced map $\tilde{\alpha}_*$ on vector fields. One assigns to the vector field F on A, the vector field $G = \tilde{\alpha}_* F$ on B defined by the equation

$$G(\mathbf{y}) = \alpha_*(F(\alpha^{-1}(\mathbf{y}))).$$

A scalar field h on A gives rise to a scalar field $k = \tilde{\alpha}_* h$ on B defined by the equation $k = h \circ \alpha^{-1}$. The map $\tilde{\alpha}_*$ is not however very natural, for it does not in general commute with the operators grad, curl, and div of vector calculus, nor with the "translation" operators α_i and β_j of §31. See the exercises.

EXERCISES

1. Prove Theorem 32.1 when ω and η have order zero and when θ has order zero.

2. Let $\alpha : \mathbf{R}^3 \to \mathbf{R}^6$ be a C^∞ map. Show directly that

$$d\alpha_1 \wedge d\alpha_3 \wedge d\alpha_5 = \big(\det D\alpha(1,3,5)\big)\, dx_1 \wedge dx_2 \wedge dx_3.$$

3. In $\mathbf{R}^3$, let

$$\omega = xy\,dx + 2z\,dy - y\,dz.$$

Let $\alpha : \mathbf{R}^2 \to \mathbf{R}^3$ be given by the equation

$$\alpha(u,v) = (uv,\ u^2,\ 3u + v).$$

Calculate $d\omega$ and $\alpha^*\omega$ and $\alpha^*(d\omega)$ and $d(\alpha^*\omega)$ directly.

4. Show that (a) of Theorem 32.2 is equivalent to the formula $\alpha^*(dy_i) = d(\alpha^* y_i)$, where $y_i : \mathbf{R}^n \to \mathbf{R}$ is the ith projection function in $\mathbf{R}^n$.

5. Prove the following formula for computing $\alpha^*\omega$ in general:

Theorem. *Let A be open in $\mathbf{R}^k$; let $\alpha : A \to \mathbf{R}^n$ be of class C^∞. Let $\mathbf{x}$ denote the general point of $\mathbf{R}^k$; let $\mathbf{y}$ denote the general point of $\mathbf{R}^n$. If $I = (i_1, \ldots, i_\ell)$ is an ascending ℓ-tuple from the set $\{1, \ldots, n\}$, then*

$$\alpha^*(dy_I) = \sum_{[J]} (\det \frac{\partial \alpha_I}{\partial x_J})\, dx_J.$$

Here $J = (j_1, \ldots, j_\ell)$ is an ascending ℓ-tuple from the set $\{1, \ldots, k\}$ and

$$\frac{\partial \alpha_I}{\partial x_J} = \frac{\partial(\alpha_{i_1}, \ldots, \alpha_{i_\ell})}{\partial(x_{j_1}, \ldots, x_{j_\ell})}.$$

*6. This exercise shows that the transformations α_i and β_j of §31 do not in general behave well with respect to the maps induced by a diffeomorphism α.

Let $\alpha : A \to B$ be a diffeomorphism of open sets in $\mathbf{R}^n$. Let $\mathbf{x}$ denote the general point of A, and let $\mathbf{y}$ denote the general point of B. If $F(\mathbf{x}) = \big(\mathbf{x}; f(\mathbf{x})\big)$ is a vector field in A, let $G(\mathbf{y}) = \alpha_*\big(F(\alpha^{-1}(\mathbf{y}))\big)$ be the corresponding vector field in B.

(a) Show that the 1-forms $\alpha_1 G$ and $\alpha_1 F$ do not in general correspond under the map α^*. Specifically, show that $\alpha^*(\alpha_1 G) = \alpha_1 F$ for all F if and only if $D\alpha(\mathbf{x})$ is an orthogonal matrix for each $\mathbf{x}$. [*Hint:* Show the equation $\alpha^*(\alpha_1 G) = \alpha_1 F$ is equivalent to the equation

$$D\alpha(\mathbf{x})^{\text{tr}} \cdot D\alpha(\mathbf{x}) \cdot f(\mathbf{x}) = f(\mathbf{x}).]$$

(b) Show that $\alpha^*(\beta_{n-1} G) = \beta_{n-1} F$ for all F if and only if $\det D\alpha = +1$. [*Hint:* Show the equation $\alpha^*(\beta_{n-1} G) = \beta_{n-1} F$ is equivalent to the equation $f(\mathbf{x}) = \big(\det D\alpha(\mathbf{x})\big) \cdot f(\mathbf{x})$.]

(c) If h is a scalar field in A, let $k = h \circ \alpha^{-1}$ be the corresponding scalar field in B. Show that $\alpha^*(\beta_n k) = \beta_n h$ for all h if and only if $\det D\alpha = +1$.

7. Use Exercise 6 to show that if α is an orientation-preserving isometry of $\mathbf{R}^n$, then the operator $\tilde{\alpha}_*$ on vector fields and scalar fields commutes with the operators grad and div, and with curl if $n = 3$. (Compare Exercise 5 of §31.)

7

Stokes' Theorem

We saw in the last chapter how k-forms provide a generalization to $\mathbf{R}^n$ of the notions of scalar and vector fields in $\mathbf{R}^3$, and how the differential operator d provides a generalization of the operators grad, curl, and div. Now we define the integral of a k-form over a k-manifold; this concept provides a generalization to $\mathbf{R}^n$ of the notions of line and surface integrals in $\mathbf{R}^3$. Just as line and surface integrals are involved in the statements of the classical Stokes' theorem and divergence theorem in $\mathbf{R}^3$, so are integrals of k-forms over k-manifolds involved in the generalized version of these theorems.

We recall here our convention that all manifolds, forms, vector fields, and scalar fields are assumed to be of class C^∞.

§33. INTEGRATING FORMS OVER PARAMETRIZED-MANIFOLDS

In Chapter 5, we defined the integral of a scalar function f over a manifold, with respect to volume. We follow a similar procedure here in defining the integral of a form of order k over a manifold of dimension k. We begin with parametrized-manifolds.

First let us consider a special case.

Definition. Let A be an open set in $\mathbf{R}^k$; let η be a k-form defined in A. Then η can be written uniquely in the form

$$\eta = f\, dx_1 \wedge \cdots \wedge dx_k.$$

We define the **integral of η over A** by the equation

$$\int_A \eta = \int_A f,$$

provided the latter integral exists.

This definition seems to be coordinate-dependent; in order to define $\int_A \eta$, we expressed η in terms of the standard elementary 1-forms dx_i, which depend on the choice of the standard basis $\mathbf{e}_1, \ldots, \mathbf{e}_k$ in $\mathbf{R}^k$. One can, however, formulate the definition in a coordinate-free fashion. Specifically, if $\mathbf{a}_1, \ldots, \mathbf{a}_k$ is *any* right-handed orthonormal basis for $\mathbf{R}^k$, then it is an elementary exercise to show that

$$\int_A \eta = \int_{\mathbf{x} \in A} \eta(\mathbf{x})\big((\mathbf{x};\mathbf{a}_1), \ldots, (\mathbf{x};\mathbf{a}_k)\big).$$

Thus the integral of η does not depend on the choice of basis in $\mathbf{R}^k$, although it does depend on the orientation of $\mathbf{R}^k$.

We now define the integral of a k-form over a parametrized-manifold of dimension k.

Definition. Let A be open in $\mathbf{R}^k$; let $\alpha : A \to \mathbf{R}^n$ be of class C^∞. The set $Y = \alpha(A)$, together with the map α, constitute the parametrized-manifold Y_α. If ω is a k-form defined in an open set of $\mathbf{R}^n$ containing Y, we define the **integral of ω over Y_α** by the equation

$$\int_{Y_\alpha} \omega = \int_A \alpha^* \omega,$$

provided the latter integral exists. Since α^* and $\int_A$ are linear, so is this integral.

We now show that the integral is invariant under reparametrization, *up to sign.*

Theorem 33.1. *Let $g : A \to B$ be a diffeomorphism of open sets in $\mathbf{R}^k$. Assume* $\det Dg$ *does not change sign on A. Let $\beta : B \to \mathbf{R}^n$ be a map of class C^∞; let $Y = \beta(B)$. Let $\alpha = \beta \circ g$; then $\alpha : A \to \mathbf{R}^n$ and $Y = \alpha(A)$. If ω is a k-form defined in an open set of $\mathbf{R}^n$ containing Y, then ω is integrable over Y_β if and only if it is integrable over Y_α; in this case,*

$$\int_{Y_\alpha} \omega = \pm \int_{Y_\beta} \omega,$$

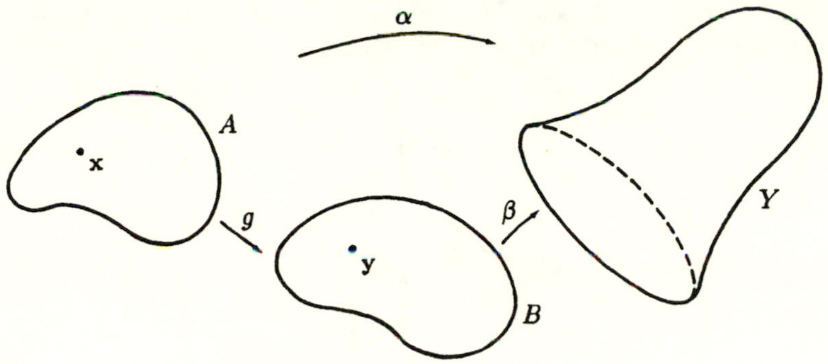

Figure 33.1

where the sign agrees with the sign of $\det Dg$.

Proof. Let $\mathbf{x}$ denote the general point of A; let $\mathbf{y}$ denote the general point of B. See Figure 33.1. We wish to show that

$$\int_A \alpha^* \omega = \epsilon \int_B \beta^* \omega,$$

where $\epsilon = \pm 1$ and agrees with the sign of $\det Dg$. If we set $\eta = \beta^* \omega$, then this equation is equivalent to the equation

$$\int_A g^* \eta = \epsilon \int_B \eta.$$

Let us write η in the form $\eta = f\ dy_1 \wedge \cdots \wedge dy_k$. Then

$$\begin{aligned} g^* \eta &= (f \circ g) g^* (dy_1 \wedge \cdots \wedge dy_k) \\ &= (f \circ g) \det(Dg)\, dx_1 \wedge \cdots \wedge dx_k. \end{aligned}$$

(Here we apply Theorem 32.2, in the case $k = n$.) Our equation then takes the form

$$\int_A (f \circ g) \det Dg = \epsilon \int_B f.$$

This equation follows at once from the change of variables theorem, since $\det Dg = \epsilon \left|\det Dg\right|$. □

We remark that if A is connected (that is, if A cannot be written as the union of two disjoint nonempty open sets), then the hypothesis that $\det Dg$ does not change sign on A is automatically satisfied. For the set of points where $\det Dg$ is positive is open, and so is the set of points where it is negative.

This integral is fairly easy to compute in practice. One has the following result:

Theorem 33.2. *Let A be open in $\mathbf{R}^k$; let $\alpha : A \to \mathbf{R}^n$ be of class C^∞; let $Y = \alpha(A)$. Let $\mathbf{x}$ denote the general point of A; and let $\mathbf{z}$ denote the general point of $\mathbf{R}^n$. If*

$$\omega = f\, dz_I$$

is a k-form defined in an open set of $\mathbf{R}^n$ containing Y, then

$$\int_{Y_\alpha} \omega = \int_A (f \circ \alpha) \det(\partial \alpha_I / \partial \mathbf{x}).$$

Proof. Applying Theorem 32.2, we have

$$\alpha^* \omega = (f \circ \alpha) \det(\partial \alpha_I / \partial \mathbf{x})\, dx_1 \wedge \cdots \wedge dx_k.$$

The theorem follows. □

The notion of a k-form is a rather abstract one; the notion of its integral over a parametrized-manifold is even more abstract. In a later section (§36) we discuss a geometric interpretation of k-forms and of their integrals that gives some insight into their intuitive meaning.

REMARK. We can now make sense of the "dx" notation commonly used in single-variable calculus. If $\eta = f\, dx$ is a 1-form defined in the open interval $A = (a, b)$ of the real line $\mathbf{R}$, then

$$\int_A \eta = \int_A f$$

by definition. That is,

$$\int_A f\, dx = \int_a^b f,$$

where the notation on the left denotes the integral of a *form;* and the notation on the right denotes the integral of a *function!* They are equal by definition. Thus the "dx" notation used in connection with single integrals in calculus makes perfect sense once one has studied differential forms.

One can also make sense of the notation commonly used in calculus to denote a *line integral.* Given a 1-form $P\, dx + Q\, dy + R\, dz$, defined in an open set A of $\mathbf{R}^3$, and given a parametrized-curve $\gamma : (a, b) \to A$, one has by the preceding theorem the formula

$$\int_{C_\gamma} P\, dx + Q\, dy + R\, dz$$

$$= \int_{(a,b)} [P(\gamma(t)) \frac{d\gamma_1}{dt} + Q(\gamma(t)) \frac{d\gamma_2}{dt} + R(\gamma(t)) \frac{d\gamma_3}{dt}]\, dt,$$

where C is the image set of γ. *This is just the formula given in calculus for evaluating the line integral* $\int_C P\,dx + Q\,dy + R\,dz$. Thus the notation used for line integrals in calculus makes perfect sense once one has studied differential forms.

It is considerably more difficult, however, to make sense of the "$dx\,dy$" notation commonly used in calculus when dealing with double integrals. If f is a continuous bounded function defined on a subset A of $\mathbf{R}^2$, it is common in calculus to denote the integral of f over A by the symbol

$$\iint_A f(x,y)\;dx\;dy.$$

Here the symbol "$dx\,dy$" has no independent meaning, since the only product operation we have defined for 1-forms is the wedge product. One justification for this notation is that it resembles the notation for the iterated integral. And indeed, if A is the interior of a rectangle $[a,b] \times [c,d]$, then we have the equation

$$\int_c^d [\int_a^b f(x,y)\;dx]\,dy = \int_A f,$$

by the Fubini theorem. Another justification for this notation is that it resembles the notation for the integral of a 2-form, and one has the equation

$$\int_A f\;dx \wedge dy = \int_A f$$

by definition. But a difficulty arises when one reverses the roles of x and y. For the iterated integral, one has the equation

$$\int_a^b [\int_c^d f(x,y)\;dy]\,dx = \int_A f,$$

and for the integral of a 2-form, one has the equation

$$\int_A f\;dy \wedge dx = -\int_A f!$$

Which rule should one follow in dealing with the symbol

$$\iint_A f(x,y)\;dy\;dx\;?$$

Which ever choice one makes, confusion is likely to occur. For this reason, the "$dx\,dy$" notation is one we shall not use.

One could, however, use the "$\mathrm{d}V$" notation introduced in Chapter 5 without ambiguity. If A is open in $\mathbf{R}^k$, then A can be considered to be a parametrized-manifold that is parametrized by the identity map $\alpha : A \to A$! Then

$$\int_{A_\alpha} f\;\mathrm{d}V = \int_A (f \circ \alpha) V\big(D(\alpha)\big) = \int_A f,$$

since $D(\alpha)$ is the identity matrix. Of course, the symbol d used here bears no relation to the differential operator d.

EXERCISES

1. Let $A = (0,1)^2$. Let $\alpha : A \to \mathbf{R}^3$ be given by the equation

$$\alpha(u,v) = (u, v, u^2 + v^2 + 1).$$

Let Y be the image set of α. Evaluate the integral over Y_α of the 2-form $x_2\, dx_2 \wedge dx_3 + x_1 x_3\, dx_1 \wedge dx_3$.

2. Let $A = (0,1)^3$. Let $\alpha : A \to \mathbf{R}^4$ be given by the equation

$$\alpha(s,t,u) = (s, u, t, (2u - t)^2).$$

Let Y be the image set of α. Evaluate the integral over Y_α of the 3-form $x_1\, dx_1 \wedge dx_4 \wedge dx_3 + 2x_2 x_3\, dx_1 \wedge dx_2 \wedge dx_3$.

3. (a) Let A be the open unit ball in $\mathbf{R}^2$. Let $\alpha : A \to \mathbf{R}^3$ be given by the equation

$$\alpha(u,v) = (u, v, [1 - u^2 - v^2]^{1/2}).$$

Let Y be the image set of α. Evaluate the integral over Y_α of the form $(1/\|\mathbf{x}\|^m)(x_1\, dx_2 \wedge dx_3 - x_2\, dx_1 \wedge dx_3 + x_3\, dx_1 \wedge dx_2)$.

(b) Repeat (a) when

$$\alpha(u,v) = (u, v, -[1 - u^2 - v^2]^{1/2}).$$

4. If η is a k-form in $\mathbf{R}^k$, and if $\mathbf{a}_1, \ldots, \mathbf{a}_k$ is a basis for $\mathbf{R}^k$, what is the relation between the integrals

$$\int_A \eta \quad \text{and} \quad \int_{\mathbf{x} \in A} \eta(\mathbf{x})\big((\mathbf{x}; \mathbf{a}_1), \ldots, (\mathbf{x}; \mathbf{a}_k)\big) \quad ?$$

Show that if the frame $(\mathbf{a}_1, \ldots, \mathbf{a}_k)$ is orthonormal and right-handed, they are equal.

§34. ORIENTABLE MANIFOLDS

We shall define the integral of a k-form ω over a k-manifold M in much the same way that we defined the integral of a scalar function over M. First, we treat the case where the support of ω lies in a single coordinate patch $\alpha : U \to V$. In this case, we define

$$\int_M \omega = \int_{\text{Int } U} \alpha^*\omega.$$

However, this integral is invariant under reparametrization *only up to sign.* Therefore, in order that the integral $\int_M \omega$ be well-defined, we need an extra condition on M. That condition is called *orientability.* We discuss it in this section.

Definition. Let $g : A \to B$ be a diffeomorphism of open sets in $\mathbf{R}^k$. We say that g is **orientation-preserving** if $\det Dg > 0$ on A. We say g is **orientation-reversing** if $\det Dg < 0$ on A.

This definition generalizes the one given in §20. Indeed, there is associated with g a linear transformation of tangent spaces,

$$g_* : \mathcal{T}_{\mathbf{x}}(\mathbf{R}^k) \to \mathcal{T}_{g(\mathbf{x})}(\mathbf{R}^k),$$

given by the equation $g_*(\mathbf{x};\mathbf{v}) = (g(\mathbf{x}); Dg(\mathbf{x}) \cdot \mathbf{v})$. Then g is orientation-preserving if and only if for each $\mathbf{x}$, the linear transformation of $\mathbf{R}^k$ whose matrix is Dg is orientation-preserving in the sense previously defined.

Definition. Let M be a k-manifold in $\mathbf{R}^n$. Given coordinate patches $\alpha_i : U_i \to V_i$ on M for $i = 0, 1$, we say they **overlap** if $V_0 \cap V_1$ is non-empty. We say they **overlap positively** if the transition function $\alpha_1^{-1} \circ \alpha_0$ is orientation-preserving. If M can be covered by a collection of coordinate patches each pair of which overlap positively (if they overlap at all), then M is said to be **orientable.** Otherwise, M is said to be **non-orientable.**

Definition. Let M be a k-manifold in $\mathbf{R}^n$. Suppose M is orientable. Given a collection of coordinate patches covering M that overlap positively, let us adjoin to this collection all other coordinate patches on M that overlap these patches positively. It is easy to see that the patches in this expanded collection overlap one another positively. This expanded collection is called an **orientation** on M. A manifold M together with an orientation of M is called an **oriented manifold.**

This discussion makes no sense for a 0-manifold, which is just a discrete collection of points. We will discuss later what one might mean by "orientation" in this case.

If V is a vector space of dimension k, then V is also a k-manifold. We thus have two different notions of what is meant by an *orientation* of V. An orientation of V was defined in §20 to be a collection of k-frames in V; it is defined here to be a collection of coordinate patches on V. The connection between these two notions is easy to describe. Given an orientation of V in the sense of §20, we specify a corresponding orientation of V in the present sense as follows: For each frame $(\mathbf{v}_1, \ldots, \mathbf{v}_k)$ belonging to the given orientation of V, the linear isomorphism $\alpha : \mathbf{R}^k \to V$ such that $\alpha(\mathbf{e}_i) = \mathbf{v}_i$ for each i is a coordinate patch on V. Two such coordinate patches overlap positively, as you can check; the collection of all such specifies an orientation of V in the present sense.

Oriented manifolds in $\mathbf{R}^n$ of dimensions 1 and n−1 and n

In certain dimensions, the notion of orientation has a geometric interpretation that is easily described. This situation occurs when k equals 1 or $n-1$ or n. In the case $k = 1$, we can picture an orientation in terms of a tangent vector field, as we now show.

Definition. Let M be an oriented 1-manifold in $\mathbf{R}^n$. We define a corresponding unit tangent vector field T on M as follows: Given $\mathbf{p} \in M$, choose a coordinate patch $\alpha : U \to V$ on M about $\mathbf{p}$ belonging to the given orientation. Define

$$T(\mathbf{p}) = (\mathbf{p}; D\alpha(t_0)/\|D\alpha(t_0)\|),$$

where t_0 is the parameter value such that $\alpha(t_0) = \mathbf{p}$. Then T is called the **unit tangent field corresponding to the orientation of** M.

Note that $(\mathbf{p}; D\alpha(t_0))$ is the velocity vector of the curve α corresponding to the parameter value $t = t_0$; then $T(\mathbf{p})$ equals this vector divided by its length.

We show T is well-defined. Let β be a second coordinate patch on M about $\mathbf{p}$ belonging to the orientation of M. Let $\mathbf{p} = \beta(t_1)$ and let $g = \beta^{-1} \circ \alpha$. Then g is a diffeomorphism of a neighborhood of t_0 with a neighborhood of t_1, and

$$\begin{aligned} D\alpha(t_0) &= D(\beta \circ g)(t_0) \\ &= D\beta(t_1) \cdot Dg(t_0). \end{aligned}$$

Now $Dg(t_0)$ is a 1 by 1 matrix; since g is orientation-preserving, $Dg(t_0) > 0$. Then

$$D\alpha(t_0)/\,\|D\alpha(t_0)\| = D\beta(t_1)/\,\|D\beta(t_1)\|.$$

It follows that the vector field T is of class C^∞, since $t_0 = \alpha^{-1}(\mathbf{p})$ is a C^∞ function of $\mathbf{p}$ and $D\alpha(t)$ is a C^∞ function of t.

EXAMPLE 1. Given an oriented 1-manifold M, with corresponding unit tangent field T, we often picture the direction of T by drawing an arrow on the curve M itself. Thus an oriented 1-manifold gives rise to what is often called in calculus a *directed curve.* See Figure 34.1.

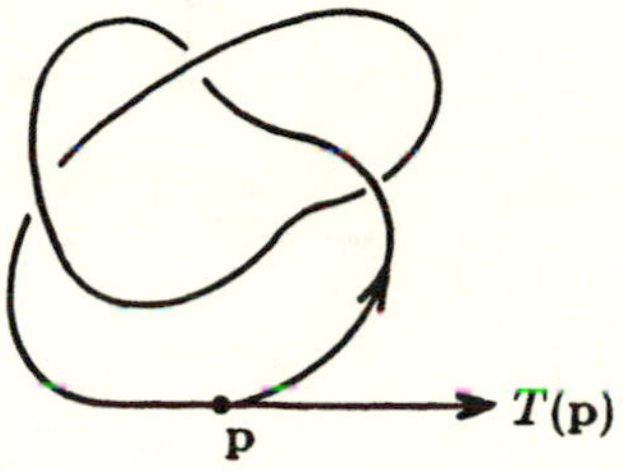

Figure 34.1

A difficulty arises if M has non-empty boundary. The problem is indicated in Figure 34.2, where ∂M consists of the two points $\mathbf{p}$ and $\mathbf{q}$. If $\alpha : U \to V$ is a coordinate patch about the boundary point $\mathbf{p}$ of M, the fact that U is open in $\mathbf{H}^1$ means that the corresponding unit tangent vector $T(\mathbf{p})$ must point *into* M from $\mathbf{p}$. Similarly, $T(\mathbf{q})$ points into M from $\mathbf{q}$. In the 1-manifold indicated, there is no way to define a unit tangent field on M that points into M at both $\mathbf{p}$ and $\mathbf{q}$. Thus it would seem that M is not orientable. Surely this is an anomaly.

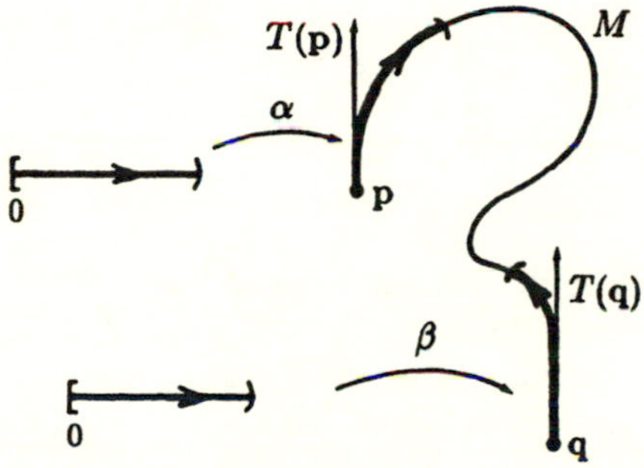

Figure 34.2

The problem disappears if we allow ourselves coordinate patches whose domains are open sets in $\mathbf{R}^1$ or $\mathbf{H}^1$ *or in the* left half-line $\mathbf{L}^1 = \{x \mid x \leq 0\}$. With this extra degree of freedom, it is easy to cover the manifold of the

previous example by coordinate patches that overlap positively. Three such patches are indicated in Figure 34.3.

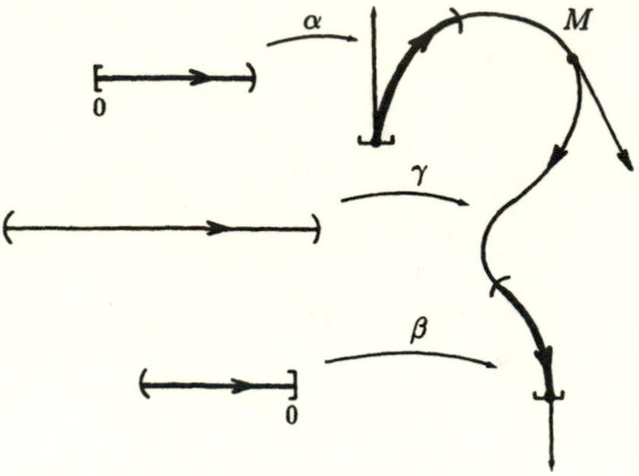

Figure 34.3

In view of the preceding example, we henceforth make the following convention:

Convention. *In the case of a 1-manifold M, we shall allow the domains of the coordinate patches on M to be open sets in* $\mathbf{R}^1$ *or in* $\mathbf{H}^1$ *or in* $\mathbf{L}^1$.

It is the case that, with this extra degree of freedom, every 1-manifold is orientable. We shall not prove this fact.

Now we consider the case where M is an $n-1$ manifold in $\mathbf{R}^n$. In this case, we can picture an orientation of M in terms of a unit *normal* vector field to M.

Definition. Let M be an $n-1$ manifold in $\mathbf{R}^n$. If $\mathbf{p} \in M$, let $(\mathbf{p};\mathbf{n})$ be a unit vector in the n-dimensional vector space $\mathcal{T}_\mathbf{p}(\mathbf{R}^n)$ that is orthogonal to the $n-1$ dimensional linear subspace $\mathcal{T}_\mathbf{p}(M)$. Then $\mathbf{n}$ is uniquely determined up to sign. Given an orientation of M, choose a coördinate patch $\alpha : U \to V$ on M about $\mathbf{p}$ belonging to this orientation; let $\alpha(\mathbf{x}) = \mathbf{p}$. Then the columns $\partial\alpha/\partial x_i$ of the matrix $D\alpha(\mathbf{x})$ give a basis

$$(\mathbf{p};\partial\alpha/\partial x_1), \ldots, (\mathbf{p};\partial\alpha/\partial x_{n-1})$$

for the tangent space to M at $\mathbf{p}$. We specify the sign of $\mathbf{n}$ by requiring that the frame

$$(\mathbf{n},\partial\alpha/\partial x_1, \ldots, \partial\alpha/\partial x_{n-1})$$

be right-handed, that is, that the matrix $[\mathbf{n}\quad D\alpha(\mathbf{x})]$ have positive determinant. We shall show in a later section that $\mathbf{n}$ is well-defined, independent of the choice of α, and that the resulting function $\mathbf{n}(\mathbf{p})$ is of class C^∞. The vector field $N(\mathbf{p}) = (\mathbf{p}; \mathbf{n}(\mathbf{p}))$ is called the **unit normal field to** M **corresponding to the orientation of** M.

EXAMPLE 2. We can now give an example of a manifold that is not orientable. The 2-manifold in $\mathbf{R}^3$ that is pictured in Figure 34.4 has no continuous unit normal vector field. You can convince yourself of this fact. This manifold is called the **Möbius band.**

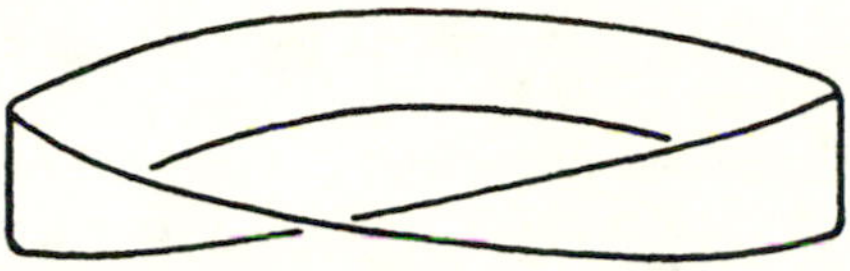

Figure 34.4

EXAMPLE 3. Another example of a non-orientable 2-manifold is the **Klein bottle.** It can be pictured in $\mathbf{R}^3$ as the self-intersecting surface of Figure 34.5. We think of K as the space swept out by a moving circle, as indicated in the figure. One can represent K as a 2-manifold *without* self-intersections in $\mathbf{R}^4$ as follows: Let the circle begin at position C_0, and move on to C_1, C_2, and so on. Begin with the circle lying in the subspace $\mathbf{R}^3 \times 0$ of $\mathbf{R}^4$; as it moves from C_0 to C_1, and on, let it remain in $\mathbf{R}^3 \times 0$. However, as the circle approaches the crucial spot where it would have to cross a part of the surface already generated, let it gradually move "up" into $\mathbf{R}^3 \times \mathbf{H}^1_+$ until it has passed the crucial spot, and then let it come back down gently into $\mathbf{R}^3 \times 0$ and continue on its way!

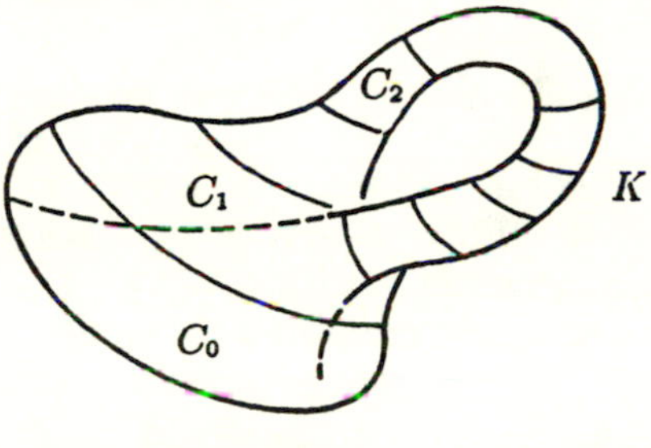

Figure 34.5

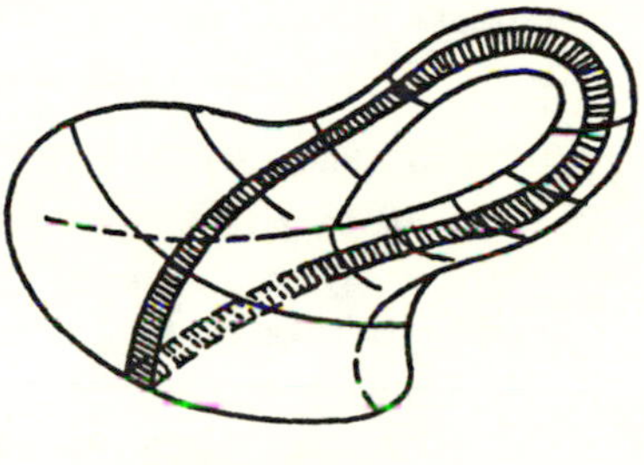

Figure 34.6

To see that K is not orientable, we need only note that K contains a copy of the Möbius band M. See Figure 34.6. If K were orientable, then M would be orientable as well. (Take all coordinate patches on M that overlap positively the coordinate patches belonging to the orientation of K.)

Finally, let us consider the case of an n-manifold M in $\mathbf{R}^n$. In this case, not only is M orientable, but it in fact has a "natural" orientation:

Definition. Let M be an n-manifold in $\mathbf{R}^n$. If $\alpha : U \to V$ is a coordinate patch on M, then $D\alpha$ is an n by n matrix. We define the **natural orientation** of M to consist of all coordinate patches on M for which $\det D\alpha > 0$. It is easy to see that two such patches overlap positively.

We must show M may be covered by such coordinate patches. Given $\mathbf{p} \in M$, let $\alpha : U \to V$ be a coordinate patch about $\mathbf{p}$. Now U is open in either $\mathbf{R}^n$ or $\mathbf{H}^n$; by shrinking U if necessary, we can assume that U is either an open ϵ-ball or the intersection with $\mathbf{H}^n$ of an open ϵ-ball. In either case, U is connected, so $\det D\alpha$ is either positive or negative on all of U. If the former, then α is our desired coordinate patch about $\mathbf{p}$; if the latter, then $\alpha \circ r$ is our desired coordinate patch about $\mathbf{p}$, where $r : \mathbf{R}^n \to \mathbf{R}^n$ is the map

$$r(x_1, x_2, \dots, x_n) = (-x_1, x_2, \dots, x_n).$$

Reversing the orientation of a manifold

Let $r : \mathbf{R}^k \to \mathbf{R}^k$ be the reflection map

$$r(x_1, x_2, \dots, x_k) = (-x_1, x_2, \dots, x_k);$$

it is its own inverse. The map r carries $\mathbf{H}^k$ to $\mathbf{H}^k$ if $k > 1$, and it carries $\mathbf{H}^1$ to the left half-line $\mathbf{L}^1$ if $k = 1$.

Definition. Let M be an oriented k-manifold in $\mathbf{R}^n$. If $\alpha_i : U_i \to V_i$ is a coordinate patch on M belonging to the orientation of M, let β_i be the coordinate patch

$$\beta_i = \alpha_i \circ r : r(U_i) \to V_i.$$

Then β_i overlaps α_i negatively, so it does not belong to the orientation of M. The coordinate patches β_i overlap each other positively, however (as you can check), so they constitute an orientation of M. It is called the **reverse**, or **opposite**, orientation to that specified by the coordinate patches α_i.

It follows that every orientable k-manifold M has at least two orientations, a given one and its opposite. If M is connected, it has only two (see

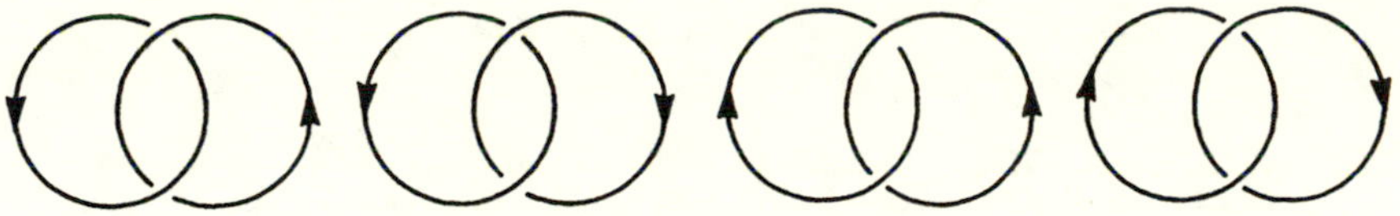

Figure 34.7

the exercises). Otherwise, it has more than two. The 1-manifold pictured in Figure 34.7, for example, has four orientations, as indicated.

We remark that if M is an oriented 1-manifold with corresponding tangent field T, then reversing the orientation of M results in replacing T by $-T$. For if $\alpha : U \to V$ is a coordinate patch belonging to the orientation of M, then $\alpha \circ r$ belongs to the opposite orientation. Now $(\alpha \circ r)(t) = \alpha(-t)$, so that $d(\alpha \circ r)/dt = -d\alpha/dt$.

Similarly, if M is an oriented $n-1$ manifold in $\mathbf{R}^n$ with corresponding normal field N, reversing the orientation of M results in replacing N by $-N$. For if $\alpha : U \to V$ belongs to the orientation of M, then $\alpha \circ r$ belongs to the opposite orientation. Now

$$\frac{\partial(\alpha \circ r)}{\partial x_1} = -\frac{\partial \alpha}{\partial x_1} \quad \text{and} \quad \frac{\partial(\alpha \circ r)}{\partial x_i} = \frac{\partial \alpha}{\partial x_i} \quad \text{if} \quad i > 1.$$

Furthermore, one of the frames

$$(\mathbf{n}, \frac{\partial \alpha}{\partial x_1}, \frac{\partial \alpha}{\partial x_2}, \dots, \frac{\partial \alpha}{\partial x_{n-1}}) \quad \text{and} \quad (-\mathbf{n}, -\frac{\partial \alpha}{\partial x_1}, \frac{\partial \alpha}{\partial x_2}, \dots, \frac{\partial \alpha}{\partial x_{n-1}})$$

is right-handed if and only if the other one is. Thus if $\mathbf{n}$ corresponds to the coordinate patch α, then $-\mathbf{n}$ corresponds to the coordinate patch $\alpha \circ r$.

The induced orientation of ∂M

Theorem 34.1. *Let $k > 1$. If M is an orientable k-manifold with non-empty boundary, then ∂M is orientable.*

Proof. Let $\mathbf{p} \in \partial M$; let $\alpha : U \to V$ be a coordinate patch about $\mathbf{p}$. There is a corresponding coordinate patch α_0 on ∂M that is said to be obtained by *restricting* α. (See §24.) Formally, if we define $b : \mathbf{R}^{k-1} \to \mathbf{R}^k$ by the equation

$$b(x_1, \dots, x_{k-1}) = (x_1, \dots, x_{k-1}, 0),$$

then $\alpha_0 = \alpha \circ b$.

We show that if α and β are coordinate patches about $\mathbf{p}$ that overlap positively, then so do their restrictions α_0 and β_0. Let $g : W_0 \to W_1$ be the

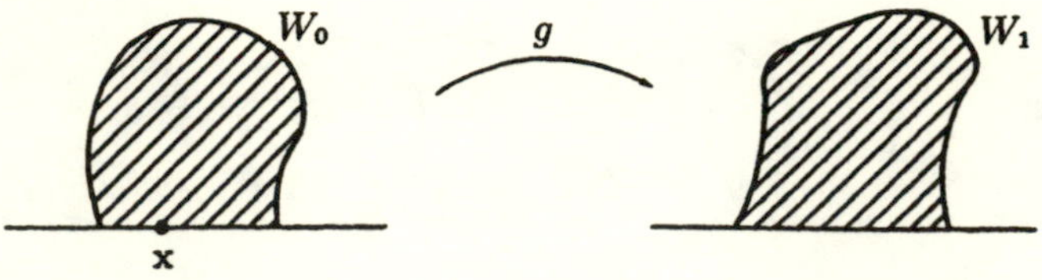

Figure 34.8

transition function $g = \beta^{-1} \circ \alpha$, where W_0 and W_1 are open in $\mathbf{H}^k$. Then $\det Dg > 0$. See Figure 34.8.

Now if $\mathbf{x} \in \partial\mathbf{H}^k$, then the derivative Dg of g at $\mathbf{x}$ has the last row

$$Dg_k = [0 \cdots 0 \quad \partial g_k/\partial x_k]$$

where $\partial g_k/\partial x_k \geq 0$. For if one begins at the point $\mathbf{x}$ and gives one of the variables $x_1, \ldots, x_{k-1}$ an increment, the value of g_k does not change, while if one gives the variable x_k a positive increment, the value of g_k increases; it follows that $\partial g_k/\partial x_j$ vanishes at $\mathbf{x}$ if $j < k$ and is non-negative if $j = k$.

Since $\det Dg \neq 0$, it follows that $\partial g_k/\partial x_k > 0$ at each point $\mathbf{x}$ of $\partial\mathbf{H}^k$. Then because $\det Dg > 0$, it follows that

$$\det \frac{\partial(g_1, \ldots, g_{k-1})}{\partial(x_1, \ldots, x_{k-1})} > 0.$$

But this matrix is just the derivative of the transition function for the coordinate patches α_0 and β_0 on ∂M. □

The proof of the preceding theorem shows that, given an orientation of M, one can obtain an orientation of ∂M by simply taking restrictions of coordinate patches that belong to the orientation of M. However, this orientation of ∂M is not always the one we prefer. We make the following definition:

Definition. Let M be an orientable k-manifold with non-empty boundary. Given an orientation of M, the corresponding **induced orientation** of ∂M is defined as follows: If k is even, it is the orientation obtained by simply restricting coordinate patches belonging to the orientation of M. If k is odd, it is the opposite of the orientation of ∂M obtained in this way.

EXAMPLE 4. The 2-sphere S^2 and the torus T are orientable 2-manifolds, since each is the boundary of a 3-manifold in $\mathbf{R}^3$, which is orientable. In general, if M is a 3-manifold in $\mathbf{R}^3$, oriented naturally, what can we say about the induced orientation of ∂M? It turns out that it is the orientation of ∂M that corresponds to the unit normal field to ∂M pointing *outwards* from the 3-manifold M. We give an informal argument here to justify this statement, reserving a formal proof until a later section.

Given M, let $\alpha : U \to V$ be a coordinate patch on M belonging to the natural orientation of M, about the point p of ∂M. Then the map

$$(\alpha \circ b)(\mathbf{x}) = \alpha(x_1, x_2, 0)$$

gives the restricted coordinate patch on ∂M about p. Since $\dim M = 3$, which is odd, the induced orientation of ∂M is opposite to the one obtained by restricting coordinate patches on M. Thus the normal field $N = (\mathbf{p};\mathbf{n})$ to ∂M corresponding to the induced orientation of M satisfies the condition that the frame $(-\mathbf{n}, \partial\alpha/\partial x_1, \partial\alpha/\partial x_2)$ is right-handed.

On the other hand, since M is oriented naturally, $\det D\alpha > 0$. It follows that $(\partial\alpha/\partial x_3, \partial\alpha/\partial x_1, \partial\alpha/\partial x_2)$ is right-handed. Thus $-\mathbf{n}$ and $\partial\alpha/\partial x_3$ lie on the same side of the tangent plane to M at p. Since $\partial\alpha/\partial x_3$ points *into* M, the vector n points outwards from M. See Figure 34.9.

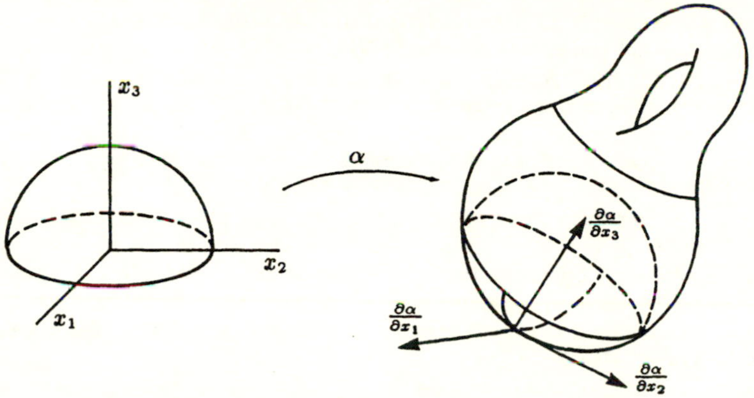

Figure 34.9

EXAMPLE 5. Let M be a 2-manifold with non-empty boundary, in $\mathbf{R}^3$. If M is oriented, let us give ∂M the induced orientation. Let N be the unit normal field to M corresponding to the orientation of M; and let T be the unit tangent field to ∂M corresponding to the induced orientation of ∂M. What is the relationship between N and T?

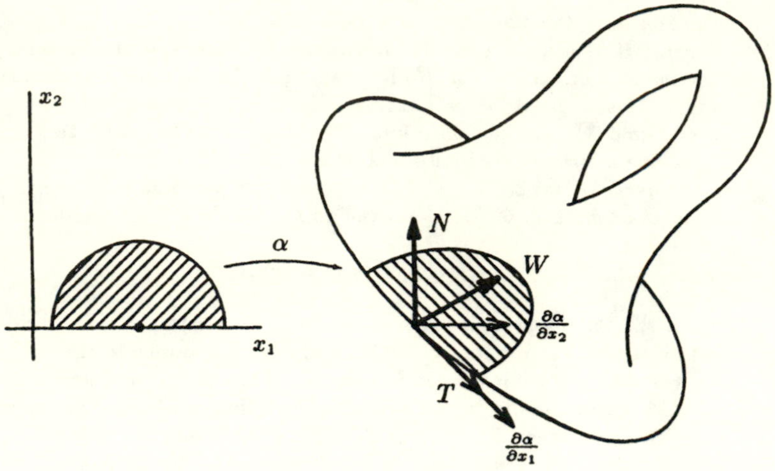

Figure 34.10

We assert the following: Given N and T, for each $\mathbf{p} \in \partial M$ let $W(\mathbf{p})$ be the unit vector that is perpendicular to both $N(\mathbf{p})$ and $T(\mathbf{p})$, chosen so that the frame $\big(N(\mathbf{p}), T(\mathbf{p}), W(\mathbf{p})\big)$ is right-handed. Then $W(\mathbf{p})$ is tangent to M at $\mathbf{p}$ and points *into* M from ∂M.

(This statement is a more precise way of formulating the description usually given in the statement of Stokes' theorem in calculus: "The relation between N and T is such that if you walk around ∂M in the direction specified by T, with your head pointing in the direction specified by N, then the manifold M is on your left." See Figure 34.10.)

To verify this assertion, let $\alpha : U \to V$ be a coordinate patch on M about the point $\mathbf{p}$ of ∂M, belonging to the orientation of M. Then the coordinate patch $\alpha \circ b$ belongs to the induced orientation of ∂M. (Note that $\dim M = 2$, which is even.) The vector $\partial\alpha/\partial x_1$ represents the velocity vector of the parametrized curve $\alpha \circ b$; hence by definition it points in the same direction as the unit tangent vector T.

The vector $\partial\alpha/\partial x_2$, on the other hand, is the velocity of a parametrized curve that begins at a point $\mathbf{p}$ of ∂M and moves *into* M as t increases. Thus, by definition, it points into M from $\mathbf{p}$. Now $\partial\alpha/\partial x_2$ need not be orthogonal to T. But we can choose a scalar λ such that the vector $\mathbf{w} = \partial\alpha/\partial x_2 + \lambda\partial\alpha/\partial x_1$ is orthogonal to $\partial\alpha/\partial x_1$ and hence to T. Then $\mathbf{w}$ also points into M; set $W(\mathbf{p}) = (\mathbf{p}; \mathbf{w}/\,\|\mathbf{w}\|\,)$.

Finally, the vector $N(\mathbf{p}) = (\mathbf{p}; \mathbf{n})$ is, by definition, the unit vector normal to M at $\mathbf{p}$ such that the frame $(\mathbf{n}, \partial\alpha/\partial x_1, \partial\alpha/\partial x_2)$ is right-handed. Now

$$\det[\mathbf{n} \quad \partial\alpha/\partial x_1 \quad \partial\alpha/\partial x_2] = \det[\mathbf{n} \quad \partial\alpha/\partial x_1 \quad \mathbf{w}],$$

by direct computation. It follows that the frame (N, T, W) is right-handed.

EXERCISES

1. Let M be an n-manifold in $\mathbf{R}^n$. Let α, β be coordinate patches on M such that $\det D\alpha > 0$ and $\det D\beta > 0$. Show that α and β overlap positively if they overlap at all.
2. Let M be a k-manifold in $\mathbf{R}^n$; let α, β be coordinate patches on M. Show that if α and β overlap positively, so do $\alpha \circ r$ and $\beta \circ r$.
3. Let M be an oriented 1-manifold in $\mathbf{R}^2$, with corresponding unit tangent vector field T. Describe the unit normal field corresponding to the orientation of M.
4. Let C be the cylinder in $\mathbf{R}^3$ given by

$$C = \{(x, y, z) \mid x^2 + y^2 = 1; 0 \leq z \leq 1\}.$$

Orient C by declaring the coordinate patch $\alpha : (0,1)^2 \to C$ given by

$$\alpha(u, v) = (\cos 2\pi u, \sin 2\pi u, v)$$

to belong to the orientation. See Figure 34.11. Describe the unit normal field corresponding to this orientation of C. Describe the unit tangent field corresponding to the induced orientation of ∂C.

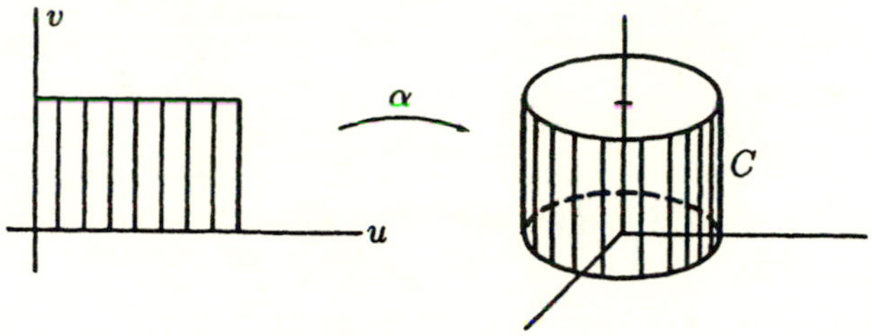

Figure 34.11

5. Let M be the 2-manifold in $\mathbf{R}^2$ pictured in Figure 34.12, oriented naturally. The induced orientation of ∂M corresponds to a unit tangent vector field; describe it. The induced orientation of ∂M also corresponds to a unit normal field; describe it.

6. Show that if M is a connected orientable k-manifold in $\mathbf{R}^n$, then M has precisely two orientations, as follows: Choose an orientation of M; it consists of a collection of coordinate patches $\{\alpha_i\}$. Let $\{\beta_j\}$ be an

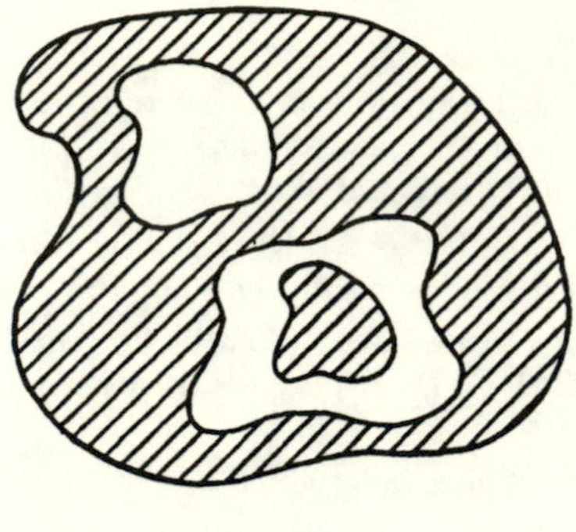

Figure 34.12

arbitrary orientation of M. Given $\mathbf{x} \in M$, choose coordinate patches α_i and β_j about $\mathbf{x}$ and define $\lambda(\mathbf{x}) = 1$ if they overlap positively at $\mathbf{x}$, and $\lambda(\mathbf{x}) = -1$ if they overlap negatively at $\mathbf{x}$.

(a) Show that $\lambda(\mathbf{x})$ is well-defined, independent of the choice of α_i and β_j.

(b) Show that λ is continuous.

(c) Show that λ is constant.

(d) Show that $\{\beta_j\}$ gives the opposite orientation to $\{\alpha_i\}$ if λ is identically -1, and the same orientation if λ is identically 1.

7. Let M be the 3-manifold in $\mathbf{R}^3$ consisting of all $\mathbf{x}$ with $1 \le \|\mathbf{x}\| \le 2$. Orient M naturally. Describe the unit normal field corresponding to the induced orientation of ∂M.

8. Let $B^n = B^n(1)$ be the unit ball in $\mathbf{R}^n$, oriented naturally. Let the unit sphere $S^{n-1} = \partial B^n$ have the induced orientation. Does the coordinate patch $\alpha : \text{Int } B^{n-1} \to S^{n-1}$ given by the equation

$$\alpha(\mathbf{u}) = (\mathbf{u}, [1 - \|\mathbf{u}\|^2]^{1/2})$$

belong to the orientation of S^{n-1}? What about the coordinate patch

$$\beta(\mathbf{u}) = (\mathbf{u}, -[1 - \|\mathbf{u}\|^2]^{1/2})?$$

§35. INTEGRATING FORMS OVER ORIENTED MANIFOLDS

Now we define the integral of a k-form ω over an oriented k-manifold. The procedure is very similar to that of §25, where we defined the integral of a scalar function over a manifold. Therefore we abbreviate some of the details.

We treat first the case where the support of ω can be covered by a single coordinate patch.

Definition. Let M be a compact oriented k-manifold in $\mathbf{R}^n$. Let ω be a k-form defined in an open set of $\mathbf{R}^n$ containing M. Let $C = M \cap (\text{Support } \omega)$; then C is compact. Suppose there is a coordinate patch $\alpha : U \to V$ on M belonging to the orientation of M such that $C \subset V$. By replacing U by a smaller open set if necessary, we can assume that U is bounded. We define the **integral of ω over M** by the equation

$$\int_M \omega = \int_{\text{Int } U} \alpha^* \omega.$$

Here Int $U = U$ if U is open in $\mathbf{R}^k$, and Int $U = U \cap \mathbf{H}^k_+$ if U is open in $\mathbf{H}^k$ but not in $\mathbf{R}^k$.

First, we note that this integral exists as an ordinary integral, and hence as an extended integral: Since α can be extended to a C^∞ map defined on a set U' open in $\mathbf{R}^k$, the form $\alpha^*\omega$ can be extended to a C^∞ form on U'. This form can be written as $h\, dx_1 \wedge \cdots \wedge dx_k$ for some C^∞ scalar function h on U'. Thus

$$\int_{\text{Int } U} \alpha^* \omega = \int_{\text{Int } U} h,$$

by definition. The function h is continuous on U and vanishes on U outside the compact set $\alpha^{-1}(C)$; hence h is bounded on U. If U is open in $\mathbf{R}^k$, then h vanishes near each point of Bd U. If U is not open in $\mathbf{R}^k$, then h vanishes near each point of Bd U not in $\partial\mathbf{H}^k$, a set that has measure zero in $\mathbf{R}^k$. In either case, h is integrable over U and hence over Int U. See Figure 35.1.

Second, we note that the integral $\int_M \omega$ is well-defined, independent of the choice of the coordinate patch α. The proof is very similar to that of Lemma 25.1; here one uses the additional fact that the transition function is orientation-preserving, so that the sign in the formula given in Theorem 33.1 is "plus."

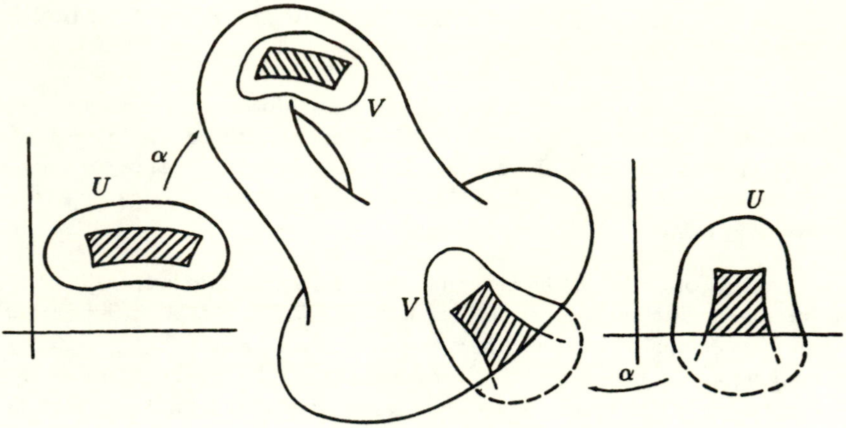

Figure 35.1

Third, we note that this integral is linear. More precisely, if ω and η have supports whose intersections with M can be covered by the single coordinate patch $\alpha : U \to V$ belonging to the orientation of M, then

$$\int_M a\omega + b\eta \quad = \quad a\int_M \omega \quad + \quad b\int_M \eta.$$

This result follows at once from the fact that α^* and $\int_{\text{Int}\, U}$ are linear.

Finally, we note that if $-M$ denotes the manifold M with the opposite orientation, then

$$\int_{-M} \omega = -\int_M \omega.$$

This result follows from Theorem 33.1.

To define $\int_M \omega$ in general, we use a partition of unity.

Definition. Let M be a compact oriented k-manifold in $\mathbf{R}^n$. Let ω be a k-form defined in an open set of $\mathbf{R}^n$ containing M. Cover M by coordinate patches belonging to the orientation of M; then choose a partition of unity $\phi_1, \ldots, \phi_\ell$ on M that is dominated by this collection of coordinate patches on M. See Lemma 25.2. We define the **integral of ω over M** by the equation

$$\int_M \omega = \sum_{i=1}^{\ell} [\int_M \phi_i \omega].$$

The fact that this definition agrees with the previous one when the support of ω is covered by a single coordinate patch follows from linearity of the earlier integral and the fact that

$$\omega(\mathbf{x}) = \sum_{i=1}^{\ell} \phi_i(\mathbf{x})\omega(\mathbf{x})$$

for each $\mathbf{x} \in M$. The fact that the integral is independent of the choice of the partition of unity follows by the same argument that was used for the integral $\int_M f \, dV$. The following is also immediate:

Theorem 35.1. *Let M be a compact oriented k-manifold in $\mathbf{R}^n$. Let ω, η be k-forms defined in an open set of $\mathbf{R}^n$ containing M. Then*

$$\int_M (a\omega + b\eta) = a\int_M \omega + b\int_M \eta.$$

If $-M$ denotes M with the opposite orientation, then

$$\int_{-M} \omega = -\int_M \omega. \quad \square$$

This definition of the integral is satisfactory for theoretical purposes, but not for computational purposes. As in the case of the integral $\int_M f \, dV$, the practical way of evaluating $\int_M \omega$ is to break M up into pieces, integrate over each piece separately, and add the results together. We state this fact formally as a theorem:

***Theorem 35.2.** *Let M be a compact oriented k-manifold in $\mathbf{R}^n$. Let ω be a k-form defined in an open set of $\mathbf{R}^n$ containing M. Suppose that $\alpha_i : A_i \to M_i$, for $i = 1, \dots, N$, is a coordinate patch on M belonging to the orientation of M, such that A_i is open in $\mathbf{R}^k$ and M is the disjoint union of the open sets $M_1, \dots, M_N$ of M and a set K of measure zero in M. Then*

$$\int_M \omega = \sum_{i=1}^{N} [\int_{A_i} \alpha_i^* \omega].$$

Proof. The proof is almost a copy of the proof of Theorem 25.4. Alternatively, it follows from Theorems 25.4 and 36.2. We leave the details to you. $\square$

EXERCISES

1. Let M be a compact oriented k-manifold in $\mathbf{R}^n$. Let ω be a k-form defined in an open set of $\mathbf{R}^n$ containing M.
   (a) Show that in the case where the set $C = M \cap (\text{Support } \omega)$ is covered by a single coordinate patch, the integral $\int_M \omega$ is well-defined.
   (b) Show that the integral $\int_M \omega$ is well-defined in general, independent of the choice of the partition of unity.
2. Prove Theorem 35.2.
3. Let S^{n-1} be the unit sphere in $\mathbf{R}^n$, oriented so that the coordinate patch $\alpha : A \to S^{n-1}$ given by

$$\alpha(\mathbf{u}) = (\mathbf{u}, [1 - \|\mathbf{u}\|^2]^{1/2})$$

belongs to the orientation, where $A = \text{Int } B^{n-1}$. Let η be the $n-1$ form

$$\eta = \sum_{i=1}^{n} (-1)^{i-1} f_i \, dx_1 \wedge \cdots \wedge \widehat{dx_i} \wedge \cdots \wedge dx_n,$$

where $f_i(\mathbf{x}) = x_i / \|\mathbf{x}\|^m$. The form η is defined on $\mathbf{R}^n - \mathbf{0}$. Show that

$$\int_{S^{n-1}} \eta \neq 0,$$

as follows:

(a) Let $\rho : \mathbf{R}^n \to \mathbf{R}^n$ be given by

$$\rho(x_1, \ldots, x_{n-1}, x_n) = (x_1, \ldots, x_{n-1}, -x_n).$$

Let $\beta = \rho \circ \alpha$. Show that $\beta : A \to S^{n-1}$ belongs to the opposite orientation of S^{n-1}. [*Hint:* The map $\rho : B^n \to B^n$ is orientation-reversing.]

(b) Show that $\beta^* \eta = -\alpha^* \eta$; conclude that

$$\int_{S^{n-1}} \eta = 2 \int_A \alpha^* \eta.$$

(c) Show that

$$\int_A \alpha^* \eta = \pm \int_A 1/[1 - \|\mathbf{u}\|^2]^{1/2} \neq 0.$$

*§36. A GEOMETRIC INTERPRETATION OF FORMS AND INTEGRALS

The notion of the integral of a k-form over an oriented k-manifold seems remarkably abstract. Can one give it any intuitive meaning? We discuss here how it is related to the integral of a scalar function over a manifold, which is a notion closer to our geometric intuition.

First, we explore the relationship between alternating tensors in $\mathbf{R}^n$ and the volume function in $\mathbf{R}^n$.

Theorem 36.1. *Let W be a k-dimensional linear subspace of $\mathbf{R}^n$; let $(\mathbf{a}_1, \dots, \mathbf{a}_k)$ be an orthonormal k-frame in W, and let f be an alternating k-tensor on W. If $(\mathbf{x}_1, \dots, \mathbf{x}_k)$ is an arbitrary k-tuple in W, then*

$$f(\mathbf{x}_1, \dots, \mathbf{x}_k) = \epsilon \; V(\mathbf{x}_1, \dots, \mathbf{x}_k) f(\mathbf{a}_1, \dots, \mathbf{a}_k),$$

where $\epsilon = \pm 1$. If the $\mathbf{x}_i$ are independent, then $\epsilon = +1$ if the frames $(\mathbf{x}_1, \dots, \mathbf{x}_k)$ and $(\mathbf{a}_1, \dots, \mathbf{a}_k)$ belong to the same orientation of W and $\epsilon = -1$ otherwise.

If the $\mathbf{x}_i$ are dependent, then $V(\mathbf{x}_1, \dots, \mathbf{x}_k) = 0$ by Theorem 21.3 and the value of ϵ does not matter.

Proof. Step 1. If $W = \mathbf{R}^k$, then the theorem holds. In that case, the k-tensor f is a multiple of the determinant function, so there is a scalar c such that for all k-tuples $(\mathbf{x}_1, \dots, \mathbf{x}_k)$ in $\mathbf{R}^k$,

$$f(\mathbf{x}_1, \dots, \mathbf{x}_k) = c \det[\mathbf{x}_1 \cdots \mathbf{x}_k].$$

If the $\mathbf{x}_i$ are dependent, it follows that f vanishes; then the theorem holds trivially. Otherwise, we have

$$f(\mathbf{x}_1, \dots, \mathbf{x}_k) = c \det[\mathbf{x}_1 \cdots \mathbf{x}_k] = c\epsilon_1 V(\mathbf{x}_1, \dots, \mathbf{x}_k),$$

where $\epsilon_1 = +1$ if $(\mathbf{x}_1, \dots, \mathbf{x}_k)$ is right-handed, and $\epsilon_1 = -1$ otherwise. Similarly,

$$f(\mathbf{a}_1, \dots, \mathbf{a}_k) = c\epsilon_2 V(\mathbf{a}_1, \dots, \mathbf{a}_k) = c\epsilon_2,$$

where $\epsilon_2 = +1$ if $(\mathbf{a}_1, \dots, \mathbf{a}_k)$ is right-handed and $\epsilon_2 = -1$ otherwise. It follows that

$$f(\mathbf{x}_1, \dots, \mathbf{x}_k) = (\epsilon_1/\epsilon_2) V(\mathbf{x}_1, \dots, \mathbf{x}_k) f(\mathbf{a}_1, \dots, \mathbf{a}_k),$$

where $\epsilon_1/\epsilon_2 = +1$ if $(\mathbf{x}_1, \dots, \mathbf{x}_k)$ and $(\mathbf{a}_1, \dots, \mathbf{a}_k)$ belong to the same orientation of $\mathbf{R}^k$, and $\epsilon_1/\epsilon_2 = -1$ otherwise.

Step 2. The theorem holds in general. Given W, choose an orthogonal transformation $h : \mathbf{R}^n \to \mathbf{R}^n$ carrying W onto $\mathbf{R}^k \times \mathbf{0}$. Let $k : \mathbf{R}^k \times \mathbf{0} \to W$ be the inverse map. Since f is an alternating tensor on W, it is mapped to an alternating tensor $k^* f$ on $\mathbf{R}^k \times \mathbf{0}$. Since $\big(h(\mathbf{x}_1), \dots, h(\mathbf{x}_k)\big)$ is a k-tuple in $\mathbf{R}^k \times \mathbf{0}$, and $\big(h(\mathbf{a}_1), \dots, h(\mathbf{a}_k)\big)$ is an orthonormal k-tuple in $\mathbf{R}^k \times \mathbf{0}$, we have by Step 1,

$$(k^* f)(h(\mathbf{x}_1), \dots, h(\mathbf{x}_k)) = \epsilon V(h(\mathbf{x}_1), \dots, h(\mathbf{x}_k))(k^* f)(h(\mathbf{a}_1), \dots, h(\mathbf{a}_k)),$$

where $\epsilon = \pm 1$. Since V is unchanged by orthogonal transformations, we can rewrite this equation as

$$f(\mathbf{x}_1, \dots, \mathbf{x}_k) = \epsilon\, V(\mathbf{x}_1, \dots, \mathbf{x}_k) f(\mathbf{a}_1, \dots, \mathbf{a}_k),$$

as desired. Now suppose the $\mathbf{x}_i$ are independent. Then the $h(\mathbf{x}_i)$ are independent, and by Step 1 we have $\epsilon = +1$ if and only if $\big(h(\mathbf{x}_1), \dots, h(\mathbf{x}_k)\big)$ and $\big(h(\mathbf{a}_1), \dots, h(\mathbf{a}_k)\big)$ belong to the same orientation of $\mathbf{R}^k \times \mathbf{0}$. By definition, this occurs if and only if $(\mathbf{x}_1, \dots, \mathbf{x}_k)$ and $(\mathbf{a}_1, \dots, \mathbf{a}_k)$ belong to the same orientation of W. □

Note that it follows from this theorem that if $(\mathbf{a}_1, \dots, \mathbf{a}_k)$ and $(\mathbf{b}_1, \dots, \mathbf{b}_k)$ are two *orthonormal* frames in W, then

$$f(\mathbf{a}_1, \dots, \mathbf{a}_k) = \pm f(\mathbf{b}_1, \dots, \mathbf{b}_k),$$

the sign depending on whether they belong to the same orientation of W or not.

Definition. Let M be a k-manifold in $\mathbf{R}^n$; let $\mathbf{p} \in M$. If M is oriented, then the tangent space to M at $\mathbf{p}$ has a natural induced orientation, defined as follows: Choose a coordinate patch $\alpha : U \to V$ belonging to the orientation of M about $\mathbf{p}$. Let $\alpha(\mathbf{x}) = \mathbf{p}$. The collection of all k-frames in $\mathcal{T}_{\mathbf{p}}(M)$ of the form

$$(\alpha_*(\mathbf{x}; \mathbf{a}_1), \dots, \alpha_*(\mathbf{x}; \mathbf{a}_k))$$

where $(\mathbf{a}_1, \dots, \mathbf{a}_k)$ is a right-handed frame in $\mathbf{R}^k$, is called the **natural orientation of $\mathcal{T}_{\mathbf{p}}(M)$**, induced by the orientation of M. It is easy to show it is well-defined, independent of the choice of α.

Theorem 36.2. *Let M be a compact oriented k-manifold in $\mathbf{R}^n$; let ω be a k-form defined in an open set of $\mathbf{R}^n$ containing M. Let λ be the scalar function on M defined by the equation*

$$\lambda(\mathbf{p}) = \omega(\mathbf{p})\,\big((\mathbf{p}; \mathbf{a}_1), \dots, (\mathbf{p}; \mathbf{a}_k)\big),$$

where $((\mathbf{p};\mathbf{a}_1),\ldots,(\mathbf{p};\mathbf{a}_k))$ *is any orthonormal frame in the linear space* $\mathcal{T}_{\mathbf{p}}(M)$ *belonging to its natural orientation. Then* λ *is continuous, and*

$$\int_M \omega = \int_M \lambda \, dV.$$

Proof. By linearity, it suffices to consider the case where the support of ω is covered by a single coordinate patch $\alpha : U \to V$ belonging to the orientation of M. We have

$$\alpha^*\omega = h\, dx_1 \wedge \cdots \wedge dx_k$$

for some scalar function h. Let $\alpha(\mathbf{x}) = \mathbf{p}$. We compute $h(\mathbf{x})$ as follows:

$$\begin{aligned} h(\mathbf{x}) &= (\alpha^*\omega)(\mathbf{x})\big((\mathbf{x};\mathbf{e}_1),\ldots,(\mathbf{x};\mathbf{e}_k)\big) \\ &= \omega(\alpha(\mathbf{x}))\big(\alpha_*(\mathbf{x};\mathbf{e}_1),\ldots,\alpha_*(\mathbf{x};\mathbf{e}_k)\big) \\ &= \omega(\mathbf{p})\big((\mathbf{p};\partial\alpha/\partial x_1),\ldots,(\mathbf{p};\partial\alpha/\partial x_k)\big) \\ &= \pm V(D\alpha(\mathbf{x}))\lambda(\mathbf{p}), \end{aligned}$$

by Theorem 36.1. The sign is "plus" because the frame

$$((\mathbf{p};\partial\alpha/\partial x_1),\ldots,(\mathbf{p};\partial\alpha/\partial x_k))$$

belongs to the natural orientation of $\mathcal{T}_{\mathbf{p}}(M)$ by definition. Now $V(D\alpha) \neq 0$ because $D\alpha$ has rank k. Then since $\mathbf{x} = \alpha^{-1}(\mathbf{p})$ is a continuous function of $\mathbf{p}$, so is

$$\lambda(\mathbf{p}) = h(\mathbf{x})/V(D\alpha(\mathbf{x})).$$

It follows that

$$\int_M \lambda \, dV = \int_{\text{Int } U} (\lambda \circ \alpha) V(D\alpha) = \int_{\text{Int } U} h.$$

On the other hand,

$$\int_M \omega = \int_{\text{Int } U} \alpha^*\omega = \int_{\text{Int } U} h,$$

by definition. The theorem follows. □

This theorem tells us that, given a k-form ω defined in an open set about the compact oriented k-manifold M in $\mathbf{R}^n$, there exists a scalar function λ (which is in fact of class C^∞) such that

$$\int_M \omega = \int_M \lambda \, dV.$$

The reverse is also true, but the proof is a good deal harder:

One first shows that there exists a k-form ω_v, defined in an open set about M, such that the value of $\omega_v(\mathrm{p})$ on any orthonormal basis for $\mathcal{T}_{\mathrm{p}}(M)$ belonging to its natural orientation is 1. Then if λ is any C^∞ function on M, we have

$$\int_M \lambda \, \mathrm{d}V = \int_M \lambda\omega_v;$$

thus the integral of λ over M can be interpreted as the integral over M of the form $\lambda\omega_v$. The form ω_v is called a **volume form** for M, since

$$\int_M \omega_v = \int_M \mathrm{d}V = v(M).$$

This argument applies, however, only if M is orientable. If M is not orientable, the integral of a scalar function is defined, but the integral of a form is not.

A remark on notation. Some mathematicians denote the volume form ω_v by the symbol $\mathrm{d}V$, or rather by the symbol dV. (See the remark on notation in §22.) While it makes the preceding equations tautologies, this practice can cause confusion to the unwary, since V is *not* a form, and d does *not* denote the differential operator in this context!

EXERCISE

1. Let M be a k-manifold in $\mathbf{R}^n$; let $\mathrm{p} \in M$. Let α and β be coordinate patches on M about p; let $\alpha(\mathrm{x}) = \mathrm{p} = \beta(\mathrm{y})$. Let $(\mathrm{a}_1, \ldots, \mathrm{a}_k)$ be a right-handed frame in $\mathbf{R}^k$. If α and β overlap positively, show that there is a right-handed frame $(\mathrm{b}_1, \ldots, \mathrm{b}_k)$ in $\mathbf{R}^k$ such that

$$\alpha_*(\mathrm{x}; \mathrm{a}_i) = \beta_*(\mathrm{y}; \mathrm{b}_i)$$

for each i. Conclude that if M is oriented, then the natural orientation of $\mathcal{T}_{\mathrm{p}}(M)$ is well-defined.

§37. THE GENERALIZED STOKES' THEOREM

Now, finally, we come to the theorem that is the culmination of all our labors. It is a general theorem about integrals of differential forms that includes the three basic theorems of vector integral calculus—Green's theorem, Stokes' theorem, and the divergence theorem—as special cases.

We begin with a lemma that is in some sense a very special case of the theorem. Let I^k denote the unit k-cube in $\mathbf{R}^k$;

$$I^k = [0,1]^k = [0,1] \times \cdots \times [0,1].$$

Then Int I^k is the open cube $(0,1)^k$, and Bd I^k equals $I^k - \text{Int } I^k$.

Lemma 37.1. *Let $k > 1$. Let η be a $k-1$ form defined in an open set U of $\mathbf{R}^k$ containing the unit k-cube I^k. Assume that η vanishes at all points of* Bd I^k *except possibly at points of the subset* (Int I^{k-1}) $\times 0$. *Then*

$$\int_{\text{Int } I^k} d\eta = (-1)^k \int_{\text{Int } I^{k-1}} b^*\eta,$$

where $b : I^{k-1} \to I^k$ is the map

$$b(u_1, \ldots, u_{k-1}) = (u_1, \ldots, u_{k-1}, 0).$$

Proof. We use $\mathbf{x}$ to denote the general point of $\mathbf{R}^k$, and $\mathbf{u}$ to denote the general point of $\mathbf{R}^{k-1}$. See Figure 37.1. Given j with $1 \leq j \leq k$, let I_j denote the $k-1$ tuple

$$I_j = (1, \ldots, \hat{\jmath}, \ldots, k).$$

Then the typical elementary $k-1$ form in $\mathbf{R}^k$ is the form

$$dx_{I_j} = dx_1 \wedge \cdots \wedge \widehat{dx_j} \wedge \cdots \wedge dx_k.$$

Because the integrals involved are linear and the operators d and b^* are linear, it suffices to prove the lemma in the special case

$$\eta = f\ dx_{I_j},$$

so we assume this value of η in the remainder of the proof.

Step 1. We compute the integral

$$\int_{\text{Int } I^k} d\eta.$$

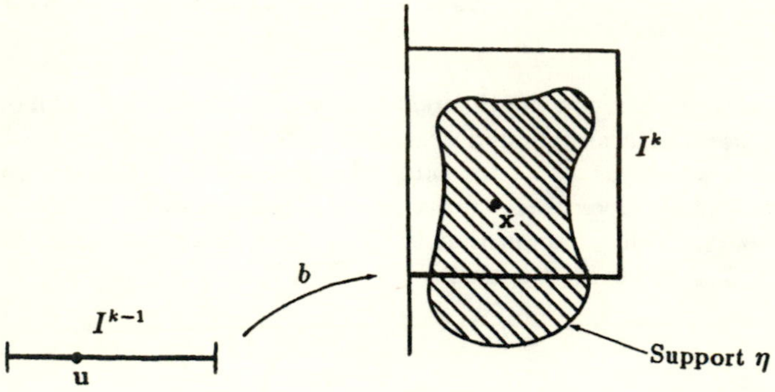

Figure 37.1

We have

$$d\eta = df \wedge dx_{I_j}$$
$$= (\sum_{i=1}^{k} D_i f \; dx_i) \wedge dx_{I_j}$$
$$= (-1)^{j-1}(D_j f) dx_1 \wedge \cdots \wedge dx_k.$$

Then we compute

$$\int_{\text{Int } I^k} d\eta = (-1)^{j-1} \int_{\text{Int } I^k} D_j f$$
$$= (-1)^{j-1} \int_{I^k} D_j f$$
$$= (-1)^{j-1} \int_{\mathbf{v} \in I^{k-1}} \int_{x_j \in I} D_j f(x_1, \ldots, x_k)$$

by the Fubini theorem, where $\mathbf{v} = (x_1, \ldots, \widehat{x}_j, \ldots, x_k)$. Using the fundamental theorem of calculus, we compute the inner integral as

$$\int_{x_j \in I} D_j f(x_1, \ldots, x_k) = f(x_1, \ldots, 1, \ldots, x_k) - f(x_1, \ldots, 0, \ldots, x_k),$$

where the 1 and the 0 appear in the jth place. Now the form η, and hence the function f, vanish at all points of Bd I^k except possibly at points of the open bottom face (Int I^{k-1}) $\times$ 0. If $j < k$, this means that the right side of this equation vanishes; while if $j = k$, it equals

$$-f(x_1, \ldots, x_{k-1}, 0).$$

We conclude the following:

$$\int_{\text{Int } I^k} d\eta = \begin{cases} 0 & \text{if } j < k, \\ (-1)^k \int_{I^{k-1}} (f \circ b) & \text{if } j = k. \end{cases}$$

Step 2. Now we compute the other integral of our theorem. The map $b : \mathbf{R}^{k-1} \to \mathbf{R}^k$ has derivative

$$Db = \begin{bmatrix} I_{k-1} \\ 0 \end{bmatrix}.$$

Therefore, by Theorem 32.2, we have

$$\begin{aligned} b^*(dx_{I_j}) &= [\det Db(1, \ldots, \hat{\jmath}, \ldots, k)]\, du_1 \wedge \cdots \wedge du_{k-1} \\ &= \begin{cases} 0 & \text{if } j < k, \\ du_1 \wedge \cdots \wedge du_{k-1} & \text{if } j = k. \end{cases} \end{aligned}$$

We conclude that

$$\int_{\text{Int } I^{k-1}} b^*\eta = \begin{cases} 0 & \text{if } j < k, \\ \int_{\text{Int } I^{k-1}} (f \circ b) & \text{if } j = k. \end{cases}$$

The theorem follows by comparing this equation with that at the end of Step 1. □

Theorem 37.2 (Stokes' theorem). *Let $k > 1$. Let M be a compact oriented k-manifold in $\mathbf{R}^n$; give ∂M the induced orientation if ∂M is not empty. Let ω be a $k-1$ form defined in an open set of $\mathbf{R}^n$ containing M. Then*

$$\int_M d\omega = \int_{\partial M} \omega$$

if ∂M is not empty; and $\int_M d\omega = 0$ if ∂M is empty.

Proof. *Step 1.* We first cover M by carefully-chosen coordinate patches. As a first case, assume that $\mathbf{p} \in M - \partial M$. Choose a coordinate patch $\alpha : U \to V$ belonging to the orientation of M, such that U is open in $\mathbf{R}^k$ and contains the unit cube I^k, and such that α carries a point of Int I^k to the point $\mathbf{p}$. (If we begin with an arbitrary coordinate patch $\alpha : U \to V$ about $\mathbf{p}$ belonging to the orientation of M, we can obtain one of the desired type by preceding α by a translation and a stretching in $\mathbf{R}^k$.) Let $W = \text{Int } I^k$,

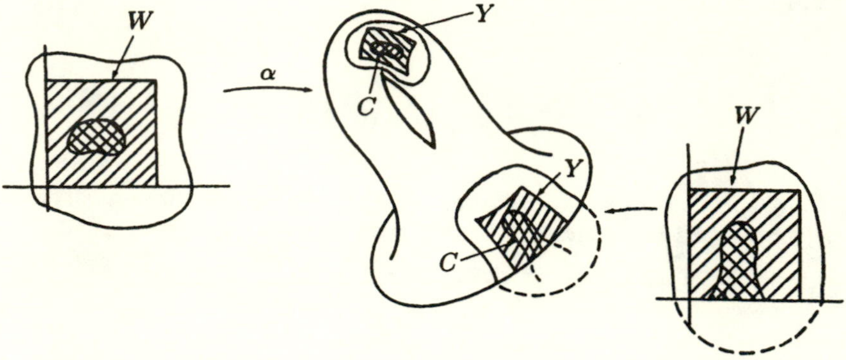

Figure 37.2

and let $Y = \alpha(W)$. Then the map $\alpha : W \to Y$ is still a coordinate patch belonging to the orientation of M about p, with $W = \text{Int } I^k$ open in $\mathbf{R}^k$. See Figure 37.2. We choose this special patch about p.

As a second case, assume that $p \in \partial M$. Choose a coordinate patch $\alpha : U \to V$ belonging to the orientation of M, such that U is open in $\mathbf{H}^k$ and U contains I^k, and such that α carries a point of $(\text{Int } I^{k-1}) \times 0$ to the point p. Let

$$W = (\text{Int } I^k) \cup ((\text{Int } I^{k-1}) \times 0),$$

and let $Y = \alpha(W)$. Then the map $\alpha : W \to Y$ is still a coordinate patch belonging to the orientation of M about p, with W open in $\mathbf{H}^k$ but not open in $\mathbf{R}^k$.

We shall use the covering of M by the coordinate patches $\alpha : W \to Y$ to compute the integrals involved in the theorem. Note that in each case, the map α can be extended if necessary to a C^∞ function defined on an open set of $\mathbf{R}^k$ containing I^k.

Step 2. Since the operator d and the integrals involved are linear, it suffices to prove the theorem in the special case where ω is a $k-1$ form such that the set

$$C = M \cap (\text{Support } \omega)$$

can be covered by a single one of the coordinate patches $\alpha : W \to Y$. Since the support of $d\omega$ is contained in the support of ω, the set $M \cap (\text{Support } d\omega)$ is contained in C, so it is covered by the same coordinate patch.

Let η denote the form $\alpha^*\omega$. The form η can be extended if necessary (without change of notation) to a C^∞ form on an open set of $\mathbf{R}^k$ containing I^k.

Furthermore, η vanishes at all points of Bd I^k except possibly at points of the bottom face (Int I^{k-1}) $\times\, 0$. Thus the hypotheses of the preceding lemma are satisfied.

Step 3. We prove the theorem when C is covered by a coordinate patch $\alpha : W \to Y$ of the type constructed in the first case. Here $W = \text{Int } I^k$ and Y is disjoint from ∂M. We compute the integrals involved. Since $\alpha^* d\omega = d\alpha^*\omega = d\eta$, we have

$$\int_M d\omega = \int_{\text{Int } I^k} \alpha^* d\omega = \int_{\text{Int } I^k} d\eta = (-1)^k \int_{\text{Int } I^{k-1}} b^*\eta.$$

Here we use the preceding lemma. In this case, the form η vanishes outside Int I^k. In particular, η vanishes on $I^{k-1} \times 0$, so that $b^*\eta = 0$. Then $\int_M d\omega = 0$.

The theorem follows. If ∂M is empty, this is the equation we wished to prove. If ∂M is non-empty, then the equation

$$\int_M d\omega = \int_{\partial M} \omega$$

holds trivially; for since the support of ω is disjoint from ∂M, the integral of ω over ∂M vanishes.

Step 4. Now we prove the theorem when C is covered by a coordinate patch $\alpha : W \to Y$ of the type constructed in the second case. Here W is open in $\mathbf{H}^k$ but not in $\mathbf{R}^k$, and Y intersects ∂M. We have Int $W = \text{Int } I^k$. We compute as before

$$\int_M d\omega = \int_{\text{Int } I^k} d\eta = (-1)^k \int_{\text{Int } I^{k-1}} b^*\eta.$$

We next compute the integral $\int_{\partial M} \omega$. The set $\partial M \cap (\text{Support } \omega)$ is covered by the coordinate patch

$$\beta = \alpha \circ b : \text{Int } I^{k-1} \to Y \cap \partial M$$

on ∂M, which is obtained by restricting α. Now β belongs to the induced orientation of ∂M if k is even, and to the opposite orientation if k is odd. If we use β to compute the integral of ω over ∂M, we must reverse the sign of the integral when k is odd. Thus we have

$$\int_{\partial M} \omega = (-1)^k \int_{\text{Int } I^{k-1}} \beta^*\omega.$$

Since $\beta^*\omega = b^*(\alpha^*\omega) = b^*\eta$, the theorem follows. $\square$

We have proved Stokes' theorem for manifolds of dimension k greater than 1. What happens if $k = 1$? If ∂M is empty, there is no problem; one proves readily that $\int_M d\omega = 0$. However, if ∂M is non-empty, we face the following questions: What does one mean by an "orientation" of a 0-manifold, and how does one "integrate" a 0-form over an oriented 0-manifold?

To see what form Stokes' theorem should take in this case, we consider first a special case.

Definition. Let M be a 1-manifold in $\mathbf{R}^n$. Suppose there is a one-to-one map $\alpha : [a,b] \to M$ of class C^∞, carrying $[a,b]$ onto M, such that $D\alpha(t) \neq 0$ for all t. Then we call M a (smooth) arc in $\mathbf{R}^n$. If M is oriented so that the coordinate patch $\alpha|(a,b)$ belongs to the orientation, we say that $\mathbf{p}$ is the **initial point** of M and $\mathbf{q}$ is the **final point** of M. See Figure 37.3.

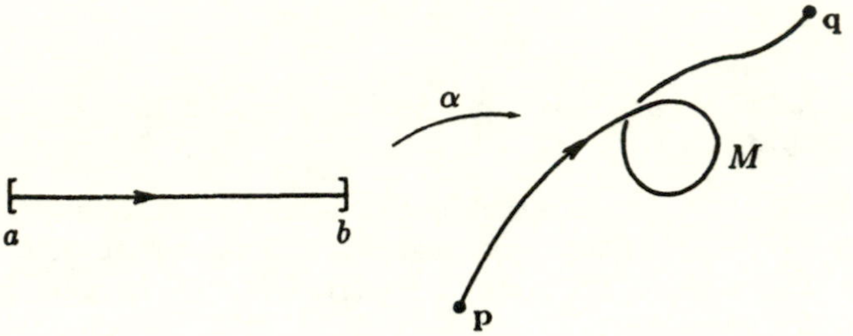

Figure 37.3

***Theorem 37.3.** *Let M be a 1-manifold in $\mathbf{R}^n$; let f be a 0-form defined in an open set about M. If M is an arc with initial point $\mathbf{p}$ and final point $\mathbf{q}$, then*

$$\int_M df = f(\mathbf{q}) - f(\mathbf{p}).$$

Proof. Let $\alpha : [a,b] \to M$ be as in the preceding definition. Then $\alpha : (a,b) \to M - \mathbf{p} - \mathbf{q}$ is a coordinate patch covering all of M except for a set of measure zero. By Theorem 35.2,

$$\int_M df = \int_{(a,b)} \alpha^*(df).$$

Now

$$\alpha^*(df) = d(f \circ \alpha) = D(f \circ \alpha)\,dt,$$

where t denotes the general point of $\mathbf{R}$. Then

$$\int_{(a,b)} \alpha^*(df) = \int_{(a,b)} D(f \circ \alpha) = f(\alpha(b)) - f(\alpha(a)),$$

by the fundamental theorem of calculus. □

This result provides a guide for formulating Stokes' theorem for 1-manifolds:

Definition. A compact 0-manifold N in $\mathbf{R}^n$ is a finite collection of points $\{\mathbf{x}_1, \dots, \mathbf{x}_m\}$ of $\mathbf{R}^n$. We define an **orientation** of N to be a function ϵ mapping N into the two-point set $\{-1, 1\}$. If f is a 0-form defined in an open set of $\mathbf{R}^n$ containing N, we define the **integral of f over the oriented manifold** N by the equation

$$\int_N f = \sum_{i=1}^{m} \epsilon(\mathbf{x}_i) f(\mathbf{x}_i).$$

Definition. If M is an oriented 1-manifold in $\mathbf{R}^n$ with non-empty boundary, we define the **induced orientation** of ∂M by setting $\epsilon(\mathbf{p}) = -1$, for $\mathbf{p} \in \partial M$, if there is a coordinate patch $\alpha : U \to V$ on M about $\mathbf{p}$ belonging to the orientation of M, with U open in $\mathbf{H}^1$. We set $\epsilon(\mathbf{p}) = +1$ otherwise. See Figure 37.4.

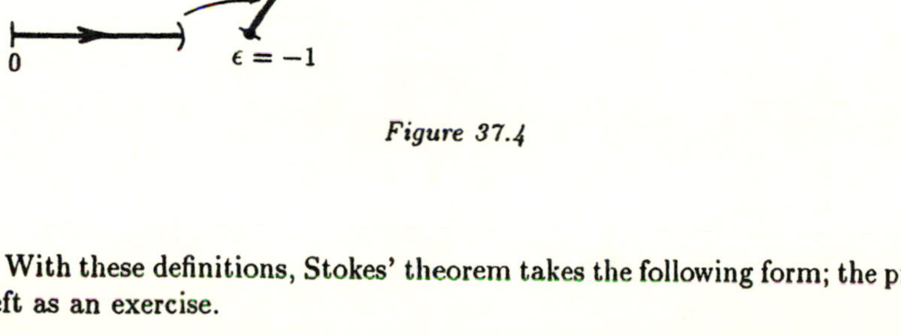

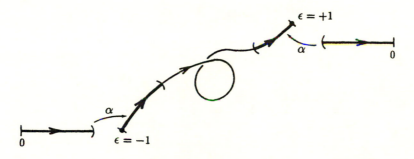

Figure 37.4

With these definitions, Stokes' theorem takes the following form; the proof is left as an exercise.

Theorem 37.4 (Stokes' theorem in dimension 1). *Let M be a compact oriented 1-manifold in $\mathbf{R}^n$; give ∂M the induced orientation if ∂M is not empty. Let f be a 0-form defined in an open set of $\mathbf{R}^n$ containing M. Then*

$$\int_M df = \int_{\partial M} f$$

if ∂M is not empty; and $\int_M df = 0$ if ∂M is empty. □

EXERCISES

1. Prove Stokes' theorem for 1-manifolds. [*Hint:* Cover M by coordinate patches, belonging to the orientation of M, of the form $\alpha : W \to Y$, where W is one of the intervals $(0,1)$ or $[0,1)$ or $(-1,0]$. Prove the theorem when the set $M \cap (\text{Support } f)$ is covered by one of these coordinate patches.]

2. Suppose there is an $n-1$ form η defined in $\mathbf{R}^n - \mathbf{0}$ such that $d\eta = 0$ and

$$\int_{S^{n-1}} \eta \neq 0.$$

Show that η is not exact. (For the existence of such an η, see the exercises of §30 and the exercises of either §35 or §38.)

3. Prove the following:

Theorem (Green's theorem). *Let M be a compact 2-manifold in $\mathbf{R}^2$, oriented naturally; give ∂M the induced orientation. Let $P\,dx + Q\,dy$ be a 1-form defined in an open set of $\mathbf{R}^2$ about M. Then*

$$\int_{\partial M} P\,dx + Q\,dy = \int_M (D_1 Q - D_2 P)\,dx \wedge dy.$$

4. Let M be the 2-manifold in $\mathbf{R}^3$ consisting of all points $\mathbf{x}$ such that

$$4(x_1)^2 + (x_2)^2 + 4(x_3)^2 = 4 \quad \text{and} \quad x_2 \geq 0.$$

Then ∂M is the circle consisting of all points such that

$$(x_1)^2 + (x_3)^2 = 1 \quad \text{and} \quad x_2 = 0.$$

See Figure 37.5. The map

$$\alpha(u,v) = (u,\ 2[1 - u^2 - v^2]^{1/2}, v),$$

for $u^2 + v^2 < 1$, is a coordinate patch on M that covers $M - \partial M$. Orient M so that α belongs to the orientation, and give ∂M the induced orientation.

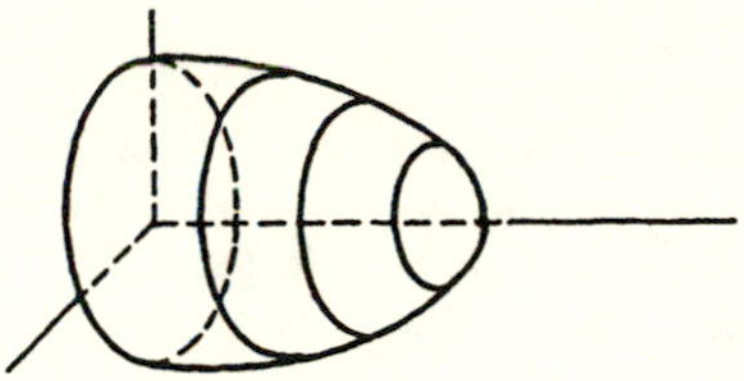

Figure 37.5

(a) What normal vector corresponds to the orientation of M? What tangent vector corresponds to the induced orientation of ∂M?

(b) Let ω be the 1-form $\omega = x_2\,dx_1 + 3x_1\,dx_3$. Evaluate $\int_{\partial M} \omega$ directly.

(c) Evaluate $\int_M d\omega$ directly, by expressing it as an integral over the unit disc in the (u, v) plane.

5. The 3-ball $B^3(r)$ is a 3-manifold in $\mathbf{R}^3$; orient it naturally and give $S^2(r)$ the induced orientation. Assume that ω is a 2-form defined in $\mathbf{R}^3 - \mathbf{0}$ such that

$$\int_{S^2(r)} \omega = a + (b/r),$$

for each $r > 0$.

(a) Given $0 < c < d$, let M be the 3-manifold in $\mathbf{R}^3$ consisting of all $\mathbf{x}$ with $c \leq \|\mathbf{x}\| \leq d$, oriented naturally. Evaluate $\int_M d\omega$.

(b) If $d\omega = 0$, what can you say about a and b?

(c) If $\omega = d\eta$ for some η in $\mathbf{R}^3 - \mathbf{0}$, what can you say about a and b?

6. Let M be an oriented $k + \ell + 1$ manifold without boundary in $\mathbf{R}^n$. Let ω be a k-form and let η be an ℓ-form, both defined in an open set of $\mathbf{R}^n$ about M. Show that

$$\int_M \omega \wedge d\eta = a \int_M d\omega \wedge \eta$$

for some a, and determine a. (Assume M is compact.)

*§38. APPLICATIONS TO VECTOR ANALYSIS

In general, we know from the discussion in §31 that differential forms of order k can be interpreted in $\mathbb{R}^n$ in certain cases as scalar or vector fields, namely when $k = 0, 1, n-1$, or n. We show here that integrals of forms can similarly be so interpreted; then Stokes' theorem can also in certain cases be interpreted in terms of scalar or vector fields. These versions of the general Stokes' theorem include the classical theorems of the vector integral calculus.

We consider the various cases one-by-one.

The gradient theorem for 1-manifolds in $\mathbb{R}^n$

First, we interpret the integral of a 1-form in terms of vector fields. If F is a vector field defined in an open set of $\mathbb{R}^n$, then F corresponds under the "translation map" α_1 to a certain 1-form ω. (See Theorem 31.1.) It turns out that the integral of ω over an oriented 1-manifold equals the integral, with respect to 1-volume, of the tangential component of the vector field F. That is the substance of the following lemma:

Lemma 38.1. *Let M be a compact oriented 1-manifold in $\mathbb{R}^n$; let T be the unit tangent vector to M corresponding to the orientation. Let*

$$F(\mathbf{x}) = (\mathbf{x}; f(\mathbf{x})) = (\mathbf{x}; \Sigma f_i(\mathbf{x})\mathbf{e}_i)$$

be a vector field defined in an open set of $\mathbb{R}^n$ containing M; it corresponds to the 1-form

$$\omega = \sum f_i\, dx_i.$$

Then

$$\int_M \omega = \int_M \langle F, T\rangle\ ds.$$

Here we use the classical notation "ds" rather than "dV" to denote the integral with respect to 1-volume (arc length), simply to make our theorems resemble more closely the classical theorems of vector integral calculus.

Note that if one replaces M by $-M$, then the integral $\int_M \omega$ changes sign. This replacement has the effect of replacing T by $-T$; thus the integral $\int_M \langle F, T\rangle\ ds$ also changes sign.

Proof. We give two proofs of this lemma. The first relies on the results of §36; the second does not.

First proof. By Theorem 36.2, we have

$$\int_M \omega = \int_M \lambda\ ds,$$

where $\lambda(\mathbf{p})$ is the value of $\omega(\mathbf{p})$ on an orthonormal basis for $\mathcal{T}_{\mathbf{p}}(M)$ that belongs to the natural orientation of this tangent space. In the present case, the tangent space is 1-dimensional, and $T(\mathbf{p})$ is such an orthonormal basis. Let $T(\mathbf{p}) = (\mathbf{p}; \mathbf{t})$. Since $\omega = \Sigma f_i \, dx_i$,

$$\omega(\mathbf{p})\,(\mathbf{p};\mathbf{t}) = \sum f_i(\mathbf{p}) t_i(\mathbf{p}).$$

Thus

$$\lambda(\mathbf{p}) = \langle F(\mathbf{p}), T(\mathbf{p})\rangle,$$

and the lemma follows.

Second proof. Since the integrals involved are linear in ω and F, respectively, it suffices to prove the lemma in the case where the set

$$C = M \cap (\text{Support } \omega)$$

lies in a single coordinate patch $\alpha : U \to V$ belonging to the orientation of M. In that case, we simply compute both integrals. Let t denote the general point of $\mathbf{R}$. Then

$$\begin{aligned}\alpha^*\omega &= \sum_{i=1}^{n} (f_i \circ \alpha)\, d\alpha_i \\ &= \sum_{i=1}^{n} (f_i \circ \alpha)\,(D\alpha_i)\, dt \\ &= \langle f \circ \alpha,\, D\alpha\rangle\, dt.\end{aligned}$$

It follows that

$$\begin{aligned}\int_M \omega &= \int_{\text{Int } U} \alpha^*\omega \\ &= \int_{\text{Int } U} \langle f \circ \alpha, D\alpha\rangle.\end{aligned}$$

On the other hand,

$$\begin{aligned}\int_M \langle F, T\rangle\, ds &= \int_{\text{Int } U} \langle F \circ \alpha, T \circ \alpha\rangle \cdot V(D\alpha) \\ &= \int_{\text{Int } U} \langle f \circ \alpha, D\alpha / \|D\alpha\|\,\rangle \cdot V(D\alpha) \\ &= \int_{\text{Int } U} \langle f \circ \alpha, D\alpha\rangle,\end{aligned}$$

since

$$V(D\alpha) = [\det(D\alpha^{\text{tr}} \cdot D\alpha)]^{1/2} = \|D\alpha\|.$$

The lemma follows. □

Theorem 38.2 (The gradient theorem). *Let M be a compact 1-manifold in $\mathbf{R}^n$; let T be a unit tangent vector field to M. Let f be a C^∞ function defined in an open set about M. If ∂M is empty, then*

$$\int_M \langle \text{grad } f, T\rangle \, ds = 0.$$

If ∂M consists of the points $\mathbf{x}_1, \ldots, \mathbf{x}_m$, let $\epsilon_i = -1$ if T points into M at $\mathbf{x}_i$ and $\epsilon_i = +1$ otherwise. Then

$$\int_M \langle \text{grad } f, T\rangle \, ds = \sum_{i=1}^{m} \epsilon_i f(\mathbf{x}_i).$$

Proof. The 1-form df corresponds to the vector field grad f, by Theorem 31.1. Therefore

$$\int_M df = \int_M \langle \text{grad } f, T\rangle \, ds,$$

by the preceding lemma. Our theorem then follows from the 1-dimensional version of Stokes' theorem. □

The divergence theorem for n−1 manifolds in $\mathbf{R}^n$

Now we interpret the integral of an $n-1$ form, over an oriented $n-1$ manifold M, in terms of vector fields. First, we must verify a result stated earlier, the fact that an orientation of M determines a unit normal vector field to M. Recall the following definition from §34:

Definition. Let M be an oriented $n-1$ manifold in $\mathbf{R}^n$. Given $\mathbf{p} \in M$, let $(\mathbf{p};\mathbf{n})$ be a unit vector in $\mathcal{T}_{\mathbf{p}}(\mathbf{R}^n)$ that is orthogonal to the $n-1$ dimensional linear subspace $\mathcal{T}_{\mathbf{p}}(M)$. If $\alpha : U \to V$ is a coordinate patch on M about $\mathbf{p}$ belonging to the orientation of M with $\alpha(\mathbf{x}) = \mathbf{p}$, choose $\mathbf{n}$ so that

$$(\mathbf{n}, \frac{\partial \alpha}{\partial x_1}(\mathbf{x}), \cdots, \frac{\partial \alpha}{\partial x_{n-1}}(\mathbf{x}))$$

is right-handed. The vector field $N(\mathbf{p}) = (\mathbf{p};\mathbf{n}(\mathbf{p}))$ is called the **unit normal field corresponding to the orientation of M.**

We show $N(\mathbf{p})$ is well-defined, and of class C^∞. To show it is well-defined, let β be another coordinate patch about $\mathbf{p}$, belonging to the orientation of M. Let $g = \beta^{-1} \circ \alpha$ be the transition function, and let $g(\mathbf{x}) = \mathbf{y}$. Since $\alpha = \beta \circ g$,

$$D\alpha(\mathbf{x}) = D\beta(\mathbf{y}) \cdot Dg(\mathbf{x}).$$

Then for any $\mathbf{v} \in \mathbf{R}^n$, we have the equation

$$[\mathbf{v} \quad D\alpha(\mathbf{x})] = [\mathbf{v} \quad D\beta(\mathbf{y})] \begin{bmatrix} 1 & 0 \\ 0 & Dg(\mathbf{x}) \end{bmatrix}.$$

(Here $D\alpha$ and $D\beta$ have size n by $n-1$, so each of these three matrices has size n by n.) It follows that

$$\det[\mathbf{v} \quad D\alpha(\mathbf{x})] = \det[\mathbf{v} \quad D\beta(\mathbf{y})] \cdot \det Dg(\mathbf{x}).$$

Since $\det Dg > 0$, we conclude that $[\mathbf{v} \quad D\alpha(\mathbf{x})]$ has positive determinant if and only if $[\mathbf{v} \quad D\beta(\mathbf{y})]$ does.

To show that N is of class C^∞, we obtain a formula for it. As motivation, let us consider the case $n = 3$:

EXAMPLE 1. Given two vectors $\mathbf{a}$ and $\mathbf{b}$ in $\mathbf{R}^3$, one learns in calculus that their cross product $\mathbf{c} = \mathbf{a} \times \mathbf{b}$ is perpendicular to both, that the frame $(\mathbf{c}, \mathbf{a}, \mathbf{b})$ is right-handed, and that $\|\mathbf{c}\|$ equals $V(\mathbf{a}, \mathbf{b})$. The vector $\mathbf{c}$ is, of course, the vector with components

$$c_1 = \det \begin{bmatrix} a_2 & b_2 \\ a_3 & b_3 \end{bmatrix}, \qquad c_2 = -\det \begin{bmatrix} a_1 & b_1 \\ a_3 & b_3 \end{bmatrix}, \qquad c_3 = \det \begin{bmatrix} a_1 & b_1 \\ a_2 & b_2 \end{bmatrix}.$$

It follows that if M is an oriented 2-manifold in $\mathbf{R}^3$, and if $\alpha : U \to V$ is a coordinate patch on M belonging to the orientation of M, and if we set

$$\mathbf{c} = \frac{\partial \alpha}{\partial x_1} \times \frac{\partial \alpha}{\partial x_2},$$

then the vector $\mathbf{n} = \mathbf{c} / \|\mathbf{c}\|$ gives the corresponding unit normal to M. See Figure 38.1.

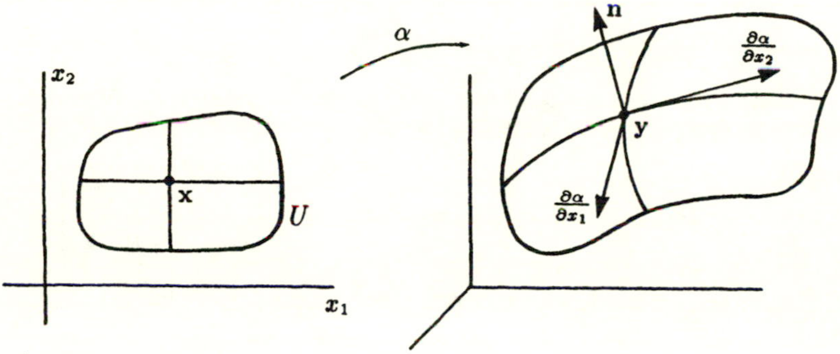

Figure 38.1

There is a formula similar to the cross product formula for determining $\mathbf{n}$ in general, as we now show.

Lemma 38.3. *Given independent vectors* $\mathbf{x}_1, \dots, \mathbf{x}_{n-1}$ *in* $\mathbf{R}^n$, *let* X *be the* n *by* $n-1$ *matrix* $X = [\mathbf{x}_1 \cdots \mathbf{x}_{n-1}]$, *and let* $\mathbf{c}$ *be the vector* $\mathbf{c} = \Sigma c_i \mathbf{e}_i$, *where*

$$c_i = (-1)^{i-1} \det X(1, \dots, \hat{\imath}, \dots, n).$$

The vector $\mathbf{c}$ *has the following properties:*

(1) $\mathbf{c}$ *is non-zero and orthogonal to each* $\mathbf{x}_i$.

(2) *The frame* $(\mathbf{c}, \mathbf{x}_1, \dots, \mathbf{x}_{n-1})$ *is right-handed.*

(3) $\|\mathbf{c}\| = V(X)$.

Proof. We begin with a preliminary calculation. Let $\mathbf{x}_1, \dots, \mathbf{x}_{n-1}$ be fixed. Given $\mathbf{a} \in \mathbf{R}^n$, we compute the following determinant; expanding by cofactors of the first column, we have:

$$\begin{aligned}\det[\mathbf{a}\ \mathbf{x}_1 \cdots \mathbf{x}_{n-1}] &= \sum_{i=1}^{n} a_i (-1)^{i-1} \det X(1, \dots, \hat{\imath}, \dots, n)\\ &= \langle \mathbf{a}, \mathbf{c} \rangle.\end{aligned}$$

This equation contains all that is needed to prove the theorem.

(1) Set $\mathbf{a} = \mathbf{x}_i$. Then the matrix $[\mathbf{a}\ \mathbf{x}_1 \cdots \mathbf{x}_{n-1}]$ has two identical columns, so its determinant vanishes. Thus $\langle \mathbf{x}_i, \mathbf{c} \rangle = 0$ for all i, so $\mathbf{c}$ is orthogonal to each $\mathbf{x}_i$. To show $\mathbf{c} \neq \mathbf{0}$, we note that since the columns of X span a space of dimension $n-1$, so do the rows of X. Hence some set consisting of $n-1$ rows of X is independent, say the set consisting of all rows but the ith. Then $c_i \neq 0$; whence $\mathbf{c} \neq \mathbf{0}$.

(2) Set $\mathbf{a} = \mathbf{c}$. Then

$$\det[\mathbf{c}\ \mathbf{x}_1 \cdots \mathbf{x}_{n-1}] = \langle \mathbf{c}, \mathbf{c} \rangle = \|\mathbf{c}\|^2 > 0.$$

Thus the frame $(\mathbf{c}, \mathbf{x}_1, \dots, \mathbf{x}_{n-1})$ is right-handed.

(3) This equation follows at once from Theorem 21.4. Alternatively, one can compute the matrix product

$$[\mathbf{c}\ \mathbf{x}_1 \cdots \mathbf{x}_{n-1}]^{\text{tr}} \cdot [\mathbf{c}\ \mathbf{x}_1 \cdots \mathbf{x}_{n-1}] = \begin{bmatrix} \|\mathbf{c}\|^2 & 0 \\ 0 & X^{\text{tr}} \cdot X \end{bmatrix}.$$

Taking determinants and using the formula in (2), we have

$$\|\mathbf{c}\|^4 = \|\mathbf{c}\|^2 V(X)^2.$$

Since $\|\mathbf{c}\| \neq 0$, we conclude that $\|\mathbf{c}\| = V(X)$. □

Corollary 38.4. *If M is an oriented $n-1$ manifold in $\mathbf{R}^n$, then the unit normal vector $N(\mathbf{p})$ corresponding to the orientation of M is a C^∞ function of* $\mathbf{p}$.

Proof. If $\alpha : U \to V$ is a coordinate patch on M about $\mathbf{p}$, let

$$c_i(\mathbf{x}) = (-1)^{i-1} \det D\alpha(1, \ldots, \hat{\imath}, \ldots, n)(\mathbf{x})$$

for $\mathbf{x} \in U$, and let $c(\mathbf{x}) = \Sigma c_i(\mathbf{x})\mathbf{e}_i$. Then for all $\mathbf{p} \in V$, we have

$$N(\mathbf{p}) = (\mathbf{p}; c(\mathbf{x})/\|c(\mathbf{x})\|),$$

where $\mathbf{x} = \alpha^{-1}(\mathbf{p})$; this function is of class C^∞ as a function of $\mathbf{p}$. □

Now we interpret the integral of an $n-1$ form in terms of vector fields. If G is a vector field in $\mathbf{R}^n$, then G corresponds under the "translation map" β_{n-1} to a certain $n-1$ form ω in $\mathbf{R}^n$. (See Theorem 31.1.) It turns out that the integral of ω over an oriented $n-1$ manifold M equals the integral over M, with respect to volume, of the normal component of the vector field G. That is the substance of the following lemma:

Lemma 38.5. *Let M be a compact oriented $n-1$ manifold in $\mathbf{R}^n$; let N be the corresponding unit normal vector field. Let G be a vector field defined in an open set U of $\mathbf{R}^n$ containing M. If we denote the general point of $\mathbf{R}^n$ by $\mathbf{y}$, this vector field has the form*

$$G(\mathbf{y}) = (\mathbf{y}; g(\mathbf{y})) = (\mathbf{y}; \Sigma g_i(\mathbf{y})\mathbf{e}_i);$$

it corresponds to the $n-1$ form

$$\omega = \sum_{i=1}^{n} (-1)^{i-1} g_i \, dy_1 \wedge \cdots \wedge \widehat{dy_i} \wedge \cdots \wedge dy_n.$$

Then

$$\int_M \omega = \int_M \langle G, N \rangle \; dV.$$

Note that if we replace M by $-M$, then the integral $\int_M \omega$ changes sign. This replacement has the effect of replacing N by $-N$, so that the integral $\int_M \langle G, N \rangle \; dV$ also changes sign.

Proof. We give two proofs of this theorem. The first relies on the results of §36 and the second does not.

First proof. By Theorem 36.2, we have

$$\int_M \omega = \int_M \lambda \, dV,$$

where $\lambda(\mathbf{p})$ is the value of $\omega(\mathbf{p})$ on an orthonormal basis for $\mathcal{T}_\mathbf{p}(M)$ that belongs to the natural orientation of this tangent space. We show that $\lambda = \langle G, N \rangle$, and the proof is complete.

Let $(\mathbf{p};\mathbf{a}_1), \dots, (\mathbf{p};\mathbf{a}_{n-1})$ be an orthonormal basis for $\mathcal{T}_\mathbf{p}(M)$ that belongs to its natural orientation. Let A be the matrix $A = [\mathbf{a}_1 \cdots \mathbf{a}_{n-1}]$; and let $\mathbf{c}$ be the vector $\mathbf{c} = \Sigma c_i \mathbf{e}_i$, where

$$c_i = (-1)^{i-1} \det A(1, \dots, \hat{\imath}, \dots, n).$$

By the preceding lemma, the vector $\mathbf{c}$ is orthogonal to each $\mathbf{a}_i$, and the frame $(\mathbf{c}, \mathbf{a}_1, \dots, \mathbf{a}_{n-1})$ is right-handed, and

$$\|\mathbf{c}\| = V(A) = [\det(A^{\text{tr}} \cdot A)]^{1/2} = [\det I_{n-1}]^{1/2} = 1.$$

Then $N = (\mathbf{p};\mathbf{c})$ is the unit normal to M at $\mathbf{p}$ corresponding to the orientation of M. Now by Theorem 27.7, we have

$$dy_1 \wedge \cdots \wedge \widehat{dy_i} \wedge \cdots \wedge dy_n((\mathbf{p};\mathbf{a}_1), \dots, (\mathbf{p};\mathbf{a}_{n-1})) = \det A(1, \dots, \hat{\imath}, \dots, n).$$

Then

$$\begin{aligned} \lambda(\mathbf{p}) &= \sum_{i=1}^{n} (-1)^{i-1} g_i(\mathbf{p}) \det A(1, \dots, \hat{\imath}, \dots, n) \\ &= \sum_{i=1}^{n} g_i(p) \cdot c_i. \end{aligned}$$

Thus $\lambda = \langle G, N \rangle$, as desired.

Second proof. Since the integrals involved in the statement of the theorem are linear in ω and G, respectively, it suffices to prove the theorem in the case where the set

$$C = M \cap (\text{Support } \omega)$$

lies in a single coordinate patch $\alpha : U \to V$ belonging to the orientation of M. We compute the first integral as follows:

$$\begin{aligned} \int_M \omega &= \int_{\text{Int } U} \alpha^* \omega \\ &= \int_{\text{Int } U} [\sum_{i=1}^{n} (-1)^{i-1} (g_i \circ \alpha) \det D\alpha(1, \dots, \hat{\imath}, \dots, n)], \end{aligned}$$

by Theorem 32.2. To compute the second integral, set $c = \Sigma c_i \mathbf{e}_i$, where

$$c_i = (-1)^{i-1} \det D\alpha(1, \ldots, \hat{\imath}, \ldots, n).$$

If N is the unit normal corresponding to the orientation, then as in the preceding corollary, $N(\alpha(\mathbf{x})) = (\alpha(\mathbf{x}); c(\mathbf{x})/\|c(\mathbf{x})\|)$. We compute

$$\begin{aligned}\int_M \langle G, N\rangle \, dV &= \int_{\text{Int } U} \langle G \circ \alpha, N \circ \alpha\rangle \cdot V(D\alpha) \\ &= \int_{\text{Int } U} \langle g \circ \alpha, c\rangle \quad \text{since} \quad \|c\| = V(D\alpha), \\ &= \int_{\text{Int } U} [\sum_{i=1}^{n} (g_i \circ \alpha)(-1)^{i-1} \det D\alpha(1, \ldots, \hat{\imath}, \ldots, n)].\end{aligned}$$

The lemma follows. □

Now we interpret the integral of an n-form in terms of scalar fields. The interpretation is just what one might expect:

Lemma 38.6. *Let M be a compact n-manifold in $\mathbf{R}^n$, oriented naturally. Let $\omega = h\, dx_1 \wedge \cdots \wedge dx_n$ be an n-form defined in an open set of $\mathbf{R}^n$ containing M. Then h is the corresponding scalar field, and*

$$\int_M \omega = \int_M h \, dV.$$

Proof. *First proof.* We use the results of §36. We have

$$\int_M \omega = \int_M \lambda \, dV,$$

where λ is obtained by evaluating ω on an orthonormal basis for $\mathcal{T}_\mathbf{p}(M)$ that belongs to its natural orientation. Now α belongs to the orientation of M if $\det D\alpha > 0$; thus the natural orientation of $\mathcal{T}_\mathbf{p}(M)$ consists of the right-handed frames. The usual basis for $\mathcal{T}_\mathbf{p}(M) = \mathcal{T}_\mathbf{p}(\mathbf{R}^n)$ is one such frame, and the value of ω on this frame is h.

Second proof. It suffices to consider the case where the set $M \cap (\text{Support } \omega)$ is covered by a coordinate patch $\alpha : U \to V$ belonging to the orientation of M. We have by definition

$$\int_M \omega = \int_{\text{Int } U} \alpha^* \omega = \int_{\text{Int } U} (h \circ \alpha) \det D\alpha,$$

$$\int_M h \, dV = \int_{\text{Int } U} (h \circ \alpha) V(D\alpha).$$

Now $V(D\alpha) = |\det D\alpha| = \det D\alpha$, since α belongs to the natural orientation of M. □

We note that the integral $\int_M h \, dV$ in fact equals the ordinary integral of h over the bounded subset M of $\mathbb{R}^n$. For if $A = M - \partial M$, then A is open in $\mathbb{R}^n$, and the identity map $i : A \to A$ is a coordinate patch on M, belonging to its natural orientation, that covers all of M except for a set of measure zero in M. By Theorem 25.4,

$$\int_M h \, dV = \int_A (h \circ i) V(Di) = \int_A h.$$

The latter is an ordinary integral; it equals $\int_M h$ because ∂M has measure zero in $\mathbb{R}^n$.

We now examine, for an n-manifold M in $\mathbb{R}^n$, naturally oriented, what the induced orientation of ∂M looks like. We considered the case $n = 3$ in Example 4 of §34. A result similar to that one holds in general:

Lemma 38.7. *Let M be an n-manifold in $\mathbb{R}^n$. If M is oriented naturally, then the induced orientation of ∂M corresponds to the unit normal field N to ∂M that points outwards from M at each point of ∂M.*

The **inward normal** to ∂M at $\mathbf{p}$ is the velocity vector of a curve that begins at $\mathbf{p}$ and moves into M as the parameter value increases. The **outward normal** is its negative.

Proof. Let $\alpha : U \to V$ be a coordinate patch on M about $\mathbf{p}$ belonging to the orientation of M. Then $\det D\alpha > 0$. Let $b : \mathbb{R}^{n-1} \to \mathbb{R}^n$ be the map

$$b(x_1, \ldots, x_{n-1}) = (x_1, \ldots, x_{n-1}, 0).$$

The map $\alpha_0 = \alpha \circ b$ is a coordinate patch on ∂M about $\mathbf{p}$. It belongs to the induced orientation of ∂M if n is even, and to its opposite if n is odd. Let N be the unit normal field to ∂M corresponding to the induced orientation of ∂M; let $N(\mathbf{p}) = (\mathbf{p}; \mathbf{n}(\mathbf{p}))$. Then

$$\det[(-1)^n \mathbf{n} \quad D\alpha_0] > 0,$$

which implies that

$$\det[D\alpha_0 \quad \mathbf{n}] = \det[\frac{\partial \alpha}{\partial x_1} \quad \cdots \quad \frac{\partial \alpha}{\partial x_{n-1}} \quad \mathbf{n}] < 0.$$

On the other hand, we have

$$\det D\alpha = \det[\frac{\partial \alpha}{\partial x_1} \quad \cdots \quad \frac{\partial \alpha}{\partial x_{n-1}} \quad \frac{\partial \alpha}{\partial x_n}] > 0.$$

The vector $\partial\alpha/\partial x_n$ is the velocity vector of a curve that begins at a point of ∂M and moves into M as the parameter increases. Thus $\mathbf{n}$ is the outward normal to ∂M at p. □

Theorem 38.8 (The divergence theorem). *Let M be a compact n-manifold in $\mathbf{R}^n$. Let N be the unit normal vector field to ∂M that points outwards from M. If G is a vector field defined in an open set of $\mathbf{R}^n$ containing M, then*

$$\int_M (\text{div } G)\, dV = \int_{\partial M} \langle G, N\rangle\, dV.$$

Here the left-hand integral involves integration with respect to n-volume, and the right-hand integral involves integration with respect to $n-1$ volume.

Proof. Given G, let $\omega = \beta_{n-1}G$ be the corresponding $n-1$ form. Orient M naturally and give ∂M the induced orientation. Then the normal field N corresponds to the orientation of ∂M, by Lemma 38.7, so that

$$\int_{\partial M} \omega = \int_{\partial M} \langle G, N\rangle\, dV,$$

by Lemma 38.5. According to Theorem 31.1, the scalar field div G corresponds to the n-form $d\omega$; that is, $d\omega = (\text{div } G) dx_1 \wedge \cdots \wedge dx_n$. Then Lemma 38.6 implies that

$$\int_M d\omega = \int_M (\text{div } G)\, dV.$$

The theorem follows from Stokes' theorem. □

In $\mathbf{R}^3$, the divergence theorem is sometimes called *Gauss' theorem.*

Stokes' theorem for 2-manifolds in $\mathbf{R}^3$

There is one more situation in which we can translate the general Stokes' theorem into a theorem about vector fields. It occurs when M is an oriented 2-manifold in $\mathbf{R}^3$.

Theorem 38.9 (Stokes' theorem—classical version). *Let M be a compact orientable 2-manifold in $\mathbf{R}^3$. Let N be a unit normal field to M. Let F be a C^∞ vector field defined in an open set about M. If ∂M is empty, then*

$$\int_M \langle \text{curl } F, N\rangle\, dV = 0.$$

If ∂M is non-empty, let T be the unit tangent vector field to ∂M chosen so that the vector $W(\mathbf{p}) = N(\mathbf{p}) \times T(\mathbf{p})$ points into M from ∂M. Then

$$\int_M \langle \text{curl } F, N \rangle \, dV = \int_{\partial M} \langle F, T \rangle \, ds.$$

Proof. Given F, let $\omega = \alpha_1 F$ be the corresponding 1-form. Then according to Theorem 31.2, the vector field curl F corresponds to the 2-form $d\omega$. Orient M so that N is the corresponding unit normal field. Then by Lemma 38.5,

$$\int_M d\omega = \int_M \langle \text{curl } F, N \rangle \, dV.$$

On the other hand, if ∂M is non-empty, its induced orientation corresponds to the unit tangent field T. (See Example 5 of §34.) It follows from Lemma 38.1 that

$$\int_{\partial M} \omega = \int_{\partial M} \langle F, T \rangle \, ds.$$

The theorem now follows from Stokes' theorem. □

EXERCISES

1. Let G be a vector field in $\mathbf{R}^3 - \mathbf{0}$. Let $S^2(r)$ be the sphere of radius r in $\mathbf{R}^3$ centered at $\mathbf{0}$. Let N_r be the unit normal to $S^2(r)$ that points away from the origin. If div $G(\mathbf{x}) = 1/\|\mathbf{x}\|$, and if $0 < c < d$, what can you say about the relation between the values of the integral

$$\int_{S^2(r)} \langle G, N_r \rangle \, dV$$

for $r = c$ and $r = d$?

2. Let G be a vector field defined in $A = \mathbf{R}^n - \mathbf{0}$ with div $G = 0$ in A.

   (a) Let M_1 and M_2 be compact n-manifolds in $\mathbf{R}^n$, such that the origin is contained in both $M_1 - \partial M_1$ and $M_2 - \partial M_2$. Let N_i be the unit outward normal vector field to ∂M_i, for $i = 1, 2$. Show that

$$\int_{\partial M_1} \langle G, N_1 \rangle \, dV = \int_{\partial M_2} \langle G, N_2 \rangle \, dV.$$

   [*Hint:* Consider first the case where $M_2 = B^n(\epsilon)$ and is contained in $M_1 - \partial M_1$. See Figure 38.2.]

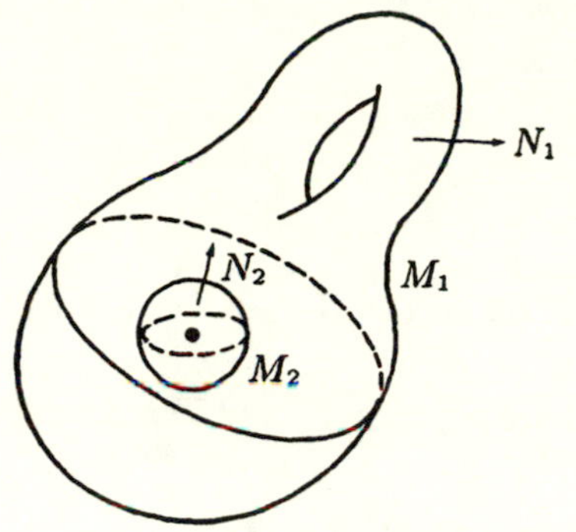

Figure 38.2

(b) Show that as M ranges over all compact n-manifolds in $\mathbb{R}^n$ for which the origin is not in ∂M, the integral

$$\int_{\partial M} \langle G, N\rangle \, dV,$$

where N is the unit normal to ∂M pointing outwards from M, has only two possible values.

3. Let G be a vector field in $B = \mathbb{R}^n - \mathbf{p} - \mathbf{q}$ with div $G = 0$ in B. As M ranges over all compact n-manifolds in $\mathbb{R}^n$ for which $\mathbf{p}$ and $\mathbf{q}$ are not in ∂M, how many possible values does the integral

$$\int_{\partial M} \langle G, N\rangle \, dV$$

have? (Here N is the unit normal to ∂M pointing outwards from M.)

4. Let η be the $n-1$ form in $A = \mathbb{R}^n - \mathbf{0}$ defined by the equation

$$\eta = \sum_{i=1}^{n} (-1)^{i-1} f_i \, dx_1 \wedge \cdots \wedge \widehat{dx_i} \wedge \cdots \wedge dx_n,$$

where $f_i(\mathbf{x}) = x_i/\|\mathbf{x}\|^m$. Orient the unit ball B^n naturally, and give $S^{n-1} = \partial B^n$ the induced orientation. Show that

$$\int_{S^{n-1}} \eta = v(S^{n-1}).$$

[*Hint:* If G is the vector field corresponding to η, and N is the unit outward normal field to S^{n-1}, then $\langle G, N\rangle = 1$.]

5. Let S be the subset of $\mathbf{R}^3$ consisting of the union of:

(i) the z-axis,

(ii) the unit circle $x^2 + y^2 = 1, z = 0$,

(iii) the points $(0, y, 0)$ with $y \geq 1$.

Let A be the open set $\mathbf{R}^3 - S$ of $\mathbf{R}^3$. Let C_1, C_2, D_1, D_2, D_3 be the oriented 1-manifolds in A that are pictured in Figure 38.3. Suppose that F is a vector field in A, with curl $F = \mathbf{0}$ in A, and that

$$\int_{C_1} \langle F, T \rangle \, ds = 3 \quad \text{and} \quad \int_{C_2} \langle F, T \rangle \, ds = 7.$$

What can you say about the integral

$$\int_{D_i} \langle F, T \rangle \, ds$$

for $i = 1, 2, 3$? Justify your answers.

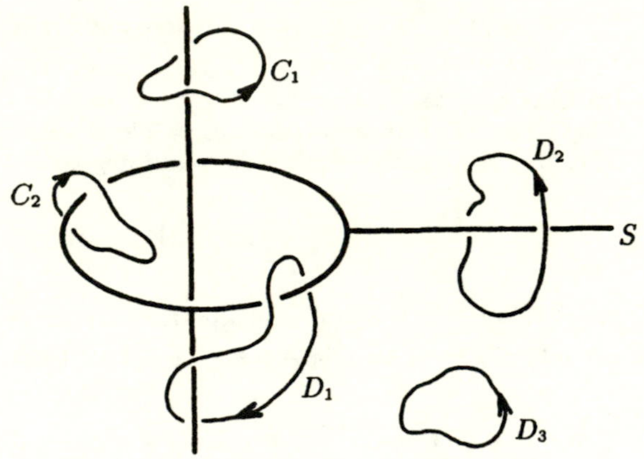

Figure 38.3

8

Closed Forms and Exact Forms

In the applications of vector analysis to physics, it is often important to know whether a given vector field F in $\mathbf{R}^3$ is the gradient of a scalar field f. If it is, F is said to be **conservative,** and the function f (or sometimes its negative) is called a **potential function** for F. Translated into the language of forms, this question is just the question whether a given 1-form ω in $\mathbf{R}^3$ is the differential of a 0-form, that is, whether ω is exact.

In other applications to physics, one wishes to know whether a given vector field G in $\mathbf{R}^3$ is the curl of another vector field F. Translated into the language of forms, this is just the question whether a given 2-form ω in $\mathbf{R}^3$ is the differential of a 1-form, that is, whether ω is exact.

We study here the analogous question in $\mathbf{R}^n$. If ω is a k-form defined in an open set A of $\mathbf{R}^n$, then a necessary condition for ω to be exact is the condition that ω be closed, i.e., that $d\omega = 0$. This condition is not in general sufficient. We explore in this chapter what additional conditions, either on A or on both A and ω, are needed in order to ensure that ω is exact.

§39. THE POINCARE LEMMA

Let A be an open set in $\mathbf{R}^n$. We show in this section that if A satisfies a certain condition called *star-convexity,* then any closed form ω on A is automatically exact. This result is a famous one called the Poincaré lemma.

We begin with a preliminary result:

Theorem 39.1 (Leibnitz's rule). *Let Q be a rectangle in $\mathbf{R}^n$; let $f : Q \times [a,b] \to \mathbf{R}$ be a continuous function. Denote f by $f(\mathbf{x},t)$ for $\mathbf{x} \in Q$ and $t \in [a,b]$. Then the function*

$$F(\mathbf{x}) = \int_{t=a}^{t=b} f(\mathbf{x},t)$$

is continuous on Q. Furthermore, if $\partial f/\partial x_j$ is continuous on $Q \times [a,b]$, then

$$\frac{\partial F}{\partial x_j}(\mathbf{x}) = \int_{t=a}^{t=b} \frac{\partial f}{\partial x_j}(\mathbf{x},t).$$

This formula is called *Leibnitz's rule for differentiating under the integral sign.*

Proof. *Step 1.* We show that F is continuous. The rectangle $Q \times [a,b]$ is compact; therefore f is uniformly continuous on $Q \times [a,b]$. That is, given $\epsilon > 0$, there is a $\delta > 0$ such that

$$|f(\mathbf{x}_1,t_1) - f(\mathbf{x}_0,t_0)| < \epsilon \quad \text{whenever} \quad |(\mathbf{x}_1,t_1) - (\mathbf{x}_0,t_0)| < \delta.$$

It follows that when $|\mathbf{x}_1 - \mathbf{x}_0| < \delta$,

$$|F(\mathbf{x}_1) - F(\mathbf{x}_0)| \leq \int_{t=a}^{t=b} |f(\mathbf{x}_1,t) - f(\mathbf{x}_0,t)| \leq \epsilon(b-a).$$

Continuity of F follows.

Step 2. In calculating the integral and derivatives involved in Leibnitz's rule, only the variables x_j and t are involved; all others are held constant. Therefore it suffices to prove the theorem in the case where $n = 1$ and Q is an interval $[c,d]$ in $\mathbf{R}$.

Let us set, for $x \in [c,d]$,

$$G(x) = \int_{t=a}^{t=b} D_1 f(x,t).$$

We wish to show that $F'(x)$ exists and equals $G(x)$. For this purpose, we apply (of all things) the Fubini theorem. We are given that $D_1 f$ is continuous on $[c,d] \times [a,b]$. Then

$$\begin{aligned}\int_{x=c}^{x=x_0} G(x) &= \int_{x=c}^{x=x_0} \int_{t=a}^{t=b} D_1 f(x,t) \\ &= \int_{t=a}^{t=b} \int_{x=c}^{x=x_0} D_1 f(x,t) \\ &= \int_{t=a}^{t=b} [f(x_0,t) - f(c,t)] \\ &= F(x_0) - F(c);\end{aligned}$$

the second equation follows from the Fubini theorem, and the third from the fundamental theorem of calculus. Then for $x \in [c,d]$, we have

$$\int_c^x G = F(x) - F(c).$$

Since G is continuous by Step 1, we may apply the fundamental theorem of calculus once more to conclude that

$$G(x) = F'(x). \quad \square$$

We now obtain a criterion for determining when two closed forms differ by an exact form. This criterion involves the notion of a *differentiable homotopy.*

Definition. Let A and B be open sets in $\mathbf{R}^n$ and $\mathbf{R}^m$, respectively; let $g, h : A \to B$ be C^∞ maps. We say that g and h are **differentiably homotopic** if there is a C^∞ map $H : A \times I \to B$ such that

$$H(\mathbf{x},0) = g(\mathbf{x}) \quad \text{and} \quad H(\mathbf{x},1) = h(\mathbf{x})$$

for $\mathbf{x} \in A$. The map H is called a **differentiable homotopy** between g and h.

For each t, the map $\mathbf{x} \to H(\mathbf{x},t)$ is a C^∞ map of A into B; if we think of t as "time," then H gives us a way of "deforming" the map g into the map h, as t goes from 0 to 1.

Theorem 39.2. *Let A and B be open sets in $\mathbf{R}^n$ and $\mathbf{R}^m$, respectively. Let $g, h : A \to B$ be C^∞ maps that are differentiably homotopic. Then there is a linear transformation*

$$\mathcal{P} : \Omega^{k+1}(B) \to \Omega^k(A),$$

defined for $k \geq 0$, such that for any form η on B of positive order,

$$d\mathcal{P}\eta + \mathcal{P}d\eta = h^*\eta - g^*\eta,$$

while for a form f of order 0,

$$\mathcal{P}df = h^*f - g^*f.$$

This theorem implies that if η is a closed form of positive order, then $h^*\eta$ and $g^*\eta$ differ by an exact form, since $h^*\eta - g^*\eta = d\mathcal{P}\eta$ if η is closed. On the other hand, if f is a closed 0-form, then $h^*f - g^*f = 0$.

Note that d raises the order of a form by 1, and $\mathcal{P}$ lowers it by 1. Thus if η has order $l > 0$, all the forms in the first equation have order l; and all the forms in the second equation have order 0. Of course, $\mathcal{P}f$ is not defined if f is a 0-form.

Proof. *Step 1.* We consider first a very special case. Given an open set A in $\mathbf{R}^n$, let U be a neighborhood of $A \times I$ in $\mathbf{R}^{n+1}$, and let $\alpha, \beta : A \to U$ be the maps given by the equations

$$\alpha(\mathbf{x}) = (\mathbf{x}, 0) \quad \text{and} \quad \beta(\mathbf{x}) = (\mathbf{x}, 1).$$

(Then α and β are differentiably homotopic.) We define, for any $k+1$ form η defined in U, a k-form $P\eta$ defined in A, such that

$$(*) \qquad \begin{aligned} dP\eta + Pd\eta &= \beta^*\eta - \alpha^*\eta \quad \text{if order} \quad \eta > 0, \\ Pdf &= \beta^*f - \alpha^*f \quad \text{if order} \quad f = 0. \end{aligned}$$

To begin, let $\mathbf{x}$ denote the general point of $\mathbf{R}^n$, and let t denote the general point of $\mathbf{R}$. Then $dx_1, \ldots, dx_n$, dt are the elementary 1-forms in $\mathbf{R}^{n+1}$. If g is any continuous scalar function in $A \times I$, we define a scalar function $\mathcal{I}g$ on A by the formula

$$(\mathcal{I}g)(\mathbf{x}) = \int_{t=0}^{t=1} g(\mathbf{x}, t).$$

Then we define P as follows: If $k \geq 0$, the general $k+1$ form η in $\mathbf{R}^{n+1}$ can be written uniquely as

$$\eta = \sum_{[I]} f_I \, dx_I + \sum_{[J]} g_J \, dx_J \wedge dt.$$

Here I denotes an ascending $k+1$ tuple, and J denotes an ascending k-tuple, from the set $\{1, \ldots, n\}$. We define P by the equation

$$P\eta = \sum_{[I]} P(f_I \; dx_I) + \sum_{[J]} P(g_J \; dx_J \wedge dt),$$

where

$$P(f_I \; dx_I) = 0 \quad \text{and} \quad P(g_J \; dx_J \wedge dt) = (-1)^k (\mathcal{I} g_J) \; dx_J .$$

Then $P\eta$ is a k-form defined on the subset A of $\mathbf{R}^n$.

Linearity of P follows at once from the uniqueness of the representation of η and linearity of the integral operator $\mathcal{I}$.

To show that $P\eta$ is of class C^∞, we need only show that the function $\mathcal{I}g$ is of class C^∞; and this result follows at once from Leibnitz's rule, since g is of class C^∞.

Note that in the special case $k = 0$, the form η is a 1-form and is written as

$$\eta = \sum_{i=1}^{n} f_i \; dx_i + g \; dt;$$

in this case, the tuple J is empty, and we have

$$P\eta = 0 + P(g \; dt) = \mathcal{I}g.$$

Although the operator P may seem rather artificial, it is in fact a rather natural one. Just as d is in some sense a "differentiation operator," the operator P is in some sense an "integration operator," one that "integrates η in the direction of the last coordinate." An alternate definition of P that makes this fact clear is given in the exercises.

Step 2. We show that the formulas

$$P(f \; dx_I) = 0 \quad \text{and} \quad P(g \; dx_J \wedge dt) = (-1)^k (\mathcal{I} g) \; dx_J$$

hold even when I is an arbitrary $k+1$ tuple, and J is an arbitrary k-tuple, from the set $\{1, \ldots, n\}$. The proof is easy. If the indices are not distinct, then these formulas hold trivially, since $dx_I = 0$ and $dx_J = 0$ in this case. If the indices are distinct and in ascending order, these formulas hold by definition. Then they hold for any sets of distinct indices, since rearranging the indices changes the values of dx_I and dx_J only by a sign.

Step 3. We verify formula $(*)$ of Step 1 for a 0-form. We have

$$\begin{aligned} P(df) &= P\Big(\sum_{j=1}^{n} \frac{\partial f}{\partial x_j} dx_j\Big) + P\Big(\frac{\partial f}{\partial t} dt\Big) \\ &= 0 + (-1)^0 \mathcal{I}\Big(\frac{\partial f}{\partial t}\Big) \\ &= f \circ \beta - f \circ \alpha \\ &= \beta^* f - \alpha^* f, \end{aligned}$$

where the third equation follows from the fundamental theorem of calculus.

Step 4. We verify formula $(*)$ for a form of positive order $k+1$. Note that because α is the map $\alpha(\mathbf{x}) = (\mathbf{x}, 0)$, then

$$\begin{aligned} \alpha^*(dx_i) &= d\alpha_i = dx_i \quad \text{for} \quad i = 1, \ldots, n, \\ \alpha^*(dt) &= d\alpha_{n+1} = 0. \end{aligned}$$

A similar remark holds for β^*.

Now because d and P and α^* and β^* are linear, it suffices to verify our formula for the forms $f\ dx_I$ and $g\ dx_J \wedge dt$. We first consider the case $\eta = f\ dx_I$. Let us compute both sides of the equation. The left side is

$$\begin{aligned} dP\eta + Pd\eta &= d(0) + P(d\eta) \\ &= \Big[\sum_{j=1}^{n} P\Big(\frac{\partial f}{\partial x_j} dx_j \wedge dx_I\Big)\Big] + P\Big(\frac{\partial f}{\partial t} dt \wedge dx_I\Big) \\ &= 0 + (-1)^{k+1} P\Big(\frac{\partial f}{\partial t} dx_I \wedge dt\Big) \quad \text{by Step 2,} \\ &= \mathcal{I}\Big(\frac{\partial f}{\partial t}\Big) dx_I \\ &= [f \circ \beta - f \circ \alpha] dx_I. \end{aligned}$$

The right side of our equation is

$$\begin{aligned} \beta^*\eta - \alpha^*\eta &= (f \circ \beta)\beta^*(dx_I) - (f \circ \alpha)\alpha^*(dx_I) \\ &= [f \circ \beta - f \circ \alpha] dx_I. \end{aligned}$$

Thus our result holds in this case.

We now consider the case when $\eta = g\, dx_J \wedge dt$. Again, we compute both sides of the equation. We have

$$(**) \qquad d(P\eta) = d[(-1)^k(\mathcal{I}g)\, dx_J)] = (-1)^k \sum_{j=1}^{n} D_j(\mathcal{I}g)\, dx_j \wedge dx_J.$$

On the other hand,

$$d\eta = \sum_{j=1}^{n} (D_j g)\, dx_j \wedge dx_J \wedge dt + (D_{n+1}g)\, dt \wedge dx_J \wedge dt,$$

so that by Step 2,

$$(***) \qquad P(d\eta) = (-1)^{k+1} \sum_{j=1}^{n} \mathcal{I}(D_j g)\, dx_j \wedge dx_J.$$

Adding (**) and (* * *) and applying Leibnitz's rule, we see that

$$d(P\eta) + P(d\eta) = 0.$$

On the other hand, the right side of the equation is

$$\beta^*(g\, dx_J \wedge dt) - \alpha^*(g\, dx_J \wedge dt) = 0,$$

since $\beta^*(dt) = 0$ and $\alpha^*(dt) = 0$. This completes the proof of the special case of the theorem.

Step 5. We now prove the theorem in general. We are given C^∞ maps $g, h : A \to B$, and a differentiable homotopy $H : A \times I \to B$ between them. Let $\alpha, \beta : A \to A \times I$ be the maps of Step 1, and let P be the linear transformation of forms whose properties are stated in Step 1. We then define our desired linear transformation $\mathcal{P} : \Omega^{k+1}(B) \to \Omega^k(A)$ by the equation

$$\mathcal{P}\eta = P(H^*\eta).$$

See Figure 39.1. Since $H^*\eta$ is a $k+1$ form defined in a neighborhood of $A \times I$, then $P(H^*\eta)$ is a k-form defined in A.

Note that since H is a differentiable homotopy between g and h,

$$H \circ \alpha = g \quad \text{and} \quad H \circ \beta = h.$$

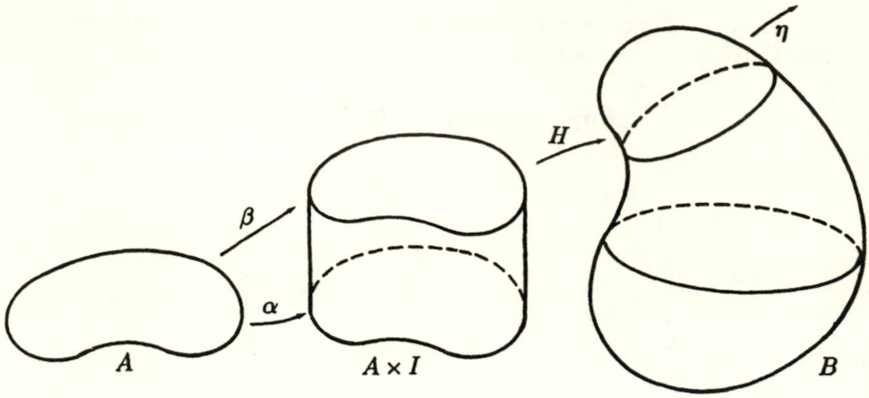

Figure 39.1

Then we compute

$$\begin{aligned} d\mathcal{P}\eta + \mathcal{P}d\eta &= dP(H^*\eta) + P(H^*d\eta) \\ &= dP(H^*\eta) + P(dH^*\eta) \\ &= \beta^*(H^*\eta) - \alpha^*(H^*\eta) \quad \text{by Step 1,} \\ &= h^*\eta - g^*\eta, \end{aligned}$$

as desired. An entirely similar computation applies for a 0-form f. □

Now we can prove the Poincaré lemma. First, a definition:

Definition. Let A be an open set in $\mathbf{R}^n$. We say that A is **star-convex** with respect to the point $\mathbf{p}$ of A if for each $\mathbf{x} \in A$, the line segment joining $\mathbf{x}$ and $\mathbf{p}$ lies in A.

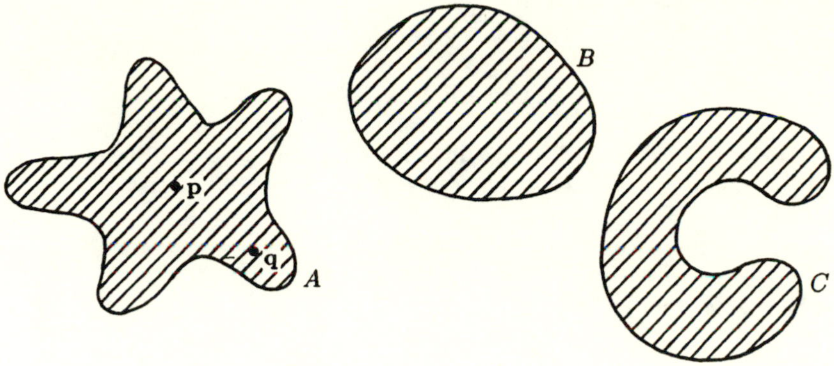

Figure 39.2

EXAMPLE 1. In Figure 39.2, the set A is star-convex with respect to the point $\mathbf{p}$, but not with respect to the point $\mathbf{q}$. The set B is star-convex with respect to each of its points; that is, it is convex. The set C is not star-convex with respect to any of its points.

Theorem 39.3 (The Poincaré lemma). *Let A be a star-convex open set in $\mathbb{R}^n$. If ω is a closed k-form on A, then ω is exact on A.*

Proof. We apply the preceding theorem. Let $\mathbf{p}$ be a point with respect to which A is star-convex. Let $h : A \to A$ be the identity map and let $g : A \to A$ be the constant map carrying each point to the point $\mathbf{p}$. Then g and h are differentiably homotopic; indeed, the map

$$H(\mathbf{x}, t) = th(\mathbf{x}) + (1-t)g(\mathbf{x})$$

carries $A \times I$ into A and is the desired differentiable homotopy. (For each t, the point $H(\mathbf{x}, t)$ lies on the line segment between $h(\mathbf{x}) = \mathbf{x}$ and $g(\mathbf{x}) = \mathbf{p}$, so that it lies in A.) We call H the **straight-line homotopy** between g and h.

Let $\mathcal{P}$ be the transformation given by the preceding theorem. If f is a 0-form on A, we have

$$\mathcal{P}(df) = h^* f - g^* f = f \circ h - f \circ g.$$

Then if $df = 0$, we have for all $\mathbf{x} \in A$,

$$0 = f(h(\mathbf{x})) - f(g(\mathbf{x})) = f(\mathbf{x}) - f(\mathbf{p}),$$

so that f is constant on A.

If ω is a k-form with $k > 0$, we have

$$d\mathcal{P}\omega + \mathcal{P}d\omega = h^*\omega - g^*\omega.$$

Now $h^*\omega = \omega$ because h is the identity map, and $g^*\omega = 0$ because g is a constant map. Then if $d\omega = 0$, we have

$$d\mathcal{P}\omega = \omega,$$

so that ω is exact on A. □

Theorem 39.4. *Let A be a star-convex open set in $\mathbf{R}^n$. Let ω be a closed k-form on A. If $k > 1$, and if η and η_0 are two $k-1$ forms on A with $d\eta = \omega = d\eta_0$, then*

$$\eta = \eta_0 + d\theta$$

for some $k-2$ form θ on A. If $k = 1$, and if f and f_0 are two 0-forms on A with $df = \omega = df_0$, then $f = f_0 + c$ for some constant c.

Proof. Since $d(\eta - \eta_0) = 0$, the form $\eta - \eta_0$ is a closed form on A. By the Poincaré lemma, it is exact. A similar comment applies to the form $f - f_0$. □

EXERCISES

1. (a) Translate the Poincaré lemma for k-forms into theorems about scalar and vector fields in $\mathbf{R}^3$. Consider the cases $k = 0, 1, 2, 3$.

   (b) Do the same for Theorem 39.4. Consider the cases $k = 1, 2, 3$.

2. (a) Let $g : A \to B$ be a diffeomorphism of open sets in $\mathbf{R}^n$, of class C^∞. Show that if A is homologically trivial in dimension k, so is B.

   (b) Find an open set in $\mathbf{R}^2$ that is not star-convex but is homologically trivial in every dimension.

3. Let A be an open set in $\mathbf{R}^n$. Show that A is homologically trivial in dimension 0 if and only if A is connected. [*Hint:* Let $\mathbf{p} \in A$. Show that if $df = 0$, and if $\mathbf{x}$ can be joined by a broken-line path in A to $\mathbf{p}$, then $f(\mathbf{x}) = f(\mathbf{p})$. Show that the set of all $\mathbf{x}$ that can be joined to $\mathbf{p}$ by a broken-line path in A is open in A.]

4. Prove the following theorem; it shows that P is in some sense an operator that integrates in the direction of the last coordinate:

   Theorem. *Let A be open in $\mathbf{R}^n$; let η be a $k+1$ form defined in an open set U of $\mathbf{R}^{n+1}$ containing $A \times I$. Given $t \in I$, let $\alpha_t : A \to U$*

be the "slice" map defined by $\alpha_t(\mathbf{x}) = (\mathbf{x}, t)$. *Given fixed vectors* $(\mathbf{x}; \mathbf{v}_1), \ldots, (\mathbf{x}; \mathbf{v}_k)$ *in* $\mathcal{T}_{\mathbf{x}}(\mathbf{R}^n)$, *let*

$$(\mathbf{y}; \mathbf{w}_i) = (\alpha_t)_*(\mathbf{x}; \mathbf{v}_i),$$

for each t. *Then* $(\mathbf{y}; \mathbf{w}_i)$ *belongs to* $\mathcal{T}_{\mathbf{y}}(\mathbf{R}^{n+1})$; *and* $\mathbf{y} = (\mathbf{x}, t)$ *is a function of* t, *but* $\mathbf{w}_i = (\mathbf{v}_i, 0)$ *is not. (See Figure 39.3.) Then*

$$(P\eta)(\mathbf{x})\big((\mathbf{x}; \mathbf{v}_1), \ldots, (\mathbf{x}; \mathbf{v}_k)\big) =$$
$$(-1)^k \int_{t=0}^{t=1} \eta(\mathbf{y})\big((\mathbf{y}; \mathbf{w}_1), \ldots, (\mathbf{y}; \mathbf{w}_k), (\mathbf{y}; \mathbf{e}_{n+1})\big).$$

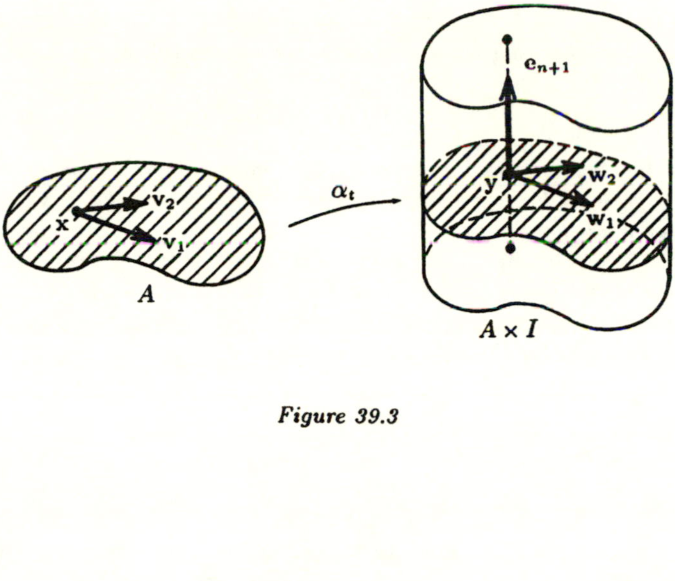

Figure 39.3

§40. THE DeRHAM GROUPS OF PUNCTURED EUCLIDEAN SPACE

We have shown that an open set A of $\mathbf{R}^n$ is homologically trivial in all dimensions if it is star-convex. We now consider some situations in which A is *not* homologically trivial in all dimensions. The simplest such situation occurs when A is the punctured euclidean space $\mathbf{R}^n - \mathbf{0}$. Earlier exercises demonstrated the existence of a closed $n-1$ form in $\mathbf{R}^n - \mathbf{0}$ that is not exact. Now we analyze the situation further, giving a definitive criterion for deciding whether or not a given closed form in $\mathbf{R}^n - \mathbf{0}$ is exact.

A convenient way to deal with this question is to define, for an open set A in $\mathbf{R}^n$, certain vector spaces $H^k(A)$ that are called the *deRham groups* of A. The condition that A be homologically trivial in dimension k is equivalent to the condition that $H^k(A)$ be the trivial vector space. We shall determine the dimensions of these spaces in the case $A = \mathbf{R}^n - \mathbf{0}$.

To begin, we consider what is meant by the *quotient* of a vector space by a subspace.

Definition. If V is a vector space, and if W is a linear subspace of V, we denote by V/W the set whose elements are the subsets of V of the form

$$\mathbf{v} + W = \{\mathbf{v} + \mathbf{w} \mid \mathbf{w} \in W\}.$$

Each such set is called a **coset** of V, determined by W. One shows readily that if $\mathbf{v}_1 - \mathbf{v}_2 \in W$, then the cosets $\mathbf{v}_1 + W$ and $\mathbf{v}_2 + W$ are equal, while if $\mathbf{v}_1 - \mathbf{v}_2 \notin W$, then they are disjoint. Thus V/W is a collection of disjoint subsets of V whose union is V. (Such a collection is called a **partition** of V.) We define vector space operations in V/W by the equations

$$(\mathbf{v}_1 + W) + (\mathbf{v}_2 + W) = (\mathbf{v}_1 + \mathbf{v}_2) + W,$$

$$c(\mathbf{v} + W) = (c\mathbf{v}) + W.$$

With these operations, V/W becomes a vector space. It is called the **quotient space** of V by W.

We must show these operations are well-defined. Suppose $\mathbf{v}_1 + W = \mathbf{v}_1' + W$ and $\mathbf{v}_2 + W = \mathbf{v}_2' + W$. Then $\mathbf{v}_1 - \mathbf{v}_1'$ and $\mathbf{v}_2 - \mathbf{v}_2'$ are in W, so that their sum, which equals $(\mathbf{v}_1 + \mathbf{v}_2) - (\mathbf{v}_1' + \mathbf{v}_2')$, is in W. Then

$$(\mathbf{v}_1 + \mathbf{v}_2) + W = (\mathbf{v}_1' + \mathbf{v}_2') + W.$$

Thus vector addition is well-defined. A similar proof shows that multiplication by a scalar is well-defined. The vector space properties are easy to check; we leave the details to you.

Now if V is finite-dimensional, then so is V/W; we shall not however need this result. On the other hand, V/W may be finite-dimensional even in cases where V and W are not.

Definition. Suppose V and V' are vector spaces, and suppose W and W' are linear subspaces of V and V', respectively. If $T : V \to V'$ is a linear transformation that carries W into W', then there is a linear transformation

$$\tilde{T} : V/W \to V'/W'$$

defined by the equation $\tilde{T}(\mathbf{v} + W) = T(\mathbf{v}) + W'$; it is said to be **induced** by T. One checks readily that $\tilde{T}$ is well-defined and linear.

Now we can define deRham groups.

Definition. Let A be an open set in $\mathbf{R}^n$. The set $\Omega^k(A)$ of all k-forms on A is a vector space. The set $C^k(A)$ of closed k-forms on A and the set $E^k(A)$ of exact k-forms on A are linear subspaces of $\Omega^k(A)$. Since every exact form is closed, $E^k(A)$ is contained in $C^k(A)$. We define the **deRham group** of A in dimension k to be the quotient vector space

$$H^k(A) = C^k(A)/E^k(A).$$

If ω is a closed k-form on A (i.e., an element of $C^k(A)$), we often denote its coset $\omega + E^k(A)$ simply by $\{\omega\}$.

It is immediate that $H^k(A)$ is the trivial vector space, consisting of the zero vector alone, if and only if A is homologically trivial in dimension k.

Now if A and B are open sets in $\mathbf{R}^n$ and $\mathbf{R}^m$, respectively, and if $g : A \to B$ is a C^∞ map, then g induces a linear transformation $g^* : \Omega^k(B) \to \Omega^k(A)$ of forms, for all k. Because g^* commutes with d, it carries closed forms to closed forms and exact forms to exact forms; thus g^* induces a linear transformation

$$g^* : H^k(B) \to H^k(A)$$

of deRham groups. (For convenience, we denote this induced transformation also by g^*, rather than by $\tilde{g}^*$.)

Studying closed forms and exact forms on a given set A now reduces to calculating the deRham groups of A. There are several tools that are used in computing these groups. We consider two of them here. One involves the notion of a *homotopy equivalence*. The other is a special case of a general theorem called the *Mayer-Vietoris theorem*. Both are standard tools in algebraic topology.

Theorem 40.1 (Homotopy equivalence theorem). *Let A and B be open sets in $\mathbf{R}^n$ and $\mathbf{R}^m$, respectively. Let $g : A \to B$ and $h : B \to A$ be C^∞ maps. If $g \circ h : B \to B$ is differentiably homotopic to the identity map i_B of B, and if $h \circ g : A \to A$ is differentiably homotopic to the identity map i_A of A, then g^* and h^* are linear isomorphisms of the deRham groups.*

If $g \circ h$ *equals* i_B and $h \circ g$ *equals* i_A, then of course g and h are diffeomorphisms. If g and h satisfy the hypotheses of this theorem, then they are called (differentiable) **homotopy equivalences.**

Proof. If η is a closed k-form on A, for $k \geq 0$, then Theorem 39.2 implies that

$$(h \circ g)^*\eta - (i_A)^*\eta$$

is exact. Then the induced maps of the deRham groups satisfy the equation

$$g^*(h^*(\{\eta\}) = \{\eta\},$$

so that $g^* \circ h^*$ is the identity map of $H^k(A)$ with itself. A similar argument shows that $h^* \circ g^*$ is the identity map of $H^k(B)$. The first fact implies that g^* maps $H^k(B)$ *onto* $H^k(A)$, since given $\{\eta\}$ in $H^k(A)$, it equals $g^*(h^*\{\eta\})$. The second fact implies that g^* is one-to-one, since the equation $g^*\{\omega\} = 0$ implies that $h^*(g^*\{\omega\}) = 0$, whence $\{\omega\} = 0$.

By symmetry, h^* is also a linear isomorphism. □

In order to prove our other major theorem, we need a technical lemma:

Lemma 40.2. *Let U and V be open sets in $\mathbf{R}^n$; let $X = U \cup V$; and suppose $A = U \cap V$ is non-empty. Then there exists a C^∞ function $\phi : X \to [0,1]$ such that ϕ is identically 0 in a neighborhood of $U - A$ and ϕ is identically 1 in a neighborhood of $V - A$.*

Proof. See Figure 40.1. Let $\{\phi_i\}$ be a partition of unity on X dominated by the open covering $\{U, V\}$. Let S_i = Support ϕ_i for each i. Divide the index set of the collection $\{\phi_i\}$ into two disjoint subsets J and K, so that for every $i \in J$, the set S_i is contained in U, and for every $i \in K$, the set S_i is contained in V. (For example, one could let J consist of all i such that $S_i \subset U$, and let K consist of the remaining i.) Then let

$$\phi(\mathbf{x}) = \sum_{i \in K} \phi_i(\mathbf{x}).$$

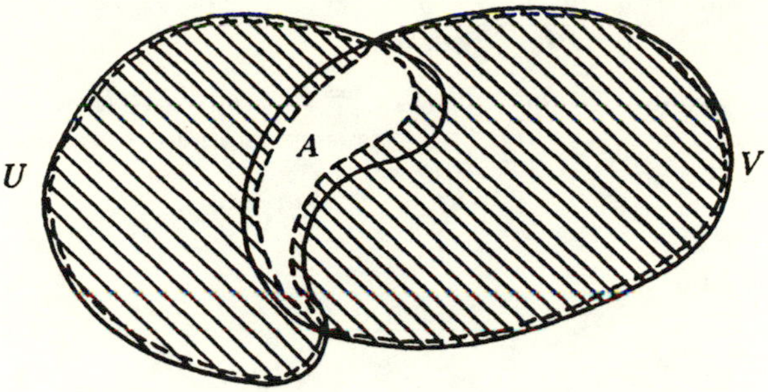

Figure 40.1

The local finiteness condition guarantees that ϕ is of class C^∞ on X, since each $\mathbf{x} \in X$ has a neighborhood on which ϕ equals a finite sum of C^∞ functions.

Let $\mathbf{a} \in U - A$; we show ϕ is identically 0 in a neighborhood of $\mathbf{a}$. First, we choose a neighborhood W of $\mathbf{a}$ that intersects only finitely many sets S_i. From among these sets S_i, take those whose indices belong to K, and let D be their union. Then D is closed, and D does not contain the point $\mathbf{a}$. The set $W - D$ is thus a neighborhood of $\mathbf{a}$, and for each $i \in K$, the function ϕ_i vanishes on $W - D$. It follows that $\phi(\mathbf{x}) = 0$ for $\mathbf{x} \in W - D$.

Since

$$1 - \phi(\mathbf{x}) = \sum_{i \in J} \phi_i(\mathbf{x}),$$

symmetry implies that the function $1 - \phi$ is identically 0 in a neighborhood of $V - A$. □

Theorem 40.3 (Mayer-Vietoris–special case). *Let U and V be open sets in $\mathbf{R}^n$ with U and V homologically trivial in all dimensions. Let $X = U \cup V$; suppose $A = U \cap V$ is non-empty. Then $H^0(X)$ is trivial, and for $k \geq 0$, the space $H^{k+1}(X)$ is linearly isomorphic to the space $H^k(A)$.*

Proof. We introduce some notation that will be convenient. If B, C are open sets of $\mathbf{R}^n$ with $B \subset C$, and if η is a k-form on C, we denote by $\eta|B$ the **restriction** of η to B. That is, $\eta|B = j^*\eta$, where j is the inclusion map $j : B \to C$. Since j^* commutes with d, it follows that the restriction of a closed or exact form is closed or exact, respectively. It also follows that if $A \subset B \subset C$, then $(\eta|B)|A = \eta|A$.

Step 1. We first show that $H^0(X)$ is trivial. Let f be a closed 0-form on X. Then $f|U$ and $f|V$ are closed forms on U and V, respectively.

Because U and V are homologically trivial in dimension 0, there are constant functions c_1 and c_2 such that $f|U = c_1$ and $f|V = c_2$. Since $U \cap V$ is non-empty, $c_1 = c_2$; thus f is constant on X.

Step 2. Let $\phi : X \to [0,1]$ be a C^∞ function such that ϕ vanishes in a neighborhood U' of $U - A$ and $1-\phi$ vanishes in a neighborhood V' of $V - A$. For $k \geq 0$, we define

$$\delta : \Omega^k(A) \to \Omega^{k+1}(X)$$

by the equation

$$\delta(\omega) = \begin{cases} d\phi \wedge \omega & \text{on } A, \\ 0 & \text{on } U' \cup V'. \end{cases}$$

Since $d\phi = 0$ on the set $U' \cup V'$, the form $\delta(\omega)$ is well-defined; since A and $U' \cup V'$ are open and their union is X, it is of class C^∞ on X. The map δ is clearly linear. It commutes with the differential operator d, up to sign, since

$$d(\delta(\omega)) = \left\{ \begin{matrix} (-1)d\phi \wedge d\omega & \text{on} & A \\ 0 & \text{on} & U' \cup V' \end{matrix} \right\} = -\delta(d\omega).$$

Then δ carries closed forms to closed forms, and exact forms to exact forms, so it induces a linear transformation

$$\tilde{\delta} : H^k(A) \to H^{k+1}(X).$$

We show that $\tilde{\delta}$ is an isomorphism.

Step 3. We first show that $\tilde{\delta}$ is one-to-one. For this purpose, it suffices to show that if ω is a closed k-form in A such that $\delta(\omega)$ is exact, then ω is itself exact.

So suppose $\delta(\omega) = d\theta$ for some k-form θ on X. We define k-forms ω_1 and ω_2 on U and V, respectively, by the equations

$$\omega_1 = \begin{cases} \phi\omega & \text{on } A, \\ 0 & \text{on } U', \end{cases} \quad \text{and} \quad \omega_2 = \begin{cases} (1-\phi)\omega & \text{on } A, \\ 0 & \text{on } V'. \end{cases}$$

Then ω_1 and ω_2 are well-defined and of class C^∞. See Figure 40.2.

We compute

$$d\omega_1 = \begin{cases} d\phi \wedge \omega + 0 & \text{on } A, \\ 0 & \text{on } U'; \end{cases}$$

the first equation follows from the fact that $d\omega = 0$. Then

$$d\omega_1 = \delta(\omega)|U = d\theta|U.$$

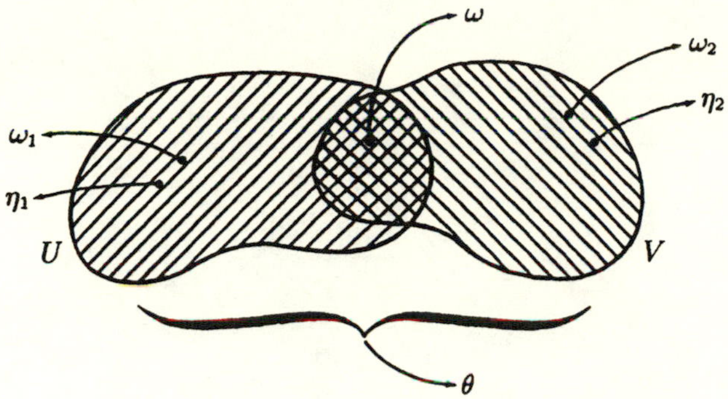

Figure 40.2

It follows that $\omega_1 - \theta|U$ is a closed k-form on U. An entirely similar proof shows that

$$d\omega_2 = -d\theta|V,$$

so that $\omega_2 + \theta|V$ is a closed k-form on V.

Now U and V are homologically trivial in all dimensions. If $k > 0$, this implies that there are $k-1$ forms η_1 and η_2 on U and V, respectively, such that

$$\omega_1 - \theta|U = d\eta_1 \quad \text{and} \quad \omega_2 + \theta|V = d\eta_2.$$

Restricting to A and adding, we have

$$\omega_1|A + \omega_2|A = d\eta_1|A + d\eta_2|A,$$

which implies that

$$\phi\omega + (1-\phi)\omega = d(\eta_1|A + \eta_2|A).$$

Thus ω is exact on A.

If $k = 0$, then there are constants c_1 and c_2 such that

$$\omega_1 - \theta|U = c_1 \quad \text{and} \quad \omega_2 + \theta|V = c_2.$$

Then

$$\phi\omega + (1-\phi)\omega = \omega_1|A + \omega_2|A = c_1 + c_2.$$

Step 4. We show $\tilde{\delta}$ maps $H^k(A)$ onto $H^{k+1}(X)$. For this purpose, it suffices to show that if η is a closed $k+1$ form in X, then there is a closed k-form ω in A such that $\eta - \delta(\omega)$ is exact.

Given η, the forms $\eta|U$ and $\eta|V$ are closed; hence there are k-forms θ_1 and θ_2 on U and V respectively, such that

$$d\theta_1 = \eta|U \quad \text{and} \quad d\theta_2 = \eta|V.$$

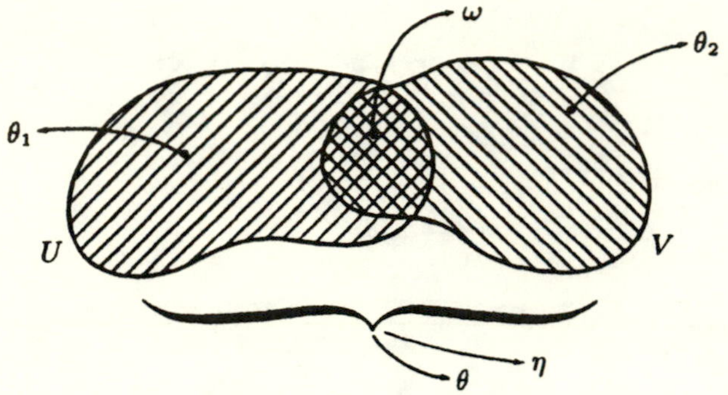

Figure 40.3

Let ω be the k-form on A defined by the equation

$$\omega = \theta_1|A - \theta_2|A;$$

then ω is closed because $d\omega = d\theta_1|A - d\theta_2|A = \eta|A - \eta|A = 0$. We define a k-form θ on X by the equation

$$\theta = \begin{cases} (1-\phi)\theta_1 + \phi\theta_2 & \text{on } A, \\ \theta_1 & \text{on } U', \\ \theta_2 & \text{on } V'. \end{cases}$$

Then θ is well-defined and of class C^∞. See Figure 40.3. We show that

$$\eta - \delta(\omega) = d\theta;$$

this completes the proof.

We compute $d\theta$ on A and U' and V' separately. Restricting to A, we have

$$\begin{aligned} d\theta|A &= [-d\phi \wedge (\theta_1|A) + (1-\phi)(d\theta_1|A)] + [d\phi \wedge (\theta_2|A) + \phi(d\theta_2|A)] \\ &= \phi\eta|A + (1-\phi)\eta|A + d\phi \wedge [\theta_2|A - \theta_1|A] \\ &= \eta|A + d\phi \wedge (-\omega) \\ &= \eta|A - \delta(\omega)|A. \end{aligned}$$

Restricting to U' and to V', we compute

$$d\theta|U' = d\theta_1|U' = \eta|U' = \eta|U' - \delta(\omega)|U',$$

$$d\theta|V' = d\theta_2|V' = \eta|V' = \eta|V' - \delta(\omega)|V',$$

since $\delta(\omega)|U' = 0$ and $\delta(\omega)|V' = 0$ by definition. It follows that

$$d\theta = \eta - \delta(\omega),$$

as desired. □

Now we can calculate the deRham groups of punctured euclidean space.

Theorem 40.4. *Let $n \geq 1$. Then*

$$\dim H^k(\mathbf{R}^n - 0) = \begin{cases} 0 & \textit{for } k \neq n-1, \\ 1 & \textit{for } k = n-1. \end{cases}$$

Proof. *Step 1.* We prove the theorem for $n = 1$. Let $A = \mathbf{R}^1 - 0$; write $A = A_0 \cup A_1$, where A_0 consists of the negative reals and A_1 consists of the positive reals. If ω is a closed k-form in A, with $k > 0$, then $\omega|A_0$ and $\omega|A_1$ are closed. Since A_0 and A_1 are star-convex, there are $k-1$ forms η_0 and η_1 on A_0 and A_1, respectively, such that $d\eta_i = \omega|A_i$ for $i = 0, 1$. Define $\eta = \eta_0$ on A_0 and $\eta = \eta_1$ on A_1. Then η is well-defined and of class C^∞, and $d\eta = \omega$.

Now let f_0 be the 0-form in A defined by setting $f_0(x) = 0$ for $x \in A_0$ and $f_0(x) = 1$ for $x \in A_1$. Then f_0 is a closed form, and f_0 is not exact. We show the coset $\{f_0\}$ forms a basis for $H^0(A)$. Given a closed 0-form f on A, the forms $f|A_0$ and $f|A_1$ are closed and thus exact. Then there are constants c_0 and c_1 such that $f|A_0 = c_0$ and $f|A_1 = c_1$. It follows that

$$f(x) = cf_0(x) + c_0$$

for $x \in A$, where $c = c_1 - c_0$. Then $\{f\} = c\{f_0\}$, as desired.

Step 2. If B is open in $\mathbf{R}^n$, then $B \times \mathbf{R}$ is open in $\mathbf{R}^{n+1}$. We show that for all k,

$$\dim H^k(B) = \dim H^k(B \times \mathbf{R}).$$

We use the homotopy equivalence theorem. Define $g : B \to B \times \mathbf{R}$ by the equation $g(\mathbf{x}) = (\mathbf{x}, 0)$, and define $h : B \times \mathbf{R} \to B$ by the equation $h(\mathbf{x}, s) = \mathbf{x}$. Then $h \circ g$ equals the identity map of B with itself. On the other hand, $g \circ h$ is differentiably homotopic to the identity map of $B \times \mathbf{R}$ with itself; the straight-line homotopy will suffice. It is given by the equation

$$H((\mathbf{x}, s), t) = t(\mathbf{x}, s) + (1-t)(\mathbf{x}, 0) = (\mathbf{x}, st).$$

Step 3. Let $n \geq 1$. We assume the theorem true for n and prove it for $n+1$.

Let U and V be the open sets in $\mathbf{R}^{n+1}$ defined by the equations

$$U = \mathbf{R}^{n+1} - \{(0, \ldots, 0, t) | t \geq 0\},$$

$$V = \mathbf{R}^{n+1} - \{(0, \ldots, 0, t) | t \leq 0\}.$$

Thus U consists of all of $\mathbf{R}^{n+1}$ except for points on the half-line $\mathbf{0} \times \mathbf{H}^1$, and V consists of all of $\mathbf{R}^{n+1}$ except for points on the half-line $\mathbf{0} \times \mathbf{L}^1$. Figure 40.4 illustrates the case $n = 3$. The set $A = U \cap V$ is non-empty; indeed, A consists of all points of $\mathbf{R}^{n+1} = \mathbf{R}^n \times \mathbf{R}$ not on the line $\mathbf{0} \times \mathbf{R}$; that is,

$$A = (\mathbf{R}^n - \mathbf{0}) \times \mathbf{R}.$$

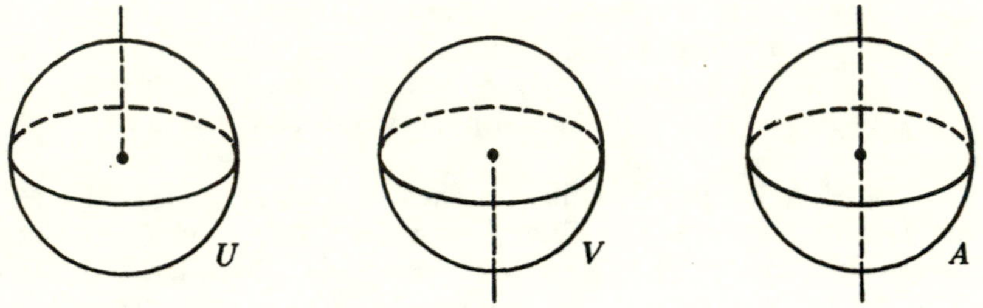

Figure 40.4

If we set $X = U \cup V$, then

$$X = \mathbf{R}^{n+1} - \mathbf{0}.$$

The set U is star-convex relative to the point $\mathbf{p} = (0, \ldots, 0, -1)$ of $\mathbf{R}^{n+1}$, and the set V is star-convex relative to the point $\mathbf{q} = (0, \ldots, 0, 1)$, as you can readily check. It follows from the preceding theorem that $H^0(X)$ is trivial, and that

$$\dim H^{k+1}(X) = \dim H^k(A) \quad \text{for} \quad k \geq 0.$$

Now Step 2 tells us that $H^k(A)$ has the same dimension as $H^k(\mathbf{R}^n - \mathbf{0})$, and the induction hypothesis implies that the latter has dimension 0 if $k \neq n - 1$, and dimension 1 if $k = n - 1$. The theorem follows. □

Let us restate this theorem in terms of forms.

Theorem 40.5. *Let $A = \mathbb{R}^n - 0$, with $n \geq 1$.*

(a) *If $k \neq n-1$, then every closed k-form on A is exact on A.*

(b) *There is a closed $n-1$ form η_0 on A that is not exact. If η is any closed $n-1$ form on A, then there is a unique scalar c such that $\eta - c\eta_0$ is exact.* □

This theorem guarantees the existence of a closed $n-1$ form in $\mathbb{R}^n - 0$ that is not exact, but it does not give us a formula for such a form. In the exercises of the last chapter, however, we obtained such a formula. If η_0 is the $n-1$ form in $\mathbb{R}^n - 0$ given by the equation

$$\eta_0 = \sum_{i=1}^{n} (-1)^{i-1} f_i \, dx_1 \wedge \cdots \wedge \widehat{dx_i} \wedge \cdots \wedge dx_n,$$

where $f_i(\mathbf{x}) = x_i / \|\mathbf{x}\|^n$, then it is easy to show by direct computation that η_0 is closed, and only somewhat more difficult to show that the integral of η_0 over S^{n-1} is non-zero, so that by Stokes' theorem it cannot be exact. (See the exercises of §35 or §38.) *Using this result,* we now derive the following criterion for a closed $n-1$ form in $\mathbb{R}^n - 0$ to be exact:

Theorem 40.6. *Let $A = \mathbb{R}^n - 0$, with $n > 1$. If η is a closed $n-1$ form in A, then η is exact in A if and only if*

$$\int_{S^{n-1}} \eta = 0.$$

Proof. If η is exact, then its integral over S^{n-1} is 0, by Stokes' theorem. On the other hand, suppose this integral is zero. Let η_0 be the form just defined. The preceding theorem tells us that there is a unique scalar c such that $\eta - c\eta_0$ is exact. Then

$$\int_{S^{n-1}} \eta = c \int_{S^{n-1}} \eta_0,$$

by Stokes' theorem. Since the integral of η_0 over S^{n-1} is not 0, we must have $c = 0$. Thus η is exact. □

EXERCISES

1. (a) Show that V/W is a vector space.

   (b) Show that the transformation $\tilde{T}$ induced by a linear transformation T is well-defined and linear.

2. Suppose $a_1, \ldots, a_n$ is a basis for V whose first k elements from a basis for the linear subspace W. Show that the cosets $a_{k+1} + W, \ldots, a_n + W$ form a basis for V/W.

3. (a) Translate Theorems 40.5 and 40.6 into theorems about vector and scalar fields in $\mathbf{R}^n - 0$, in the case $n = 2$.

   (b) Repeat for the case $n = 3$.

4. Let U and V be open sets in $\mathbf{R}^n$; let $X = U \cup V$; assume that $A = U \cap V$ is non-empty. Let $\tilde{\delta} : H^k(A) \to H^{k+1}(X)$ be the transformation constructed in the proof of Theorem 40.3. What hypotheses on $H^i(U)$ and $H^i(V)$ are needed to ensure that:

   (a) $\tilde{\delta}$ is one-to-one?

   (b) The image of $\tilde{\delta}$ is all of $H^{k+1}(X)$?

   (c) $H^0(X)$ is trivial?

5. Prove the following:

   Theorem. *Let* p *and* q *be two points of* $\mathbf{R}^n$; *let* $n \geq 1$. *Then*

$$\dim H^k(\mathbf{R}^n - \mathrm{p} - \mathrm{q}) = \begin{cases} 0 & \text{if } k \neq n-1, \\ 2 & \text{if } k = n-1. \end{cases}$$

   Proof. Let $S = \{\mathrm{p}, \mathrm{q}\}$. Use Theorem 40.3 to show that the open set $\mathbf{R}^{n+1} - S \times \mathbf{H}^1$ of $\mathbf{R}^{n+1}$ is homologically trivial in all dimensions. Then proceed by induction, as in the proof of Theorem 40.4.

6. Restate the theorem of Exercise 5 in terms of forms.

7. Derive a criterion analogous to that in Theorem 40.6 for a closed $n - 1$ form in $\mathbf{R}^n - \mathrm{p} - \mathrm{q}$ to be exact.

8. Translate results of Exercises 6 and 7 into theorems about vector and scalar fields in $\mathbf{R}^n - \mathrm{p} - \mathrm{q}$ in the cases $n = 2$ and $n = 3$.

9

Epilogue—Life Outside $\mathbf{R}^n$

§41. DIFFERENTIABLE MANIFOLDS AND RIEMANNIAN MANIFOLDS

Throughout this book, we have dealt with submanifolds of euclidean space and with forms defined in open sets of euclidean space. This approach has the advantage of conceptual simplicity; one tends to be more comfortable dealing with subspaces of $\mathbf{R}^n$ than with arbitrary metric spaces. It has the disadvantage, however, that important ideas are sometimes obscured by the familiar surroundings. That is the case here.

Furthermore, it is true that, in higher mathematics as well as in other subjects such as mathematical physics, manifolds often occur as abstract spaces rather than as subspaces of euclidean space. To treat them with the proper degree of generality requires that one move outside $\mathbf{R}^n$.

In this section, we describe briefly how this can be accomplished, and indicate how mathematicians really look at manifolds and forms.

Differentiable manifolds

Definition. Let M be a metric space. Suppose there is a collection of homeomorphisms $\alpha_i : U_i \to V_i$, where U_i is open in $\mathbf{H}^k$ or $\mathbf{R}^k$, and V_i is open in M, such that the sets V_i cover M. (To say that α_i is a **homeomorphism** is to say that α_i carries U_i onto V_i in a one-to-one fashion, and that both α_i and α_i^{-1} are continuous.) Suppose that the maps α_i overlap with class C^∞;

this means that the **transition function** $\alpha_i^{-1} \circ \alpha_j$ is of class C^∞ whenever $V_i \cap V_j$ is nonempty. The maps α_i are called **coordinate patches** on M, and so is any other homeomorphism $\alpha : U \rightarrow V$, where U is open in $\mathbf{H}^k$ or $\mathbf{R}^k$, and V is open in M, that overlaps the α_i with class C^∞. The metric space M, together with this collection of coordinate patches on M, is called a **differentiable** k**-manifold** (of class C^∞).

In the case $k = 1$, we make the special convention that the domains of the coordinate patches may be open sets in $\mathbf{L}^1$ as well as $\mathbf{R}^1$ or $\mathbf{H}^1$, just as we did before.

If there is a coordinate patch $\alpha : U \rightarrow V$ about the point p of M such that U is open in $\mathbf{R}^k$, then p is called an **interior point** of M. Otherwise, p is called a **boundary point** of M. The set of boundary points of M is denoted ∂M. If $\alpha : U \rightarrow V$ is a coordinate patch on M about p, then p belongs to ∂M if and only if U is open in $\mathbf{H}^k$ and $p = \alpha(\mathbf{x})$ for some $\mathbf{x} \in \mathbf{R}^{k-1} \times 0$. The proof is the same as that of Lemma 24.2.

Throughout this section, M will denote a differentiable k-manifold.

Definition. Given coordinate patches α_0, α_1 on M, we say they **overlap positively** if $\det D(\alpha_1^{-1} \circ \alpha_0) > 0$. If M can be covered by coordinate patches that overlap positively, then M is said to be **orientable.** An **orientation** of M consists of such a covering of M, along with all other coordinate patches that overlap these positively. An **oriented manifold** consists of a manifold M together with an orientation of M.

Given an orientation $\{\alpha_i\}$ of M, the collection $\{\alpha_i \circ r\}$, where $r : \mathbf{R}^k \rightarrow \mathbf{R}^k$ is the reflection map, gives a different orientation of M; it is called the orientation **opposite** to the given one.

Suppose M is a differentiable k-manifold with non-empty boundary. Then ∂M is a differentiable $k - 1$ manifold without boundary. The maps $\alpha \circ b$, where α is a coordinate patch on M about $p \in \partial M$ and $b : \mathbf{R}^{k-1} \rightarrow \mathbf{R}^k$ is the map

$$b(x_1, \ldots, x_{k-1}) = (x_1, \ldots, x_{k-1}, 0),$$

are coordinate patches on ∂M. The proof is the same as that of Theorem 24.3.

If the patches α_0 and α_1 on M overlap positively, so do the coordinate patches $\alpha_0 \circ b$ and $\alpha_1 \circ b$ on ∂M; the proof is that of Theorem 34.1. Thus if M is oriented and ∂M is nonempty, then ∂M can be oriented simply by taking coordinate patches on M belonging to the orientation of M about points of ∂M, and composing them with the map b. If k is even, the orientation of ∂M obtained in this way is called the **induced orientation** of ∂M; if k is odd, the opposite of this orientation is so called.

Now let us define differentiability for maps between two differentiable manifolds.

Definition. Let M and N be differentiable manifolds of dimensions k and n, respectively. Suppose A is a subset of M; and suppose $f : A \to N$. We say that f is of class C^∞ if for each $x \in A$, there is a coordinate patch $\alpha : U \to V$ on M about x, and a coordinate patch $\beta : W \to Y$ on N about $y = f(x)$, such that the composite $\beta^{-1} \circ f \circ \alpha$ is of class C^∞, as a map of a subset of $\mathbf{R}^k$ into $\mathbf{R}^n$. Because the transition functions are of class C^∞, this condition is independent of the choice of the coordinate patches. See Figure 41.1.

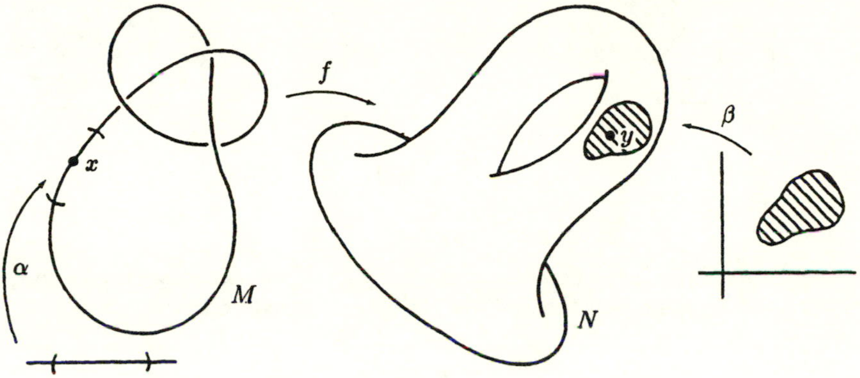

Figure 41.1

Of course, if M or N equals euclidean space, this definition simplifies, since one can take one of the coordinate patches to be the identity map of that euclidean space.

A one-to-one map $f : M \to N$ carrying M onto N is called a **diffeomorphism** if both f and f^{-1} are of class C^∞.

Now we define what we mean by a tangent vector to M. Since we have here no surrounding euclidean space to work with, it is not obvious what a tangent vector should be.

Our usual picture of a tangent vector to a manifold M in $\mathbf{R}^n$ at a point $\mathbf{p}$ of M is that it is the velocity vector of a C^∞ curve $\gamma : [a, b] \to M$ that passes through $\mathbf{p}$. This vector is just the pair $(\mathbf{p}; D\gamma(t_0))$ where $\mathbf{p} = \gamma(t_0)$ and $D\gamma$ is the derivative of γ.

Let us try to generalize this notion. If M is an arbitrary differentiable manifold, and γ is a C^∞ curve in M, what does one mean by the "derivative" of the function γ? Certainly one cannot speak of derivatives in the ordinary sense, since M does not lie in euclidean space. However, if $\alpha : U \to V$ is a coordinate patch in M about the point p, then the composite function $\alpha^{-1} \circ \gamma$ is a map from a subset of $\mathbf{R}^1$ into $\mathbf{R}^k$, so we *can* speak of its derivative. We

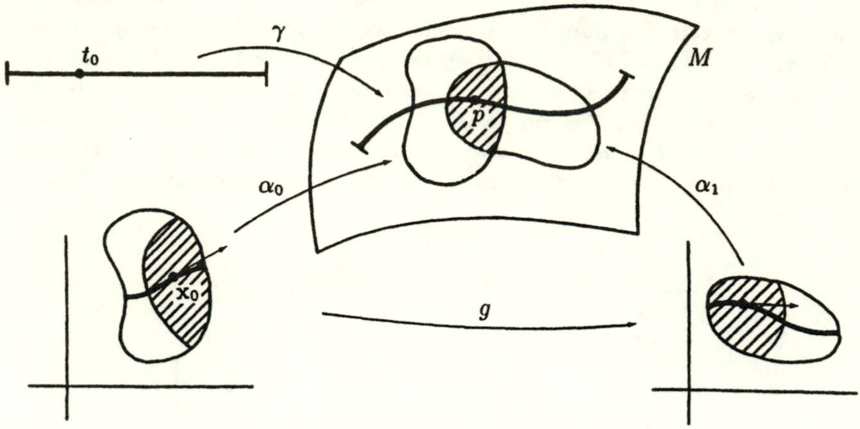

Figure 41.2

can thus think of the "derivative" of γ at t_0 as the function $\mathbf{v}$ that assigns, to each coordinate patch α about the point p, the matrix

$$\mathbf{v}(\alpha) = D(\alpha^{-1} \circ \gamma)(t_0),$$

where $p = \gamma(t_0)$.

Of course, the matrix $D(\alpha^{-1} \circ \gamma)$ depends on the particular coordinate patch chosen; if α_0 and α_1 are two coordinate patches about p, the chain rule implies that these matrices are related by the equation

$$\mathbf{v}(\alpha_1) = Dg(\mathbf{x}_0) \cdot \mathbf{v}(\alpha_0),$$

where g is the transition function $g = \alpha_1^{-1} \circ \alpha_0$, and $\mathbf{x}_0 = \alpha_0^{-1}(p)$. See Figure 41.2.

The pattern of this example suggests to us how to define a tangent vector to M in general.

Definition. Given $p \in M$, a **tangent vector to M at p** is a function $\mathbf{v}$ that assigns, to each coordinate patch $\alpha : U \to V$ in M about p, a column matrix of size k by 1 which we denote $\mathbf{v}(\alpha)$. If α_0 and α_1 are two coordinate patches about p, we require that

$$(*) \qquad \mathbf{v}(\alpha_1) = Dg(\mathbf{x}_0) \cdot \mathbf{v}(\alpha_0),$$

where $g = \alpha_1^{-1} \circ \alpha_0$ is the transition function and $\mathbf{x}_0 = \alpha_0^{-1}(p)$. The entries of the matrix $\mathbf{v}(\alpha)$ are called the **components** of $\mathbf{v}$ with respect to the coordinate patch α.

It follows from $(*)$ that a tangent vector $\mathbf{v}$ to M at p is entirely determined once its components are given with respect to a single coordinate system. It also follows from $(*)$ that if $\mathbf{v}$ and $\mathbf{w}$ are tangent vectors to M at p, then we can define $a\mathbf{v} + b\mathbf{w}$ unambiguously by setting

$$(a\mathbf{v} + b\mathbf{w})(\alpha) = a\mathbf{v}(\alpha) + b\mathbf{w}(\alpha)$$

for each α. That is, we add tangent vectors by adding their components in the usual way in each coordinate patch. And we multiply a vector $\mathbf{v}$ by a scalar similarly.

The set of tangent vectors to M at p is denoted $\mathcal{T}_p(M)$; it is called the **tangent space** to M at p. It is easy to see that it is a k-dimensional space; indeed, if α is a coordinate patch about p with $\alpha(\mathbf{x}) = p$, one checks readily that the map $\mathbf{v} \to (\mathbf{x}; \mathbf{v}(\alpha))$, which carries $\mathcal{T}_p(M)$ onto $\mathcal{T}_{\mathbf{x}}(\mathbb{R}^k)$, is a linear isomorphism. The inverse of this map is denoted by

$$\alpha_* : \mathcal{T}_{\mathbf{x}}(\mathbb{R}^k) \to \mathcal{T}_p(M).$$

It satisfies the equation $\alpha_*(\mathbf{x}; \mathbf{v}(\alpha)) = \mathbf{v}$.

Given a C^∞ curve $\gamma : [a,b] \to M$ in M, with $\gamma(t_0) = p$, we define the **velocity vector** $\mathbf{v}$ of this curve corresponding to the parameter value t_0 by the equation

$$\mathbf{v}(\alpha) = D(\alpha^{-1} \circ \gamma)(t_0);$$

then $\mathbf{v}$ is a tangent vector to M at p. One readily shows that every tangent vector to M at p is the velocity vector of some such curve.

REMARK. There is an alternate approach to defining tangent vectors that is quite common. We describe it here.

Suppose $\mathbf{v}$ is a tangent vector to M at the point p of M. There is associated with $\mathbf{v}$ a certain operator $X_{\mathbf{v}}$ on real-valued C^∞ functions defined near p. This operator is called the **derivative with respect to v**; it arises from the following considerations:

Suppose f is a C^∞ function on M defined in a neighborhood of p, and suppose $\mathbf{v}$ is the velocity vector of the curve $\gamma : [a,b] \to M$ corresponding to the parameter value t_0, where $\gamma(t_0) = p$. Then the derivative $d(f \circ \gamma)/dt$ measures the rate of change of f with respect to the parameter t of the curve. If $\alpha : U \to V$ is a coordinate patch about p, with $\alpha(\mathbf{x}) = p$, we can express this derivative as follows: We write $f \circ \gamma = (f \circ \alpha) \circ (\alpha^{-1} \circ \gamma)$, and compute

$$\begin{aligned}\frac{d(f \circ \gamma)}{dt}(t_0) &= D(f \circ \alpha)(\mathbf{x}) \cdot D(\alpha^{-1} \circ \gamma)(t_0), \\ &= D(f \circ \alpha)(\mathbf{x}) \cdot \mathbf{v}(\alpha).\end{aligned}$$

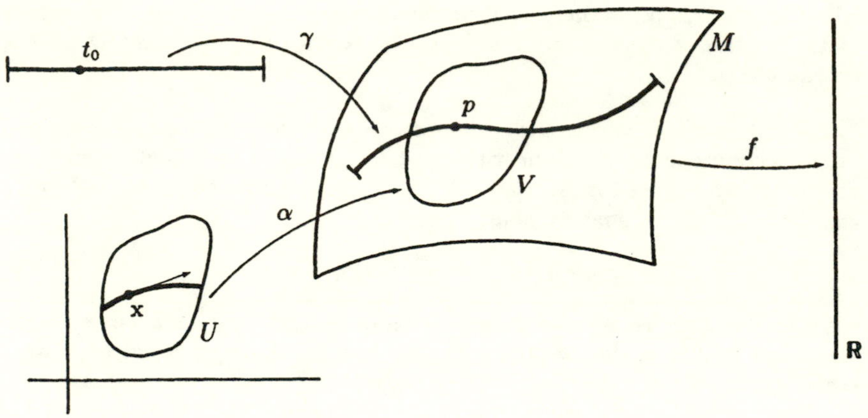

Figure 41.3

See Figure 41.3. Note that this derivative depends only on f and the velocity vector $\mathbf{v}$, not on the particular curve γ.

This formula leads us to define the operator $X_{\mathbf{v}}$ as follows:

If $\mathbf{v}$ is a tangent vector to M at p, and if f is a C^∞ real-valued function defined near p, choose a coordinate patch $\alpha : U \to V$ about p with $\alpha(\mathbf{x}) = p$, and define the **derivative of f with respect to v** by the equation

$$X_{\mathbf{v}}(f) = D(f \circ \alpha)(\mathbf{x}) \cdot \mathbf{v}(\alpha).$$

One checks readily that this number is independent of the choice of α. One checks also that $X_{\mathbf{v}+\mathbf{w}} = X_{\mathbf{v}} + X_{\mathbf{w}}$ and $X_{c\mathbf{v}} = cX_{\mathbf{v}}$. Thus the sum of vectors corresponds to the sum of the corresponding operations, and similarly for a scalar multiple of a vector.

Note that if $M = \mathbf{R}^k$, then the operator $X_{\mathbf{v}}$ is just the directional derivative of f with respect to the vector $\mathbf{v}$.

The operator $X_{\mathbf{v}}$ satisfies the following properties, which are easy to check:

(1) (Locality). If f and g agree in a neighborhood of p, then $X_{\mathbf{v}}(f) = X_{\mathbf{v}}(g)$.

(2) (Linearity). $X_{\mathbf{v}}(af + bg) = aX_{\mathbf{v}}(f) + bX_{\mathbf{v}}(g)$.

(3) (Product rule). $X_{\mathbf{v}}(f \cdot g) = X_{\mathbf{v}}(f)g(p) + f(p)X_{\mathbf{v}}(g)$.

These properties in fact characterize the operator $X_{\mathbf{v}}$. One has the following theorem: Let X be an operator that assigns to each C^∞ real-valued function f defined near p a number denoted $X(f)$, such that X satisfies conditions (1)–(3). Then there is a unique tangent vector $\mathbf{v}$ to M at p such that $X = X_{\mathbf{v}}$. The proof requires some effort; it is outlined in the exercises.

This theorem suggests an alternative approach to defining tangent vectors. One could define a tangent vector to M at p to be simply an operator X

satisfying conditions (1)–(3). The set of these operators is a linear space if we add operators in the usual way and multiply by scalars in the usual way, and thus it can be identified with the tangent space to M at p.

Many authors prefer to use this definition of tangent vector. It has the appeal that it is "intrinsic"; that is, it does not involve coordinate patches explicitly.

Now we define forms on M.

Definition. An **ℓ-form** on M is a function ω assigning to each $p \in M$, an alternating ℓ-tensor on the vector space $\mathcal{T}_p(M)$. That is,

$$\omega(p) \in \mathcal{A}^\ell(\mathcal{T}_p(M))$$

for each $p \in M$.

We require ω to be of class C^∞ in the following sense: If $\alpha : U \to V$ is a coordinate patch on M about p, with $\alpha(\mathbf{x}) = p$, one has the linear transformation

$$T = \alpha_* : \mathcal{T}_{\mathbf{x}}(\mathbf{R}^k) \to \mathcal{T}_p(M)$$

and the dual transformation

$$T^* : \mathcal{A}^\ell(\mathcal{T}_p(M)) \to \mathcal{A}^\ell(\mathcal{T}_{\mathbf{x}}(\mathbf{R}^k)).$$

If ω is an ℓ-form on M, then the ℓ-form $\alpha^*\omega$ is defined as usual by setting

$$(\alpha^*\omega)(\mathbf{x}) = T^*(\omega(p)).$$

We say that ω is **of class C^∞ near p** if $\alpha^*\omega$ is of class C^∞ near $\mathbf{x}$ in the usual sense. This condition is independent of the choice of coordinate patch. Thus ω is of class C^∞ if for every coordinate patch α on M, the form $\alpha^*\omega$ is of class C^∞ in the sense defined earlier.

Henceforth, we assume all our our forms are of class C^∞.

Let $\Omega^\ell(M)$ denote the space of ℓ-forms on M. Note that there are no elementary forms on M that would enable us to write ω in canonical form, as there were in $\mathbf{R}^n$. However, one can write $\alpha^*\omega$ in canonical form as

$$\alpha^*\omega = \sum_{[I]} f_I \, dx_I,$$

where the dx_I are the elementary forms in $\mathbf{R}^k$. We call the functions f_I the **components** of ω with respect to the coordinate patch α. They are of course of class C^∞.

Definition. If ω is an ℓ-form on M, we define the **differential** of ω as follows: Given $p \in M$, and given tangent vectors $\mathbf{v}_1, \ldots, \mathbf{v}_{\ell+1}$ to M at p, choose a coordinate patch $\alpha : U \to V$ on M about p with $\alpha(\mathbf{x}) = p$. Then define

$$d\omega(p)(\mathbf{v}_1, \ldots, \mathbf{v}_{\ell+1}) = d(\alpha^*\omega)(\mathbf{x})\big((\mathbf{x}; \mathbf{v}_1(\alpha)), \ldots, (\mathbf{x}; \mathbf{v}_{\ell+1}(\alpha))\big).$$

That is, we define $d\omega$ by choosing a coordinate patch α, pulling ω back to a form $\alpha^*\omega$ in $\mathbf{R}^k$, pulling $\mathbf{v}_1, \ldots, \mathbf{v}_{\ell+1}$ back to tangent vectors in $\mathbf{R}^k$, and then applying the operator d in $\mathbf{R}^k$. One checks that this definition is independent of the choice of the patch α. Then $d\omega$ is of class C^∞.

We can rewrite this equation as follows: Let $\mathbf{a}_i = \mathbf{v}_i(\alpha)$. The preceding equation can be written in the form

$$d\omega(p)\big(\alpha_*(\mathbf{x}; \mathbf{a}_1), \ldots, \alpha_*(\mathbf{x}; \mathbf{a}_{\ell+1})\big) = d(\alpha^*\omega)(\mathbf{x})\big((\mathbf{x}; \mathbf{a}_1), \ldots, (\mathbf{x}; \mathbf{a}_{\ell+1})\big).$$

This equation says simply that $\alpha^*(d\omega) = d(\alpha^*\omega)$. Thus one has an alternate version of the preceding definition:

Definition. If ω is an ℓ-form on M, then $d\omega$ is defined to be the unique $\ell + 1$ form on M such that for every coordinate patch α on M,

$$\alpha^*(d\omega) = d(\alpha^*\omega).$$

Here the "d" on the right side of the equation is the usual differential operator d in $\mathbf{R}^k$, and the "d" on the left is our new differential operator in M.

Now we define the integral of a k-form over M. We need first to discuss partitions of unity. Because we assume M is compact, matters are especially simple.

Theorem 41.1. *Let M be a compact differentiable manifold. Given a covering of M by coordinate patches, there exist functions $\phi_i : M \to \mathbf{R}$ of class C^∞, for $i = 1, \ldots, \ell$, such that:*

(1) *$\phi_i(p) \geq 0$ for each $p \in M$.*

(2) *For each i, the set* Support ϕ_i *is covered by one of the given coordinate patches.*

(3) *$\Sigma\, \phi_i(p) = 1$ for each $p \in M$.*

Proof. Given $p \in M$, choose a coordinate patch $\alpha : U \to V$ about p. Let $\alpha(\mathbf{x}) = p$; choose a non-negative C^∞ function $f : U \to \mathbf{R}$ whose support is compact and is contained in U, such that f is positive at the point $\mathbf{x}$. Define $\psi_p : M \to \mathbf{R}$ by setting

$$\psi_p(y) = \begin{cases} f(\alpha^{-1}(y)) & \text{if } y \in V, \\ 0 & \text{otherwise.} \end{cases}$$

Because $f(\alpha^{-1}(y))$ vanishes outside a compact subset of V, the function ψ_p is of class C^∞ on M.

Now ψ_p is positive on an open set U_p about p. Cover M by finitely many of the open sets U_p, say for $p = p_1, \dots, p_\ell$. Then set

$$\lambda = \sum_{j=1}^{\ell} \psi_{p_j} \quad \text{and} \quad \phi_i = (1/\lambda)\psi_{p_i}. \quad \square$$

Definition. Let M be a compact, oriented differentiable k-manifold. Let ω be a k-form on M. If the support of ω lies in a single coordinate patch $\alpha : U \to V$ belonging to the orientation of M, define

$$\int_M \omega = \int_{\text{Int } U} \alpha^* \omega.$$

In general, choose $\phi_1, \dots, \phi_\ell$ as in the preceding theorem and define

$$\int_M \omega = \sum_{i=1}^{\ell} [\int_M \phi_i \omega].$$

The usual argument shows this integral is well-defined and linear.

Finally, we have:

Theorem 41.2 (Stokes' theorem). *Let M be a compact, oriented differentiable k-manifold. Let ω be a $k-1$ form on M. If ∂M is non-empty, give ∂M the induced orientation; then*

$$\int_M d\omega = \int_{\partial M} \omega.$$

If ∂M is empty, then $\int_M d\omega = 0$.

Proof. The proof given earlier goes through *verbatim*. Since all the computations were carried out by working within coordinate patches, no changes are necessary. The special conventions involved when $k = 1$ and ∂M is a 0-manifold are handled exactly as before. $\square$

Not only does Stokes' theorem generalize to abstract differentiable manifolds, but the results in Chapter 8 concerning closed forms and exact forms generalize as well. Given M, one defines the deRham group $H^k(M)$ of M in dimension k to be the quotient of the space of closed k-forms on M by the space of exact k-forms. One has various methods for computing the dimensions of these spaces, including a general *Mayer-Vietoris theorem.* If M is written as the union of the two open sets U and V in M, it gives relations between the deRham groups of M and U and V and $U \cap V$. These topics are explored in [B–T].

The vector space $H^k(M)$ is obviously a diffeomorphism invariant of M. It is an unexpected and striking fact that it is also a topological invariant of M. This means that if there is a homeomorphism of M with N, then the vector spaces $H^k(M)$ and $H^k(N)$ are linearly isomorphic. This fact is a consequence of a celebrated theorem called *deRham's theorem,* which states that the algebra of closed forms on M modulo exact forms is isomorphic to a certain algebra, defined in algebraic topology for an arbitrary topological space, called the "cohomology algebra of M with real coefficients."

Riemannian manifolds

We have indicated how Stokes' theorem and the deRham groups generalize to abstract differentiable manifolds. Now we consider some of the other topics we have treated. Surprisingly, many of these do not generalize as readily.

Consider for instance the notions of the volume of a manifold M, and of the integral $\int_M f \, dV$ of a scalar function over M with respect to volume. These notions do not generalize to abstract differentiable manifolds.

Why should this be so? One way of answering this question is to note that, according to the discussion in §36, one can define the volume of a compact oriented k-manifold M in $\mathbf{R}^n$ by the formula

$$v(M) = \int_M \omega_v,$$

where ω_v is a "volume form" for M, that is, ω_v is a k-form whose value is 1 *on any orthonormal basis* for $\mathcal{T}_{\mathbf{p}}(M)$ belonging to the natural orientation of this tangent space. In this case, $\mathcal{T}_{\mathbf{p}}(M)$ is a linear subspace of $\mathcal{T}_{\mathbf{p}}(\mathbf{R}^n) = \mathbf{p} \times \mathbf{R}^n$, so $\mathcal{T}_{\mathbf{p}}(M)$ has a natural inner product derived from the dot product in $\mathbf{R}^n$. This notion of a volume form cannot be generalized to an arbitrary differentiable manifold M because we have no inner product on $\mathcal{T}_p(M)$ in general, so we do not know what it means for a set of vectors to be orthonormal.

In order to generalize our definition of volume to a differentiable manifold M, we need to have an inner product on each tangent space $\mathcal{T}_p(M)$:

Definition. Let M be a differentiable k-manifold. A **Riemannian metric** on M is an inner product $\langle \mathbf{v}, \mathbf{w} \rangle$ defined on each tangent space $\mathcal{T}_p(M)$;

it is required to be of class C^∞ as a 2-tensor field on M. A **Riemannian manifold** consists of a differentiable manifold M along with a Riemannian metric on M.

(Note that the word "metric" in this context has nothing to do with the use of the same word in the phrase "metric space.")

Now it is true that for any differentiable manifold M, there exists a Riemannian metric on M. The proof is not particularly difficult; one uses a partition of unity. But the Riemannian metric is certainly not unique.

Given a Riemannian metric on M, one has a corresponding volume function $V(\mathbf{v}_1, \ldots, \mathbf{v}_k)$ defined for k-tuples of vectors of $\mathcal{T}_p(M)$. (See the exercises of §21.) Then one can define the integral of a scalar function just as before:

Definition. Let M be a compact Riemannian manifold of dimension k. Let $f : M \to \mathbf{R}$ be a continuous function. If the support of f is covered by a single coordinate patch $\alpha : U \to V$, we define the **integral of f over M** by the equation

$$\int_M f \, dV = \int_{\text{Int } U} (f \circ \alpha) V(\alpha_*(\mathbf{x};\mathbf{e}_1), \ldots, \alpha_*(\mathbf{x};\mathbf{e}_k)).$$

The integral of f over M is defined in general by using a partition of unity, just as in §25. The **volume** of M is defined by the equation

$$v(M) = \int_M dV.$$

If M is a compact oriented Riemannian manifold, one can interpret the integral $\int_M \omega$ of a k-form over M as the integral $\int_M \lambda \, dV$ of a certain scalar function, just as we did before, where $\lambda(p)$ is the value of $\omega(p)$ on an orthonormal k-tuple of tangent vectors to M at p that belongs to the natural orientation of $\mathcal{T}_p(M)$ (derived from the orientation of M). If $\lambda(p)$ is identically 1, then ω is called the **volume form** of the Riemannian manifold M, and is denoted by ω_v. Then

$$v(M) = \int_M \omega_v.$$

For a Riemannian manifold M, a host of interesting questions arise. For instance, one can define what one means by the **length** of a smooth parametrized curve $\gamma : [a,b] \to M$; it is just the integral

$$\int_{t=a}^{t=b} \|\gamma_*(t;\mathbf{e}_1)\|.$$

The integrand is the norm of the velocity vector of the curve γ, defined of course by using the inner product on $\mathcal{T}_p(M)$. Then one can discuss "geodesics," which are "curves of minimal length" joining two points of M. One goes on to discuss such matters as "curvature." All this is dealt with in a subject called Riemannian geometry, which I hope you are tempted to investigate!

One final comment. As we have indicated, most of what is done in this book can be generalized, either to abstract differentiable manifolds or to Riemannian manifolds. One aspect that does *not* generalize is the interpretation of Stokes' theorem in terms of scalar and vector fields given in §38. The reason is clear. The "translation functions" of §31, which interpret k-forms in $\mathbf{R}^n$ as scalar fields or vector fields in $\mathbf{R}^n$ for certain values of k, depend crucially on having forms that are defined in $\mathbf{R}^n$, not on some abstract manifold M. Furthermore, the operators grad and div apply only to scalar and vector fields in $\mathbf{R}^n$; and curl applies only in $\mathbf{R}^3$. Even the notion of a "normal vector" to a manifold M depends on the surrounding space, not just on M.

Said differently, while manifolds and differential forms and Stokes' theorem have meaning outside euclidean space, classical vector analysis does not.

EXERCISES

1. Show that if $\mathbf{v} \in \mathcal{T}_p(M)$, then $\mathbf{v}$ is the velocity vector of some C^∞ curve γ in M passing through p.
2. (a) Let $\mathbf{v} \in \mathcal{T}_p(M)$. Show that the operator $X_\mathbf{v}$ is well-defined.

   (b) Verify properties (1)–(3) of the operator $X_\mathbf{v}$.
3. If ω is an ℓ-form on M, show that $d\omega$ is well-defined (independent of the choice of the coordinate patch α).
4. Verify that the proof of Stokes' theorem holds for an arbitrary differentiable manifold.
5. Show that any compact differentiable manifold has a Riemannian metric.

*6. Let M be a differentiable k-manifold; let $p \in M$. Let X be an operator on C^∞ real-valued functions defined near p, satisfying locality, linearity, and the product rule. Show there is exactly one tangent vector $\mathbf{v}$ to M at p such that $X = X_\mathbf{v}$, as follows:

   (a) Let F be a C^∞ function defined on the open cube U in $\mathbf{R}^k$ consisting of all $\mathbf{x}$ with $|\mathbf{x}| < \epsilon$. Show there are C^∞ functions $g_1, \ldots, g_k$ defined on U such that

$$F(\mathbf{x}) - F(\mathbf{0}) = \sum_j x_j g_j(\mathbf{x})$$

   for $\mathbf{x} \in U$. [*Hint:* Set

$$g_j(\mathbf{x}) = \int_{u=0}^{u=1} D_j F(x_1, \ldots, x_{j-1}, u x_j, 0, \ldots, 0).$$

Then g_j is of class C^∞ and

$$x_j g_j(\mathbf{x}) = \int_{t=0}^{t=x_j} D_j F(x_1, \ldots, x_{j-1}, t, 0, \ldots, 0).]$$

(b) If F and g_j are as in (a), show that

$$D_j F(\mathbf{0}) = g_j(\mathbf{0}).$$

(c) Show that if c is a constant function, then $X(c) = 0$. [*Hint:* Show that $X(1 \cdot 1) = 0$.]

(d) Given X, show there is at most one $\mathbf{v}$ such that $X = X_{\mathbf{v}}$. [*Hint:* Let α be a coordinate patch about p; let $h = \alpha^{-1}$. If $X = X_{\mathbf{v}}$, show that the components of $\mathbf{v}(\alpha)$ are the numbers $X(h_i)$.]

(e) Given X, show there exists a $\mathbf{v}$ such that $X = X_{\mathbf{v}}$. [*Hint:* Let α be a coordinate patch with $\alpha(\mathbf{0}) = p$; let $h = \alpha^{-1}$. Set $v_i = X(h_i)$, and let $\mathbf{v}$ be the tangent vector at p such that $\mathbf{v}(\alpha)$ has components $v_1, \ldots, v_k$. Given f defined near p, set $F = f \circ \alpha$. Then

$$X_{\mathbf{v}}(f) = \sum_j D_j F(\mathbf{0}) \cdot v_j.$$

Write $F(\mathbf{x}) = \Sigma_j x_j g_j(\mathbf{x}) + F(\mathbf{0})$ for $\mathbf{x}$ near $\mathbf{0}$, as in (a). Then

$$f = \sum_j h_j \cdot (g_j \circ h) + F(\mathbf{0})$$

in a neighborhood of p. Calculate $X(f)$ using the three properties of X.]

Bibliography

[A] Apostol, T.M., *Mathematical Analysis, 2nd edition,* Addison-Wesley, 1974.

[A-M-R] Abraham, R., Mardsen, J.E., and Ratiu, T., *Manifolds, Tensor Analysis, and Applications,* Addison-Wesley, 1983, Springer-Verlag, 1988.

[B] Boothby, W.M., *An Introduction to Differentiable Manifolds and Riemannian Geometry,* Academic Press, 1975.

[B-G] Berger, M., and Gostiaux, B., *Differential Geometry: Manifolds, Curves, and Surfaces,* Springer-Verlag, 1988.

[B-T] Bott, R., and Tu, L.W., *Differential Forms in Algebraic Topology,* Springer-Verlag, 1982.

[D] Devinatz, A., *Advanced Calculus,* Holt, Rinehart and Winston, 1968.

[F] Fleming, W., *Functions of Several Variables,* Addison-Wesley, 1965, Springer-Verlag, 1977.

[Go] Goldberg, R.P., *Methods of Real Analysis,* Wiley, 1976.

[G-P] Guillemin, V., and Pollack, A., *Differential Topology,* Prentice-Hall, 1974.

[Gr] Greub, W.H., *Multilinear Algebra, 2nd edition,* Springer-Verlag, 1978.

[M] Munkres, J.R., *Topology, A First Course,* Prentice-Hall, 1975.

[N] Northcott, D.G., *Multilinear Algebra,* Cambridge U. Press, 1984.

[N-S-S] Nickerson, H.K., Spencer, D.C., and Steenrod, N.E., *Advanced Calculus,* Van Nostrand, 1959.

[Ro] Royden, H., *Real Analysis, 3rd edition,* Macmillan, 1988.

[Ru] Rudin, W., *Principles of Mathematical Analysis, 3rd edition,* McGraw-Hill, 1976.

[S] Spivak, M., *Calculus on Manifolds,* Addison-Wesley, 1965.

Index

Printed in the United States
96039LV00003B/70-105/A

9 780201 315967